CW01160618

Human-based Systems for Translational Research

RSC Drug Discovery Series

Editor-in-Chief:
Professor David Thurston, *King's College, London, UK*

Series Editors:
Professor David Rotella, *Montclair State University, USA*
Professor Ana Martinez, *Medicinal Chemistry Institute-CSIC, Madrid, Spain*
Dr David Fox, *Vulpine Science and Learning, UK*

Advisor to the Board:
Professor Robin Ganellin, *University College London, UK*

Titles in the Series:
 1: Metabolism, Pharmacokinetics and Toxicity of Functional Groups
 2: Emerging Drugs and Targets for Alzheimer's Disease; Volume 1
 3: Emerging Drugs and Targets for Alzheimer's Disease; Volume 2
 4: Accounts in Drug Discovery
 5: New Frontiers in Chemical Biology
 6: Animal Models for Neurodegenerative Disease
 7: Neurodegeneration
 8: G Protein-Coupled Receptors
 9: Pharmaceutical Process Development
10: Extracellular and Intracellular Signaling
11: New Synthetic Technologies in Medicinal Chemistry
12: New Horizons in Predictive Toxicology
13: Drug Design Strategies: Quantitative Approaches
14: Neglected Diseases and Drug Discovery
15: Biomedical Imaging
16: Pharmaceutical Salts and Cocrystals
17: Polyamine Drug Discovery
18: Proteinases as Drug Targets
19: Kinase Drug Discovery
20: Drug Design Strategies: Computational Techniques and Applications
21: Designing Multi-Target Drugs
22: Nanostructured Biomaterials for Overcoming Biological Barriers
23: Physico-Chemical and Computational Approaches to Drug Discovery
24: Biomarkers for Traumatic Brain Injury
25: Drug Discovery from Natural Products
26: Anti-Inflammatory Drug Discovery
27: New Therapeutic Strategies for Type 2 Diabetes: Small Molecules
28: Drug Discovery for Psychiatric Disorders
29: Organic Chemistry of Drug Degradation
30: Computational Approaches to Nuclear Receptors
31: Traditional Chinese Medicine

32: Successful Strategies for the Discovery of Antiviral Drugs
33: Comprehensive Biomarker Discovery and Validation for Clinical Application
34: Emerging Drugs and Targets for Parkinson's Disease
35: Pain Therapeutics; Current and Future Treatment Paradigms
36: Biotherapeutics: Recent Developments using Chemical and Molecular Biology
37: Inhibitors of Molecular Chaperones as Therapeutic Agents
38: Orphan Drugs and Rare Diseases
39: Ion Channel Drug Discovery
40: Macrocycles in Drug Discovery
41: Human-based Systems for Translational Research

How to obtain future titles on publication:
A standing order plan is available for this series. A standing order will bring delivery of each new volume immediately on publication.

For further information please contact:
Book Sales Department, Royal Society of Chemistry, Thomas Graham House, Science Park, Milton Road, Cambridge, CB4 0WF, UK
Telephone: +44 (0)1223 420066, Fax: +44 (0)1223 420247
Email: booksales@rsc.org
Visit our website at www.rsc.org/books

Human-based Systems for Translational Research

Edited by

Robert Coleman
Drug Discovery Consultant, Falmouth, UK
Email: robt.coleman@btinternet.com

RSC Drug Discovery Series No. 41

Print ISBN: 978-1-84973-825-5
PDF eISBN: 978-1-78262-013-6
ISSN: 2041-3203

A catalogue record for this book is available from the British Library

© The Royal Society of Chemistry 2015

All rights reserved

Apart from fair dealing for the purposes of research for non-commercial purposes or for private study, criticism or review, as permitted under the Copyright, Designs and Patents Act 1988 and the Copyright and Related Rights Regulations 2003, this publication may not be reproduced, stored or transmitted, in any form or by any means, without the prior permission in writing of The Royal Society of Chemistry or the copyright owner, or in the case of reproduction in accordance with the terms of licences issued by the Copyright Licensing Agency in the UK, or in accordance with the terms of the licences issued by the appropriate Reproduction Rights Organization outside the UK. Enquiries concerning reproduction outside the terms stated here should be sent to The Royal Society of Chemistry at the address printed on this page.

The RSC is not responsible for individual opinions expressed in this work.

The authors have sought to locate owners of all reproduced material not in their own possession and trust that no copyrights have been inadvertently infringed.

Published by The Royal Society of Chemistry,
Thomas Graham House, Science Park, Milton Road,
Cambridge CB4 0WF, UK

Registered Charity Number 207890

For further information see our web site at www.rsc.org

Printed and bound by CPI Group (UK) Ltd, Croydon, CR0 4YY

Preface

Experimental animals have been a mainstay in the discovery and development of new medicines for human diseases for over half a century, so why do we now need to consider the use of humanised systems?

Historically, the pharmaceutical industry has been hugely successful, bringing a wide range of powerful and effective medicines to the market. Medicines to treat a wide range of disorders for which there had previously been, at best, inadequate treatment have been discovered and added to the clinician's armamentarium. It seemed that nothing could stop the almost exponential increase in the pharmaceutical industry's output. However, the last two to three decades have seen a marked decline in the productivity of this once prolific industry,[1,2] the effects of which have been stark, leading to a complete restructuring of the sector. Unfortunately, however dramatic such restructuring has been, it has not resulted in the hoped for increases in output of new safe and effective medicines per unit investment.[3] Increasingly, when potential medicines developed using animal models are tested in human subjects, they are being found to fail for reasons of either lack of efficacy or associated use-limiting side effects and frank toxicity.

While lack of efficacy is of course a real problem, even more serious is the capacity of new medicines to do harm. In the early part of the 20th century, human medicines could be introduced onto the market with no legal requirement for the manufacturer to explore and report their safety profiles, but this changed following the Elixir sulfanilamide scandal. In 1937, a novel preparation of the antibacterial drug, sulfanilamide, was introduced to the US market, and was responsible for in excess of 100 fatalities.[4] This led to Congress passing the 1938 Food, Drug, and Cosmetic Act, requiring companies to present safety data obtained in experimental animals on any proposed new drug, and to submit the data for approval by the FDA before

RSC Drug Discovery Series No. 41
Human-based Systems for Translational Research
Edited by Robert Coleman
© The Royal Society of Chemistry 2015
Published by the Royal Society of Chemistry, www.rsc.org

being allowed to proceed to market. And so the use of animals to identify potential human hazards became routine within the pharmaceutical industry. The idea of using animals as human surrogates was based on the physiological parallels between humans and other animals, particularly those most closely related in evolutionary terms, *i.e.* mammalian species. While this was a clearly rational approach, it did not totally resolve the issue, and this was most clearly demonstrated in the 1950s with the thalidomide disaster.[5] Thalidomide had been introduced as a mild sedative, particularly useful in reducing morning sickness in pregnancy, but was subsequently identified as the causative agent in the sudden explosion of cases of serious birth defects in the offspring of women who had been prescribed the drug. It is not that thalidomide had not been tested for side effects in experimental animals, it was simply the case that none had been performed on pregnant females to assess effects of the drug on the offspring, and interestingly, even if it had, because of species differences in this effect, the problem would probably not have been picked up. However, it was the thalidomide affair that consolidated the need for the rigorous safety evaluation of new potential medicines, and led to the increased use of animals as patient substitutes that persists to the present day.

As first-line testing for safety issues in human patients or volunteers was and remains ethically unacceptable, the logic behind using experimental animals on the basis of overall physiological similarity was uncontroversial. However, as time has passed, it has become increasingly clear that non-human species do not always respond to drugs in the same way as the humans that they are meant to represent, and their ability to reflect both clinical efficacy and safety issues has been shown to be highly variable, depending on particular drug type and disease target. Despite this, there has been a general acceptance by the medical establishment, the pharmaceutical industry and the regulators that overall, despite their shortcomings, they are doing a fair job, and with the lack of any obvious alternative, we have continued to rely on the animal-based status quo to assure the public of the safety of new medicines. But in the light of ever-increasing numbers of high-profile failures, including, but by no means restricted to, phen-fen,[6] various statins[7,8] and COX2 inhibitors,[9,10] this situation has become untenable, and it has become essential to do something to improve our ability to ensure the lack of safety problems with new medicines.

Much effort has been expended in exploring alternative approaches to new medicines R&D, and official organisations such as ECVAM (Europe) and ICCVAM (USA) have been established to determine their value. Unfortunately the output from these organisations is disappointingly low,[11] and relatively few alternatives have received the necessary level of validation required for regulatory authorities to accept these tests as viable contributors to preclinical safety testing. Furthermore, even in cases where tests have been approved for use in drug submissions, there remains the issue of persuading the pharmaceutical industry to abandon their established animal-based methods in favour of the validated alternative.

Preface

Although a rigorous validation process is applied to new technologies, it is a fact that few if any of the animal-based tests that form the basis of our current safety testing paradigm have themselves been subjected to such scrutiny; their value has been taken as a given, despite the wealth of evidence of their shortcomings. Indeed, if one looks for peer-reviewed publications providing support for the status quo, there is really only one, and that merely reports an overall modest (approximately 70%) concordance between toxicity seen in animal species and in human subjects.[12] In contrast, there is a wealth of publications providing data on the lack of value of animal data in identifying potential clinical safety issues.[13-16] Indeed, a recent publication has explored the value of dogs as predictors of human safety profiles using more sophisticated measures of predictive power (likelihood ratios) than simple concordance, and have concluded that while demonstration of toxicity can in some cases reflect toxicity in humans (in line with Olson's findings), absence of toxicity in dogs provides no useful indication of absence in man.[17]

If testing in animals is an unreliable indicator of potential clinical safety issues, then why are they still demanded and relied upon? A comprehensive answer to this question is outside the scope of this volume, but one key factor is the widespread belief that while flawed, they are superior to whatever else is available.[18] While there have long been proponents of a wider use of *in vitro* human systems, most moves in this direction have been rebuffed by claims that it is impossible to reflect the complexity of a whole integrated organism by looking at its parts in isolation, thus necessitating the use of animal surrogates. Such an attitude may be justified if there really are no viable alternatives available, *and* if the level of inaccuracy can be regarded as acceptable. For a long time, this attitude was (and in some quarters still is) the case, but technology moves on, and the seemingly impossible of yesterday becomes the possible of today, and the commonplace of tomorrow. With this in mind, together with the increasingly obvious inadequacies of the status quo, the exploration of human-based approaches seems unavoidable. Indeed, even the FDA, the primary regulatory authority in the USA, has come to this conclusion, stating on its website: "Consideration should be given to the use of appropriate *in vitro* alternative methods for safety evaluation. These methods, if accepted by all ICH regulatory authorities, can be used to replace current standard methods."[19]

It is fair to say that at this moment, we are not in a position to replace the current *in vivo* animal-based paradigm of assessing the potential safety and efficacy of new medicines with methods based on human *in vitro* models; there is much more work to do. However, we do now have greatly improved access to voluntarily donated viable human cells and tissues, an ever-increasing ability to generate a wide variety of tissue types from induced stem cells, and huge advances being made in a wide range of other technologies to apply to these materials, including those presented in this volume. With all this, we are now ready to perform proper evaluations of approaches which alone and in combination may offer superior means of establishing safety and efficacy of new medicines, consistent with 21st century science.

The purpose of this book therefore is not to present a list of established assays that can reliably identify the potential efficacies and safety issues associated with new medicines, but rather to illustrate what is currently possible in humanising medicines R&D, which with sufficient commitment of time, effort and imagination will provide a basis on which to establish more rational and reliable means of developing safe and effective drugs for the future than those on which we currently rely.

Robert Coleman

References

1. J. W. Scannell, A. Blanckley, H. Boldon and B. Warrington, *Nat. Rev. Drug Discov.*, 2012, **11**, 191.
2. F. Sams-Dodd, *Drug Discov. Today*, 2005, **10**, 139.
3. M. D. Breyer, *Expert Opin. Drug Dis*, 2014, **9**, 115.
4. J. C. Geiger, *Cal. West. Med.*, 1937, **47**, 353.
5. A. E. Rodin, L. A. Koller and J. D. Taylor, *J. Can. Med. Assoc.*, 1962, **86**, 744.
6. H. M. Connolly, J. L. Crary, M. D. McGoon, D. D. Hensrud, B. S. Edwards, W. D. Edwards and H. V. Schaff, *N. Engl. J. Med.*, 1997, **337**, 581.
7. C. D. Furberg and B. Pitt, *Curr Contr. Trials C*, 2001, **2**, 205.
8. R. S. Rosenson, *Am. J. Med.*, 2004, **116**, 408.
9. E. M. Antman, J. S. Bennett, A. Daugherty, C. Furberg, H. Roberts and K. A. Taubert, *Circulation*, 2007, **115**, 1634.
10. H. M. Krumholz, J. S. Ross, A. H. Presler and D. S. Egilman, *Br. Med. J.*, 2007, **334**, 120.
11. M. Leist, N. Hasiwa, M. Daneshian and T. Hartung, *Toxicol. Res.*, 2012, **1**, 8.
12. H. Olson, G. Betton, D. Robinson, K. Thomas, A. Monro, G. Kolaja, P. Lilly, J. Sanders, G. Sipes, W. Bracken, M. Dorato, K. Van Deun, P. Smith, B. Berger and A. Heller, *Regul. Toxicol. Pharm.*, 2000, **32**, 56.
13. P. Pound, S. Ebrahim, P. Sandercock, M. B. Bracken and I. Roberts, *Br. Med. J.*, 2004, **328**, 514.
14. A. Knight, *Rev. Recent Clin. Trials*, 2008, **3**, 89.
15. R. A. J. Matthews, *J. R. Soc. Med.*, 2008, **101**, 95.
16. P. J. van Meer, M. Kooijman, C. C. Gispen-de Wied, H. Moors and H. Schellekens, *Regul. Toxicol. Pharmacol.*, 2012, **64**, 345.
17. J. Bailey, M. Thew and M. Balls, *Altern. Lab. Anim.*, 2013, **41**, 335.
18. http://www.animalresearch.info/ (accessed 12 May 2014).
19. http://www.fda.gov/downloads/Drugs/GuidanceComplianceRegulatory Information/Guidances/UCM194490.pdf (accessed 7 August 2014).

Contents

Chapter 1	**Access to Human Cells and Tissues** *Gerry Thomas*	1
	1.1 Ethics and Law Regarding Collection of Human Samples	1
	1.1.1 The Four Cornerstones of Ethics and Biological Samples	1
	1.1.2 Legal Issues	2
	1.2 Practical Issues Regarding Human Tissue Collection and Annotation	8
	1.2.1 Obtaining Tissue with Consent for Research	9
	1.2.2 Accessing Material from a Diagnostic Archive	11
	1.3 Improving Access and Annotation	12
	1.3.1 Access Policies	12
	1.3.2 Annotation	13
	1.4 Summary	14
	References	14
Chapter 2	**Functional Studies with Human Isolated Tissues to Better Predict Clinical Safety and Efficacy** *David C. Bunton*	17
	2.1 Introduction	17
	2.2 Sourcing, Storing and Transporting Human Fresh Tissues: The First Step is the Most Important	21

RSC Drug Discovery Series No. 41
Human-based Systems for Translational Research
Edited by Robert Coleman
© The Royal Society of Chemistry 2015
Published by the Royal Society of Chemistry, www.rsc.org

2.3	Common Experimental Approaches to Investigations in Isolated Fresh Human Tissues	23
	2.3.1 Tissue Baths and Wire Myographs	23
	2.3.2 Perfusion Myographs	25
	2.3.3 Organoculture Systems and Precision-Cut Tissue Slices	25
	2.3.4 Membrane Transport and Ussing Chambers: Skin, Lung Mucosa, Gastrointestinal Tract and Glandular Tissues	28
2.4	Applications of Functional Tissues in Drug Development: Safety, Efficacy and Personalised Medicines	29
	2.4.1 Predicting Efficacy in Clinically Relevant Human Tissues	30
	2.4.2 Predicting Safety and Toxicology Risks Using Functional Tissues	31
	2.4.3 Functional Tissues and the Development of Personalised Medicines	32
2.5	Summary	35
Acknowledgements		35
References		35

Chapter 3 Translational Research in Pharmacology and Toxicology Using Precision-Cut Tissue Slices   38
G. M. M. Groothuis, A. Casini, H. Meurs, and P. Olinga

3.1	Introduction	38
	3.1.1 Precision-Cut Tissue Slices	38
	3.1.2 Availability of Different Sources of Human Organs and Tissues	40
	3.1.3 Techniques to Prepare and Incubate Precision-Cut Tissue Slices from Various Organs and Tissues	41
3.2	Human Precision-Cut Tissue Slices in ADME and Toxicology	43
3.3	Application of Human Precision-Cut Tissue Slices in Disease Models	47
	3.3.1 Fibrosis in Liver and Intestine	47
	3.3.2 Obstructive Lung Diseases	50
	3.3.3 Cancer Research	54
	3.3.4 Viral Infection Research	57
3.4	Concluding Remarks	58
References		58

Contents

Chapter 4	Modelling the Human Respiratory System: Approaches for *in Vitro* Safety Testing and Drug Discovery		66
	Human-Derived Lung Models; The Future of Toxicology Safety Assessment		
	Zoë Prytherch and Kelly BéruBé		
	4.1	Introduction	66
		4.1.1 Overview of the Respiratory System	66
		4.1.2 Inhalation Toxicology	67
		4.1.3 *Status Quo* in Safety Assessment	68
		4.1.4 *In Vitro* Toxicology	68
		4.1.5 Trends and Technology Uptake	69
	4.2	Current Human-Based Models of the Respiratory System: An Overview	70
		4.2.1 Cells and Tissues	70
		4.2.2 General Culture Conditions	76
	4.3	Discussion	80
	References		81
Chapter 5	Complex Primary Human Cell Systems for Drug Discovery		88
	Ellen L. Berg and Alison O'Mahony		
	5.1	Introduction	88
		5.1.1 Challenges of Target-Based Drug Discovery	88
		5.1.2 Kinase Inhibitors – An Example Target Class	89
		5.1.3 Complex Primary Human Cell Systems for Translational Biology	90
	5.2	Case Studies – Anti-Inflammatory Kinase Inhibitors	95
		5.2.1 JAK Kinase Inhibitors	95
		5.2.2 SYK Kinase Inhibitors	101
	5.3	Conclusions	105
	Acknowledgements		106
	References		106
Chapter 6	Human *in Vitro* ADMET and Prediction of Human Pharmacokinetics and Toxicity Liabilities at the Discovery Stage		110
	Katya Tsaioun		
	6.1	Introduction	110
	6.2	The Science of ADMET	112
	6.3	The ADMET Optimisation Loop	113
	6.4	Impact of Early Human Pharmacokinetics Prediction	116

6.5	New Human ADMET Prediction Tools	117
	6.5.1 The Blood-Brain Barrier Challenge	117
	6.5.2 Mechanisms of Human Toxicity	121
	6.5.3 The Power of Multiparametric Screening in Toxicology	125
6.6	Bridging the Gap between *in Vitro* and *in Vivo*	126
6.7	Conclusions and Future Directions	128
References		129

Chapter 7 'Body-on-a-Chip' Technology and Supporting Microfluidics — 132
A. S. T. Smith, C. J. Long, C. McAleer, X. Guo, M. Esch, J. M. Prot, M. L. Shuler, and J. J. Hickman

7.1	Introduction	132
7.2	Multi-Organ *in Vitro* Models	133
7.3	Functional Single-Organ *in Vitro* Models	136
	7.3.1 Heart/Cardiac Tissue	137
	7.3.2 Lung	137
	7.3.3 Gastrointestinal (GI) Tract	139
	7.3.4 Liver	141
	7.3.5 Kidney	143
	7.3.6 Adipose Tissue	143
	7.3.7 Central Nervous System (CNS)/Peripheral Nervous System (PNS)	144
	7.3.8 Skeletal Muscle	150
	7.3.9 Blood Surrogate	150
7.4	Microfluidics	151
7.5	Concluding Remarks	153
Acknowledgements		153
References		153

Chapter 8 Utility of Human Stem Cells for Drug Discovery — 162
Satyan Chintawar, Martin Graf, and Zameel Cader

8.1	Introduction	162
8.2	Existing Approaches to Drug Discovery	164
8.3	Advances in Human Stem Cell Technology	165
	8.3.1 Embryonic Stem Cells	165
	8.3.2 Reprogramming Somatic Cells	166
8.4	iPSC-Based Disease Models	167
	8.4.1 iPSC-Based Neurological and Psychiatric Disease Models	167
	8.4.2 iPSC-Based Cardiovascular Disease Models	169
	8.4.3 Other Examples of iPSC-Based Disease Models	172
	8.4.4 Genome Editing Tools	173

8.5	Use of iPSCs in Drug Efficacy Assessment	173
	8.5.1 Target-Based Screening *versus* Phenotypic Screening	173
	8.5.2 Examples of Drug Testing to Validate iPSC-Based Models	175
	8.5.3 Drug Screens on iPSC-Based Models	176
8.6	Toxicity Testing Using iPSC-Based Models	177
	8.6.1 Cardiotoxicity	177
	8.6.2 Hepatotoxicity	178
	8.6.3 Neurotoxicity	178
8.7	Integration of iPSCs in Drug Discovery	179
	8.7.1 Challenges	179
	8.7.2 Future Directions	180
8.8	Emerging Resources of Diseased iPS Cell Lines	181
	8.8.1 StemBANCC (Stem Cells for Biological Assays of Novel Drugs and Predictive Toxicology)	181
	8.8.2 EBiSC (European Bank for Induced Pluripotent Stem Cells)	182
	8.8.3 HipSci (Human Induced Pluripotent Stem Cells Initiative)	183
8.9	Summary	183
	References	183

Chapter 9 *In Silico* Solutions for Predicting Efficacy and Toxicity   194
Glenn J. Myatt and Kevin P. Cross

9.1	Introduction	194
9.2	Representing Chemical Structures and Associated Data	196
	9.2.1 Overview	196
	9.2.2 Chemical Representation	196
	9.2.3 Biological Data Representation	199
	9.2.4 Calculated Data	199
9.3	Searching Chemical Databases	199
9.4	Calculating Molecular Fragment Descriptors	201
	9.4.1 Overview of Fragment Types	201
	9.4.2 Predefined Features	202
	9.4.3 Structural Alerts	203
	9.4.4 Common Chemical Scaffolds	204
	9.4.5 Scaffolds Associated with Data	204
9.5	Understanding Structure–Activity or Toxicity Relationships	205
	9.5.1 Overview	205
	9.5.2 Classification Based on Substructures	205

	9.5.3 Clustering	205
	9.5.4 Decision Trees	207
9.6	Quantitative Structure–Activity Relationship (QSAR) Modelling	208
	9.6.1 Overview	208
	9.6.2 Building Models	208
	9.6.3 Applying Models	210
	9.6.4 Model Validation	211
	9.6.5 Explaining the Results	211
9.7	Regulatory *in Silico* Case Study	212
9.8	Prediction of Human Adverse Events Case Study	215
9.9	Conclusions	216
	References	217

Chapter 10 *In Silico* Organ Modelling in Predicting Efficacy and Safety of New Medicines 219
Blanca Rodriguez

10.1	Introduction	219
10.2	State-of-the-Art Computational Cardiac Electrophysiology	220
	10.2.1 Single Cardiomyocyte Models	220
	10.2.2 Whole Organ Heart Models	223
	10.2.3 Simulating Interactions of Medicines with Ionic Channels	226
10.3	Investigating Variability: Population of Models Approach	230
10.4	Validation of *in Silico* Models and Simulations	234
	References	236

Chapter 11 Human Microdosing/Phase 0 Studies to Accelerate Drug Development 241
R. Colin Garner

11.1	Introduction	241
11.2	What is Microdosing (Human Phase 0 Studies) and Where Does it Add Value?	243
11.3	Scientific, Regulatory and Ethical Aspects of Microdosing	244
	11.3.1 Scientific Considerations	244
	11.3.2 Regulatory	245
	11.3.3 Ethical Considerations in Conducting Microdose Studies	247
11.4	Analytical Technologies used for Microdosing Studies	247
11.5	Applications of Microdosing	249

		11.5.1	Example of a Microdosing Study with Several Molecules and Practical Implications	249
		11.5.2	Microdosing Use for Clarification of Conflicts in Metabolism between *in Vitro* and Animal Models for Human PK Prediction	251
		11.5.3	Microdosing to Assist in Determining the Starting Dose in a Phase 1 Study	251
	11.6	Microdosing and Drug Targeting		252
	11.7	Special Applications of Microdosing		255
		11.7.1	Use in Paediatric Studies	255
		11.7.2	Use of AMS in Fluxomics	257
		11.7.3	Microdosing of Protein Therapeutics	257
		11.7.4	Drug–Drug Interaction Studies	258
		11.7.5	Absolute Bioavailability Determination	259
	11.8	Conclusions		261
	References			262

Subject Index   267

CHAPTER 1

Access to Human Cells and Tissues

GERRY THOMAS

Department of Surgery and Cancer, Imperial College London, Room 11L05, Charing Cross Hospital, Fulham Palace Road, London W6 8RF, UK
E-mail: gerry.thomas@imperial.ac.uk

1.1 Ethics and Law Regarding Collection of Human Samples

The human body is a treasure trove of biological material that can potentially be used to further our understanding of how the human body functions in both health and disease. In addition to providing specimens that help doctors to decide whether the donor themselves is healthy and, if not, what drugs or treatment we may require to improve the donor's health, biological material left over from diagnosis has the potential to provide information on biomarkers that may be useful in developing better diagnostic procedures for use in other individuals. However, the way in which human samples are used in biomedical research raises a host of ethical and legal issues that have concerned lawyers, policy makers, clinicians and scientists for a great many years.

1.1.1 The Four Cornerstones of Ethics and Biological Samples

The four important ethical principles that guide the collection and use of human biological material for research are autonomy, beneficence, non-maleficence and justice. Briefly, in relation to human material, autonomy

RSC Drug Discovery Series No. 41
Human-based Systems for Translational Research
Edited by Robert Coleman
© The Royal Society of Chemistry 2015
Published by the Royal Society of Chemistry, www.rsc.org

relates to the ability of the donor to consent or not to consent to collection and use of their material for research. Beneficence and non-maleficence are considered together and mean the minimisation of risk to the donor, either from the physical effect of providing the sample, or protection from knowledge of medical information relating to the donor being provided to an inappropriate person, for example the identification of an individual donor to someone without the appropriate clinical duty of care. This is weighed against the possible benefit to the patient, or to future patients. In many cases, providing there is appropriate protection relating to donor identification, the risk to the donor is very small, particularly in the case of obtaining material that is left over from a clinical procedure, such as an operation. When material from a patient is used for diagnosis there is a clear and immediate benefit to the donor. However, with regard to research, the benefit to the individual donor is likely to be minimal, if there is a benefit at all. However, there is potentially a much greater benefit to the wider patient community, in particular the future patient. With respect to biological samples it is perhaps more difficult to see where justice comes into the picture. It could be argued that it would be ethically wrong to collect biological samples only from one population, therefore providing the benefits of better treatment to only a limited population of patients. This argument could be used to support wider collection of samples from donors from different ethnic groups within a society, or to encourage collection of samples from rarer disease types. The element of justice could also be used to encourage generalised access to samples once collected, rather than use being restricted to a single 'owner' of a collection for his or her research.

1.1.2 Legal Issues

There are two aspects to the law relating to human tissue. One aspect is how material is obtained and the information provided to research participants to allow them to exert their autonomy with regard to participation. The second relates to the use of data related to individuals or tissue samples. Human tissue usually requires some degree of annotation to make it useful for research projects. The question that is being posed by the researcher usually dictates the degree of annotation required. For example, it may be sufficient in some circumstances to know that the material comes from a female of age 42. In other circumstances, where the aim is to identify a biomarker that associates with drug response, it may be necessary to have details on the patient themselves (age, gender, race), clinical presentation, treatment and outcome. In general, the larger the degree of annotation required, the greater the risk of accidental identification of the donor, especially where the disease type is rare. The avoidance of accidental identification is achieved by tighter regulation of who has access to data, and in particular, the need for anonymisation of that data. Where longitudinal data is required but is not available at the time at which the sample is provided to the researcher, it will be necessary to pseudo-anonymise the sample. This means that the researcher is unable to identify the donor, but a trusted intermediary, such as a tissue bank, can hold an identifier that can be linked to personal identifiers on, for example, a clinical database, can used to acquire more data on the donor as time progresses.

1.1.2.1 Legal Regulation of Human Participation in Research

The laws governing the participation of humans in research, including the use of material derived from humans, have their origin in the Declaration of Helsinki, which was adopted by the World Medical Association as a set of ethical principles in 1964.[1] Although the Declaration itself was not a legally binding instrument, it has influenced the development of national laws and guidelines on the use of humans and human tissue in research. In the USA, the regulation that addresses the protection of human research participants is referred to as the Common Rule.[2] The Common Rule is interpreted and enforced by the Office of Human Research Protection (OHRP), which is a division of the Department of Health and Human Services (HHS). Institutions both within and outside the USA that are engaged in research which is supported financially by the HHS must submit a Federal-wide Assurance to the OHRP. This assurance states that the institution will comply with human research participant regulations of all federal agencies. In the USA, most universities have agreed to apply the Common Rule to all human research, not just to that supported financially by the US Government.

The Common Rule sets out the composition, function and role of Institutional Review Boards (IRBs) in protecting human participants in research activities. This includes requirements for obtaining informed consent from humans involved in research, and the need for special protection for vulnerable populations, *e.g.* pregnant patients, children, prisoners, *etc*. One thing that the Common Rule does not do is address the issue of who owns human tissue. It does not apply to material obtained post-mortem; the Health Insurance Portability and Accountability Act (HIPPA) of 1996 does regulate the use of information related to post-mortem material.

In the UK, the Human Tissue Act 2004 repealed and replaced the Human Tissue Act 1961, the Anatomy Act 1984 and the Human Organ Transplants Act 1989 as they relate to England and Wales, and the corresponding Orders in Northern Ireland. The Act makes consent the fundamental principle underpinning the lawful storage and use of body parts, organs and tissue from the living or the deceased for specified health-related purposes and public display. It lists the purposes for which consent is required (the so-called 'scheduled purposes'). The Act notably prohibited private individuals from covertly collecting biological samples (hair, fingernails, *etc.*) for DNA analysis, but excluded medical and criminal investigations from the offence.[3] One important exemption that affects the research community is that explicit consent is not required for the analysis of DNA when this is related to research into the working of the human body, providing that the sample is anonymised. The Act came into force on 1 September 2006. Anyone wishing to hold collections of human samples containing cells must do so under a licence issued by the Human Tissue Authority, who are responsible for administering the Act. However, there is an exemption that relates to holding anonymised tissue samples that are covered for a specific use with approval from the National Research Ethics Service (NRES) in England and Wales. Where consent includes the future use of samples in undefined research (*e.g.* for 'research into cancer'), samples must be held under a licence and can only be issued for research following approval by a review board to ensure ethical use of samples (see Section 1.3.1 on access policies). In contrast to the legislation in the USA, this law does apply to material taken post-mortem.

The Act covers a number of different sectors, including research, and the Human Tissue Authority provides detailed guidance for each of the sectors.[4]

The Act also includes responsibility for licensing the public display of whole bodies, body parts and human tissue from those who were less than 100 years old on the date the Act came into place. This effectively means that pathology museum collections that display materials taken from the living and deceased, usually in formalin, are required to hold an appropriate licence, whereas museums that house archaeological material are not.

1.1.2.2 Ownership of Human Samples

One area that often sparks discussion and is responsible for at least some of the restriction on access to human samples is the ownership of human material. Although it is difficult to pin a monetary value to an individual human sample, processing of that tissue to a cell line, or the use of multiple samples to identify novel drug targets can lead to considerable monetary gain. For this reason, courts have been asked to consider the question of whether an individual retains ownership of their excised tissue. Retention of ownership might authorise an individual to share in the profits of any commercialisation of research results and dictate who controls access to the samples for any future research. The arguments centre on the difference between 'ownership' and 'guardianship' (or 'bailment' in the USA) of samples. Guardianship is different from the sale (which implies 'ownership') or donation or gift of a specimen, which would result in a relinquishment of ownership, in that it relates to transfer of possession of a specimen. In the UK the current HTA guidance suggests that donors retain some rights of ownership in that they can withdraw their consent for their samples to be used at any time. Property owners generally have the right to use, sell, transfer, exchange or destroy their property at any time.

The majority of case law in this area comes from the USA, where in most cases involving tissue excised for clinical purposes and tissue donated for research, the courts have concluded that patients and other research study participants do not retain ownership of their samples. Allen *et al.*[5] provide a summary of the legal decisions in a number of important US cases. The important lessons that come from these cases are that where tissue has been given voluntarily for research and there is no expectation of return, the donors have no rights of ownership in the tissue itself or research conducted on it. Another important case, which involved a dispute between a researcher who wished to transfer to his new employer a substantial collection of samples from patients obtained over a 20 year period and his former employer (Washington University v Catalona, 490F 3d 667 2007: referenced in Allen et al.[5]), supported this viewpoint. In this case, the researcher wrote to his donors and sought their permission to release their samples to him in order to transfer their material to his new employer. The courts ruled that, despite the wording in the consent form that the donors retained the right to request that their samples be destroyed, or to withdraw their consent, that they could not request that their samples should be transferred to a different institution. The courts also ruled that the samples could not be legally returned to the donors because of the laws that govern proper handling of clinical waste – this significantly undermined the donors' claims to ownership of their samples. Interestingly, in the case of Moore v Regents of California in 1990, where a patient's cells were used for research without his knowledge, although the courts did not support the patient's property rights, the patient could have pursued a claim relating to the physician's failure to obtain consent for research use of his material.

In the case of material that is left over from the diagnostic process, *e.g.* formalin-fixed, paraffin-embedded tissue, there are currently no laws that exist regarding ownership of these samples. These samples, particularly so in the case of paraffin blocks, are often retained by hospitals for long periods of time, often in off-site storage facilities. Many bioethicists regard this as 'abandoned' material, and as such can be used for research subject to adherence to appropriate ethical and legal guidelines. Both the Common Rule in the USA and the HTA regulations in the UK permit the use of this material, providing that information relating to the specimen is recorded in a manner that does not permit identification of the donor. This material, which was not obtained with the primary intent of research, can be used in research with the appropriate ethics approvals in the absence of patient consent in both the USA and UK.

Although the cases reported above indicate that donors may not 'own' their donated specimens, other case law indicates that donors may retain rights to determine the use of their material. In one of the most recent cases, members of the Havasupai tribe in Arizona sued Arizona State University when blood samples that had been consented for use in diabetes research were used in studies on mental health, inbreeding and population migration.[6] In the end, the University Board of Regents agreed to pay $700,000 to the tribe members, as well as providing other forms of assistance to the tribe, and return the blood samples. This and a number of other cases, including that of Moore v Regents of California in 1990, all highlight the need for transparency of use and the importance of informed consent.

1.1.2.3 Informed Consent for Research Use

Obtaining appropriate consent for current and future use of human samples in research is clearly of paramount importance if researchers are to stay on the right side of the law. The HTA in the UK makes consent paramount and reflects the 'autonomy' element of ethics. However, there are many situations in which consent may be difficult or impractical to obtain. Here it may be possible to obtain permission from an ethics committee to carry out research on anonymised samples without consent from the original donor. The wording of the patient information sheet and the consent form that the patient signs is very important with respect to the use to which the material collected from the donor is later used. There remains (a sometimes heated) debate on what constitutes adequate informed consent and what form of consent is appropriate in different situations.[7,8] Generic or broad consent is recommended by both the Human Tissue Authority[9] and the National Research Ethics Service in the UK,[10] and is endorsed by the UK research funders,[11] and the Nuffield Council on Bioethics.[12] In a review reporting the results of studies comprising some 33 000 people in a number of different countries over a 10 year period, 80% of people were willing to donate a sample with generic consent if asked. This rose to 93–99% of people when the participants were asked specifically about donation of leftover samples.[13] This approach provides the greatest flexibility in use of the samples. One common criticism is that this approach does not provide 'informed' consent as future research uses are not defined at the time of consent.[14] It also seems that public opinion remains divided on this issue.[15–19] Public awareness of the important role that donated human specimens play in research needs to be increased, and it will be extremely important to maintain public trust and confidence in the consent procedures used to increase public support for research using human samples.

There are a number of different consent models that can be adopted (reviewed in Lewis *et al.*[20]). Initial consent is either opt-in, where the donor actively agrees to the storage and use of their samples in research, or the opt-out model where samples may be collected and used unless the donor specifically expresses a preference that their samples should not be used. There then follows a number of refinements to the opt-in consent model, whereby donors can consent once for life, consent may be obtained at certain points for use of residual samples in research, *e.g.* at each specified time point, *e.g.* every 5 years, or at the beginning of an additional episode of care (*e.g.* consent once for blood samples to be donated at each follow-up visit for cancer treatment), or consent every time an extra sample from that patient becomes available (*e.g.* the patient is consented at each follow-up visit). The second element of consent relates to consent for use of samples. Generic consent is a one-off consent, where the donor agrees to a trusted third party (*e.g.* ethics committee or a project approval board) making a decision on their behalf as to whether their samples are used appropriately in a specific project. This type of consent can cover any type of study in any category of research. A tiered consent enables the donor to specify that they do or do not wish their samples to be used in specific research areas (*e.g.* genetic research), while still permitting decisions to be made on their behalf as to which particular projects their samples can be used for. These are the most flexible approaches, but with tiered consent, particular care must be taken in sample management to ensure that the donor's wishes are respected.

Two other forms of consent that have been used, particularly in the past, are a once-only specific consent, where the donor consents only for use in a very specific study (*e.g.* study of protein A only in their sample), and a model where the donor is re-contacted for specific consent for every new study using their sample. These are particularly restrictive in terms of access to samples and lead to delays due to the requirement for re-contact, and wastage of valuable residual samples (once-only specific consent model). The last model is often used for large-scale population banks where longitudinal lifestyle questionnaire data and access to medical records are required. While the annotation of samples in such collections is very extensive and valuable, such collections are composed usually only of the type of biological sample that is easy to obtain (*e.g.* blood or urine) from healthy volunteers. The first type of consent (opt-in and generic) is favoured by banks that collect leftover clinical samples from patients. This is partly because patients are often approached at a time when they have just received their diagnosis, and because individuals may only be undergoing one course of treatment in the establishment where their sample is being collected. Repeated contact is therefore unlikely, and further annotation of the samples can be obtained from linkage to medical records, either electronically, or more often *via* the donor's hospital notes.

A recent project report from the Strategic Tissue Repository Alliance Through Unified Methods (STRATUM), which was supported by the UK's Technology Strategy Board and the pharmaceutical industry, provides information on the preferences of UK patients for the consent models given above.[21] Two approaches were used to assess public opinion – qualitative focus groups and a quantitative online survey. Their results indicated that an opt-in approach was preferred by survey responders, and the group speculated that this may be the approach used for organ donation in the UK. Interestingly, a preference for opt-out amongst focus group participants, particularly in those who were aged under 65, from higher social economic groups and with a higher education level, and the significant proportion (27%) of survey responders who preferred this approach, suggests that opt-out might be worth consideration. Interestingly, Wales has just adopted an opt-out model for organ donation, although elsewhere in the UK opt-in is still required. It will be interesting to see whether communities in Wales would now be more in favour of an opt-out model for donation of human biospecimens for research. Fewer than half of the participants wanted to be re-consented each time a biological sample was taken, although 30% wished to be consented at more than one time point. In addition, the STRATUM study found that the majority of people were happy to donate biosamples *via* the least restrictive model, generic consent. This finding reinforces others (see Wendler[13] for review) that have found the generic approach to be higher amongst the general public (79%) and particular patient groups (95%). While encouraging, the STRATUM study indicates that there is still a requirement for continued public engagement in order to encourage more of the population to be willing to donate biological material using the model that facilitates greatest access to them.

1.1.2.4 What Should Consent Cover?

There are a number of key areas that donors must be aware of to ensure that both their rights and the rights of researchers are protected. Donors have a right to withdraw their permission for use of their stored samples at any point. At the time of taking consent, they must be made aware that if they withdraw their consent sometime after the sample was taken, some or all of it may already have been used for research, and it may be impossible to withdraw data related to their sample. To facilitate best use of samples, they should be made available to all types of researchers. The information provided to the patient should therefore also state that their sample may be used by researchers within or outside the institute where their sample was taken, and therefore also by researchers based in commercial enterprises. However, no matter who uses their sample, the donors themselves will not benefit financially from the use of their sample. Donors should be made aware that their sample could be used to identify genetic components to disease that may be heritable. Perhaps the most difficult issue at the current time is the return of research results. There appears to be little consensus regarding this issue[22] and there are many elements that need clarification (see Wolff[23] for a review on return of genetic results). These issues are even more complicated where results from tissue samples are concerned, especially given that patients may be much further along their disease pathway by the time the relevance of the research findings is known. Perhaps the best advice is that those taking consent should state the biobank's policy on return of research results at the time of consent.

1.2 Practical Issues Regarding Human Tissue Collection and Annotation

Streamlined access to high-quality, highly annotated human material is essential for clinical science to advance. The largest hurdle to overcome is that the majority of tissue-based research is carried out outside hospitals whereas tissue acquisition is carried out inside the healthcare system. Our healthcare systems are usually risk averse (in many ways rightly so) and are largely undermanned and demoralised in the key areas dealing with tissue acquisition, namely pathology departments. The welfare of the patient is of prime concern, and although it is recognised that better treatments can only be developed from good research, the priority of hospital staff is to ensure that patient safety is paramount. Given the appropriate resources and institutional support, there is ample opportunity to integrate biosample acquisition into clinical care pathways. However, in order to be successful in this approach, a thorough understanding of the clinical pathway is necessary. This is particularly important when running multi-site studies – practice is often different in different hospitals.

Population banks that house collections usually of either blood, urine or both, are by comparison somewhat easier to run, but they share the issues of managing consent, and collation of data. However, these collections are often obtained outside the healthcare system and run as standalone projects.

1.2.1 Obtaining Tissue with Consent for Research

There is an opportunity to obtain biological material at the same time that similar material is being removed from the patient for diagnostic purposes (see Figure 1.1); either material that is left over after the diagnostic procedure has been completed, or extra samples may be taken specifically for research. This can only be achieved with the consent of the patient. Obtaining that consent can either be integrated into consent for the procedure, or a separate consent may be required. While written consent may be obtained for a procedure that involves anaesthesia, often those that require the cooperation of the patient (for example taking a blood sample) have no requirement for written consent to be taken by the person taking the diagnostic sample. Documenting consent and ensuring that consent is informed, particularly in the latter scenario, may therefore require additional staff or staff time to that required for the diagnostic process.

Once a sample has been taken, it requires further processing in the laboratory. The nature and, therefore, the cost of this processing (freezing, processing to paraffin, extracting analytes (*e.g.* serum from blood or RNA/DNA from tissue) depends on the individual collection site. However, all processes require annotation that is unlikely to be included in routine practice, and therefore are likely to be provided at an additional cost.

All of these processes require the cooperation of an interdisciplinary team of nurses, phlebotomists, radiologists (*e.g.* for biopsy samples), surgeons, pathologists, laboratory technical staff and, where longitudinal studies are involved, physicians or oncologists. Quality assurance throughout the chain of custody is important for both data and specimen quality. Additional clinical annotation throughout the patient pathway may be possible *via* integration with the electronic patient record. However, it is not uncommon for the patient pathway to involve multiple database systems dependent on the time point in the patient pathway, which can lead to a considerable headache in tying all the information together. Often the only way to solve this issue, and to collate items of data that do not get recorded in busy clinics, is to resort to the patient's paper notes. This again requires specialist staff time.

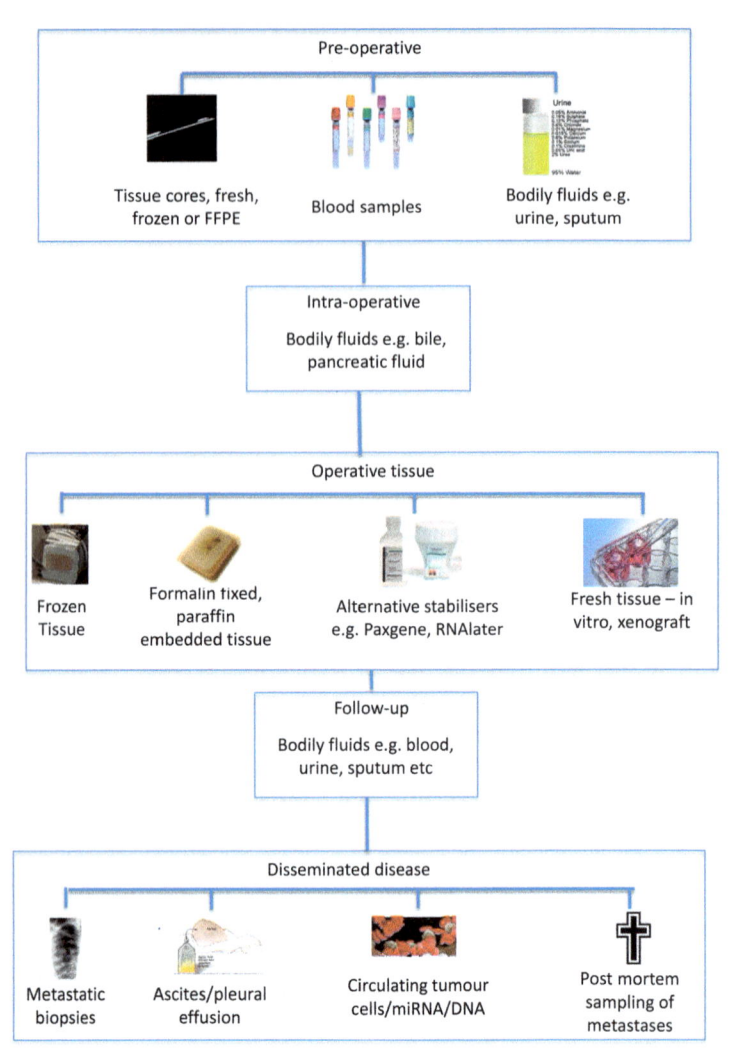

Figure 1.1 Opportunities for obtaining samples during a cancer patient pathway.

In addition, there should be a considerable investment in a database that is capable of storing the clinical annotation, information on sample pre-analytical variables (*e.g.* time taken between blood separation for serum, or cold ischaemic time for surgical tissue), and tracking samples taken at different time points from the same patient. Although many off-the-shelf versions of these databases exist, they often require adaptation for the specific ways in which patient pathways and laboratory practices operate in specific institutions. This adaptation can be costly. The alternative is to design a bespoke database from scratch, but this is not a trivial undertaking.

1.2.2 Accessing Material from a Diagnostic Archive

In the majority of hospitals, there is no facility to obtain generic consent from all patients undergoing operative procedures. However, material that has been obtained with diagnostic intent is a valuable source of research material, and comes with rich pathological annotation. Following completion of the diagnostic procedure there is often a reasonable amount of tissue remaining, but it comes without the consent of the patient for research use.

The HTA and UK NRES have agreed a position whereby NHS Research Ethics Committees (RECs) can give generic ethical approval for a research tissue bank's arrangements for collection, storage and release of tissue, providing the tissue in the bank is stored on HTA-licensed premises. This approval can extend to specific projects receiving non-identifiable tissue from the bank. Whenever identifiable tissue is released for research from a diagnostic archive, it must only be released in accordance with the donor's consent; unless it was stored prior to implementation of the HTA Act on 1 September 2006, in which case consent is not required, as it is regarded as an 'existing holding'.

Tissue that has not been consented for research (other than existing holdings) can only be released if it is from a living person, and the researcher is not in possession, and not likely to come into possession of information that identifies the person from whom it has come; and where the material is released by a research tissue bank with generic ethical approval from an REC for research within the terms of the approval; or it is to be used for a specific research project approved by an REC.[24]

In short, this means that any hospital archive may be registered as a research tissue bank *via* NRES, providing the material is held under an HTA licence, with all the aspects of traceability this requires. Material can be issued to researchers that have either an REC approved project or, if the tissue bank has sought approval from the REC of a mechanism of review to provide 'deemed' ethics approval (see the websites of the Wales Cancer Bank[25] and the Imperial College Healthcare Tissue Bank[26] for examples of this). In all cases, the material used for research must be anonymised.

Clinical annotation of these samples requires access to secure databases on hospital servers and as many research projects involve clinicians, there may be occasions when a clinician involved in research has access to a secure database that would permit identification of a sample and the identity of the patient whose material is being used. Providing the research material is not identifiable to the researcher (*e.g.* coded by a laboratory accession number) and the researcher does not seek to link the tissue to the patient, the sample will still be regarded as non-identifiable and the research will be permissible without consent if it is given ethical approval by an REC.

All of the above should serve to illustrate that obtaining human clinical biological samples is fraught with difficulty. However, the problems are not insoluble and considerable effort is being put into extended frameworks for access to human tissue across Europe. The Biobanking and BioMolecular

Resources Infrastructure (BBMRI: http://bbmri.eu/en_GB), funded initially under the EC's Framework 7, seeks to integrate large biobanks across Europe, with the aim of facilitating across-border access to human samples. This is a very ambitious project that will require considerable support from national hubs in the majority of member states to fulfil its potential. The project completed its preparatory and information gathering stage, and was recently re-funded as a European Research Infrastructure Consortium.

There certainly appears to be no shortage of tissue banks, but accessing the right type of sample from the right type of patient with adequate annotation for specific research projects still remains an issue.

1.3 Improving Access and Annotation

1.3.1 Access Policies

An enormous amount of time and effort goes into the collection and documentation of human tissue for research. By taking the consent of patients for use of their material in research, the biobanker in effect gives an undertaking that the material will be put to good use. Usage depends on the visibility of the sample to the research community, and the procedure to access the sample. There are many online directories of biobanks that collect human tissue, *e.g.* the German Biobank registry[27] or the National Cancer Research Institute (NCRI) biosample directory,[28] but few provide the ability to query the collection online to look for specific samples, let alone enable cross-tissue bank searches for rarer disease types. A few banks are starting to provide online searchable databases,[29-31] but these enable searches only of single tissue banks. In reality, and particularly in the case of rarer types of tissue, a single tissue bank will not hold sufficient samples. The ability to search across different banks and to have one portal for access to samples would greatly facilitate research. The BBMRI have the first of this type of search facility[32] but progress still needs to be made on a single access procedure to obtain material. All of the search portals require an account with a user name and password before searching of the database can commence – this is usually so that appropriate metrics can be collected and so that access is restricted to bona fide researchers.

Access procedures are necessary to strike an appropriate balance between protecting the rights of the biobank, and its donors, and making material from the biobank available to researchers. Access committees are usually established either from the individual biobank's steering committee, or an independent committee comprised of scientists with an appropriate knowledge of the uses of clinical material in research. A member of the lay community is usually either included as a member of the committee or to chair the committee. The lay member serves effectively as the patient or donor voice. In order to function appropriately, access committees should be provided with a clear mandate and specific criteria that inform their decision-making process and procedures that allow them to conduct a fair evaluation of all access requests. A policy for dealing with situations that arise from applicants contesting the decision of the access committee also need to be in place. Transparency of the decision-making process is paramount. A useful document that sets out the principles of access to human biological samples has been produced by the STRATUM project,[33] and an access policy template is available from the website of the NCRI.[34] There is a fine line between access policies being seen as obstructive to research and facilitating research. The population biobanking community has devoted a lot of time to developing appropriate strategies and these have been recently reviewed.[35]

The mechanism for ensuring compliance with the rules of access stipulated by a particular biobank are Material and Data Transfer Agreements (MDTAs). MDTAs should state the conditions that are imposed by the biobank on the use of the samples. Typically these include that the researcher should make no attempt to identify the donors of the samples they receive, that they accept the risks associated with handling human biological material, that they will store the material in a way appropriate to local laws and ethics regulations, that they will only use the samples they receive for the purpose approved in their application for use, and that they will acknowledge the biobank as the supplier of their samples in an appropriate manner. MDTAs may also include a requirement to return research data generated from the samples to the biobank, and whether unused samples should be returned to the biobank or destroyed by the researcher. The MDTA should be signed by the researcher prior to release of samples, and a list of samples provided should accompany the MDTA. This enables tracking of material out of the biobank and provides a way of linking data generated from different samples to different research groups back to the individual patient. This is particularly important when the biobank is requesting return of research data.

1.3.2 Annotation

Use of human biological samples can be maximised by the addition of data. For a clinical sample this can include data on the pathological nature of the sample, the treatment the patient subsequently undergoes, and the outcome of that treatment. For population biobanks the annotation can include any related clinical history, plus further information on lifestyle factors. In addition, information on pre-analytical variables[36] can be invaluable in defining what elements of the research data are related to this rather than to any other variable of the study. Annotation of samples is time-consuming. Where possible information should be obtained from clinical electronic records, but in practice it is difficult to persuade risk-averse and cash-strapped NHS hospitals to provide downloads of the information held on their secure networks and often the data elements useful for research are not recorded on the relevant hospital databases. Many of the hospital databases are designed more to monitor the flow of patients through the system than to record data relevant to researchers. Full annotation of samples by biobank staff is therefore likely to be a costly exercise and alternative partnerships to enable access to longitudinal data need to be sought. One possibility is to facilitate linkage of biobank information to large datasets that are collected nationally to enable strategic planning. One such example, currently in pilot phase, is to link information on clinical treatment and outcome from data collected by the National Cancer Registry Service (NCRS) with data on sample availability from tissue banks. NHS hospitals in England are obliged to make a comprehensive dataset available to the NCRS on all of the cancer patients they are treating.[37] Patients donating samples to hospital-based tissue banks are asked for their consent for their medical notes to be accessed to provide annotation of any clinical samples they may donate for research. Providing a secure mechanism is in place to protect the personal identity of the donor, provision of this facility would greatly enhance annotation of cancer biobanks.

1.4 Summary

Access to human biological samples is a priority in order to understand biological mechanisms in the species of choice, man. Considerable effort has gone into public engagement in this area, and needs to continue. Models of good practice exist that show access to material is possible, and can be done in a legal and ethical manner. However, researchers must understand that we cannot treat man as we do animals in research, and that restrictions on the type of material available may need to be in force to minimise risk to the donor. Funders need to understand that it is not only the collection of the material but the annotation of the sample that is important, and that research takes time to bear fruit. The cost of collecting and annotating human material is considerable, but the cost in terms of lack of improvement in targeting health care appropriately in future is likely to be considerably greater. It is the responsibility of tissue banks to ensure that there are improvements in access to biological samples already obtained from donors. At the time of taking consent an agreement was made with the donor that their material would be used in research. There is an obligation therefore to ensure that maximum use is made of their samples to benefit those who will come after us.

References

1. World Medical Association Declaration of Helsinki: Ethical Principles for Medical Research Involving Human Subjects [online]. Available at: www.wma.net/en/30publications/10policies/b3/17c.pdf (accessed 8 August 2014).
2. US HHS. Code of Federal Regulations. Title 21. http://www.hhs.gov/ohrp/humansubjects/commonrule/ (accessed 11 September 2014).

3. Department of Health. Human Tissue Act 2004 [online]. Available at: www.legislation.gov.uk/ukpga/2004/30/pdfs/ukpga_20040030_en.pdf (accessed 8 August 2014).
4. Human Tissue Authority. Licensing and inspections [online]. Available at: www.hta.gov.uk/licensingandinspections.cfm (accessed 8 August 2014).
5. M. J. Allen, M. L. E. Powers, K. S. Gronowski and A. M. Gronowski, *Clin. Chem.*, 2010, **56**, 1675.
6. L. Andrews, The Havasupai Case: Research without Patient Consent [online]. Available at: www.whoownsyourbody.org/havasupai.html (accessed 8 August 2014).
7. M. G. Hansson, J. Dillner, C. R. Bartram, *et al.*, *Lancet Oncol.*, 2006, **7**, 266.
8. K. Hoeyer, *Public Health Genomics,* 2010, **13**, 345.
9. Human Tissue Authority. Consent exemptions from the Human Tissue Act 2004 [online]. Available at: http://www.hta.gov.uk/legislationpolicies andcodesofpractice/definitionofrelevantmaterial/consentexemptions.cfm (accessed 11 September 2014).
10. Integrated Research Application System. Guidance document for researchers [online]. Available at: https://www.myresearchproject.org.uk/Help/Help%20Documents/PdfDocuments/researchtissuebank.pdf (accessed 9 September 2014).
11. Medical Research Council (MRC) and National Cancer Research Institute (NCRI). UK Funders' Vision for Human Tissue Resources [online]. Available at: http://www.ukcrc.org/wp-content/uploads/2014/03/Vision+for+human+tissue+resources.pdf (accessed 8 August 2014).
12. Nuffield Council on Bioethics. *Human bodies: donations for medicine and research*. 2011 http://nuffieldbioethics.org/wp-content/uploads/2014/07/Donation_full_report.pdf (accessed 11 September 2014).
13. D. Wendler, *Br. Med. J.*, 2006, **332**, 544.
14. C. Petrini, *Soc. Sci. Med.*, 2010, **70**, 217.
15. A. Tupasela, S. Sihvo, K. Snell, P. Jallinoja, A. R. Aro and E. Hemminki, *Scand. J. Public Health,* 2010, **38**, 46.
16. D. Kaufman, J. Bollinger, R. Dvoskin and J. Scott, *Genet. Med.*, 2012, **14**, 787.
17. C. M. Simon, J. L'Heureux, J. C. Murray, P. Winokur, G. Weiner, E. Newbury, L. Shinkunas and B. Zimmerman, *Genet. Med.,* 2011, **13**, 821.
18. D. J. Willison, M. Swinton, L. Schwartz, J. Abelson, C. Charles, D. Northrup, J. Cheng and L. Thabane, *BMC Med. Ethics,* 2008, **9**, 18.
19. Å. Kettis-Lindblad, L. Ring, E. Viberth and M. G. Hansson, *Scand. J. Public Health,* 2007, **35**, 148.
20. C. Lewis, M. Clotworthy, S. Hilton, C. Magee, M. J. Robertson, L. J. Stubbins and J. Corfield, *BMJ Open,* 2013, **3**, e003022.
21. C. Lewis, M. Clotworthy, S. Hilton, C. Magee, M. J. Robertson, L. J. Stubbins and J. Corfield, *BMJ Open,* 2013, **3**, e003056.
22. M. H. Zawati, B. Van Ness and B. M. Knoppers, *GenEdit,* 2011, **9**, 1.
23. S. M. Wolff, *Annu. Rev. Genomics Hum. Genet.*, 2013, **14**, 557.

24. Human Tissue Authority. Position statement on diagnostic archives [online]. Available at: www.hta.gov.uk/legislationpoliciesandcodesofpractice/positionstatementondiagnosticarchivesreleasingtissueforresearch.cfm (accessed 8 August 2014).
25. Wales Cancer Bank. Available at: http://www.walescancerbank.com/conditions-of-sample-supply.htm (accessed 11 September 2014).
26. Imperial College London. Imperial Tissue Bank: Accessing materials for use in research [online]. Available at: www.imperial.ac.uk/tissuebank/usingtissue/accessingmaterials (accessed 8 August 2014).
27. German Biobank Registry. Available at www.tmf-ev.de/BiobankenRegisterEN/Registry.aspx (accessed 8 August 2014).
28. NCRI Cancer Biosample Directory. Search biosample collections [online]. Available at: http://biosampledirectory.ncri.org.uk (accessed 8 August 2014).
29. G. Thomas, K. Unger, M. Krznaric, A. Galpine, J. Bethel, C. Tomlinson, M. Woodbridge and S. Butcher, *Genes*, 2012, **3**, 278.
30. Wales Cancer Bank Search Facility. Available at: https://walescancerbank.swan.ac.uk/StartForm.ashx (accessed 9 Nov 2013).
31. Breast Cancer Campaign Tissue Bank. Available at: https://breastcancertissuebank.org/bcc/tissueBank (accessed 9 Nov 2013).
32. Biobanking and BioMolecular Resources Infrastructure. Available at: https://www.bbmriportal.eu/bbmri2.0/jsp/bbmri/coresearch.jsf (accessed 9 Nov 2013).
33. Strategic Tissue Repository Alliances Through Unified Methods (STRATUM). Principles of Access to Human Biological Sample Resources in the UK [online]. Available at: www.stratumbiobanking.org/docs/STRATUM%20Access%20Policy%20Principles%20Final%20June%2024%202013.pdf (accessed 8 August 2014).
34. National Cancer Research Institute Template for Access. Available at: http://www.ncri.org.uk/wp-content/uploads/2013/09/Initiatives-Biobanking-2-Access-template.pdf (accessed 11 September 2014).
35. S. Fortin, S. Pathmasiri, R. Grintuch and M. Deschesnes, *Public Health Genomics*, 2011, **14**, 104.
36. F. Betsou, S. Lehmann, G. Ashton, M. Barnes, E. E. Benson, D. Coppola, Y. DeSouza, J. Eliason, B. Glazer, F. Guadagni, K. Harding, D. J. Horsfall, C. Kleeberger, U. Nanni, A. Prasad, K. Shea, A. Skubitz, S. Somiari and E. Gunter, *Cancer Epidemiol., Biomarkers Prev.*, 2010, **19**, 1004.
37. Health & Social Care Information Centre. Data sets [online]. Available at: http://www.hscic.gov.uk/cancerdataset (accessed 11 September 2014).

CHAPTER 2

Functional Studies with Human Isolated Tissues to Better Predict Clinical Safety and Efficacy

DAVID C. BUNTON

Biopta Ltd, Weipers Centre, Bearsden Road, Glasgow G61 1QH, UK
E-mail: davidbunton@biopta.com

2.1 Introduction

Research using human fresh tissue represents one of the fastest growing areas of drug discovery and development. There are two key drivers in the use of human tissue: firstly, the failure of the current approach to drug development which demands new approaches to reduce clinical attrition, and secondly, the drive towards biomarkers for personalised medicine.

The dominant approach to drug development, based on primary screening in high-throughput models and secondary screening in animals, has previously produced numerous 'blockbuster' drugs, but clinical attrition rates of 95% are no longer viewed as sustainable. Human disease-relevant tissue is increasingly viewed as a way to decrease clinical failures, particularly during phase II and III where poor efficacy has been partly attributed to an over-reliance on animal models. The second major factor is the drive towards the use of biomarkers and personalised medicines; as the search for blockbusters diminishes, the need for targeted therapies based on predictive non-clinical and clinical human data increases (Table 2.1).

Fresh, intact, functional human tissue assays aim to bridge the gaps between *in vitro* cell-based studies, *in vivo* animal studies and clinical trials. Such tissues offer advantages over simpler cell-based models, avoid species differences and truly reflect the diverse patient population. For example, cell-based assays lose functional relevance and do not retain important cell-to-cell relationships in a 3-D structure. Reconstructed or engineered 3-D tissues produced from stem cells or as reconstructed organoids fail to reflect the actual disease phenotype and diversity of responses found in healthy and diseased tissues obtained directly from patients.

Functional tissue studies also offer advantages over animal experiments. Clearly, animal *in vivo* experiments provide important information on drug behaviour at a system level, and the industry will continue to rely on these models to investigate central nervous system control of the cardiovascular, respiratory and gastrointestinal systems. Animal models do, however, present one major problem. Few, if any, can truly be considered human disease models that accurately and comprehensively reflect human pathophysiology. At best they are complex models representing some mechanistic features of a disease process, typically with artificial initiators of the 'disease' process (*e.g.* chemical irritation) and certainly not the chronic timeframe of most major diseases. Fresh human tissues, obtained directly from the target patient population, avoid many of these problems and while there are some limitations in the range of experiments that can be conducted, there appears to be an urgent need for human tissue studies to bridge the gap between cell-based approaches, *in vivo* animal experiments and clinical trials.

Human tissue studies are also of value for non-clinical safety pharmacology studies in support of ICHS7A and ICHS6 guidelines on non-clinical safety studies for pharmaceuticals and biotechnology products, respectively. Information generated by human tissues may even be included in a clinical trial application or marketing authorisation application. For example, the FDA review of gastrointestinal drugs used studies in fresh human coronary arteries to evaluate the safety of 5-HT$_4$ agonists (Table 2.2).[1]

Despite these encouraging developments, functional human tissue assays represent something of a conundrum: fresh intact human tissues are generally accepted as the gold standard non-clinical models; however, the tests are not yet deemed an essential part of the non-clinical development of most drugs.

The present chapter describes the wide range of uses of human isolated fresh tissues throughout drug development, including their traditional role in target discovery and validation, through to their increasing use as a functional model during lead optimisation, safety pharmacology and preclinical strategies for stratified medicines. The chapter also describes some of the specific challenges in acquiring and using fresh tissues.

Functional Studies with Human Isolated Tissues

Table 2.1 Sources of human fresh isolated tissue for research

Sources of human fresh isolated tissue	Organisation/Provider	Typical applications	Advantages	Disadvantages
Post mortem	Pathology department or biobank	Target identification and validation rather than functional studies	A wide range of tissues are available which are difficult or impossible to obtain from surgery	Post mortem interval (PMI) for collection of tissue is typically hours rather than minutes; tissue quality may be compromised. Not suitable for most functional assessments but may be the only available access route
Residual surgical	Surgical department, pathology or biobank	Target identification and validation; studies in target disease tissue; functional studies of safety and efficacy	Rapid collection of tissues; wide range of diseased tissues available which are a valuable model for drug	Main purpose of collection is diagnosis, aligning collection procedure with requirements for research can be difficult – often obtain leftover tissue sample following pathologist's analysis
Non-transplantable organs	Organ procurement organisation (OPO) or biobank that liaises with OPO	Functional studies of safety and efficacy; may also be used for development of surgical implants or medical devices	Freshness and volume of tissue; ability to receive healthy control tissues	Frequency of tissue is much less than for residual material and costs are much greater
Clinical biopsies	Clinical research organisation; clinical research department of some hospitals	Biomarker studies as part of a clinical trial; collection of diseased tissues from specific patient groups, *e.g.* psoriasis or atopic dermatitis	Collection conditions can be specified exactly. Allows more extensive investigations of safety and efficacy in patients during a clinical trial, *e.g.* measurement of surrogate markers of clinical effects	Time and costs involved in organising and planning the study, very small amounts of tissue may be available and may not be practical to retrieve the tissue of interest

Table 2.2 Human tissue tests used in support of safety pharmacology submissions

Human tissue	End point	Regulatory test	Regulatory document
Cardiac muscle/ Purkinje fibres	Action potential measurements, especially duration and rhythm	Cardiovascular telemetry core battery *in vivo* animal	ICH S7A Section 2.7.2
Coronary artery	Coronary artery vasospasm leading to myocardial infarction	Cardiovascular telemetry core battery *in vivo* animal – but would only pick this up indirectly through change in blood pressure and heart rate	ICH S7A Section 2.7.2
Resistance arteries	Vasoconstriction/dilatation Control of organ blood flow and main determinant of peripheral vascular resistance (mean arterial blood pressure = cardiac output × total peripheral resistance)	Cardiovascular telemetry core battery *in vivo* animal – measures of blood pressure and blood flow (by plethysmography)	ICH S7A Section 2.7.2
Bronchi	Bronchoconstriction/dilatation	Respiratory core battery *in vivo* animal	ICH S7A Section 2.7.3
Stomach or intestinal smooth muscle	GI motility – detection of undesirable side effects on transit time such as diarrhoea and constipation	Recommended as a supplementary study as part of ICHS7A Section 2.8.2.3 – transit time	ICH S7A Section 2.8.2.3
Skin inflammation tests	Measurement of cytokine production	Not part of pharmaceutical regulations but similar assays recommended for many biologicals using *in vivo* models of immune system	REACH EU directive
Gut inflammation tests	Measurement of cytokine production	Recommended as a supplementary study as part of ICHS7A Section 2.8.2.3 – gastrointestinal injury potential	ICHS7A Section 2.8.2.3

2.2 Sourcing, Storing and Transporting Human Fresh Tissues: The First Step is the Most Important

Human fresh, intact isolated tissues appear an obvious choice of test system, yet the use of such tissues has until recently remained the preserve of a small number of specialist academic centres willing and able to build their lab activities around the irregular and often unpredictable availability of fresh tissue specimens from surgery or transplants. In this chapter 'human tissue' is defined as those samples which contain groups of different cell types that together form a working tissue, *e.g.* an isolated blood vessel, a blood sample, a section of gastrointestinal mucosa or an isolated airway.

Functional studies are almost always carried out in fresh tissue samples and only rarely in samples that have been frozen, allowing biological mechanisms to be challenged by test compounds. This type of study is technically challenging and with current preservation techniques, the studies are often of short duration, typically hours or days. For non-clinical functional studies to be of translational value to clinical effects, the relevance of the tissue is dependent on the quality of the collection, storage and preparation of the tissue prior to experiments. This is carried out in such a way as to minimise functional and/or structural changes within the tissue with the aim of retaining the *in vivo* properties. Extensive characterisation of the tissue is often carried out prior to or during experiments to ensure that the relevance has been retained, for example a RIN score (RNA Integrity Number) of 7.0 or higher may be appropriate to ensure the quality of RNA isolated prior to any investigation into gene expression levels in healthy and diseased tissue samples. Sample quality has been found to vary quite markedly, in some cases with a low percentage of samples being considered suitable for use in research.[2]

In many instances, the rapid turnaround of tests using fresh tissue means that some sort of functional qualification of tissue viability must be used as the primary inclusion/exclusion criterion, with follow-up by conventional means such as RIN scores or histology being conducted at a later time. For example, functional pharmacology experiments in blood vessels may include an assessment of the responses from cell types most sensitive to damage such as endothelial release of nitric oxide.

The provision of human tissue samples for research is now much easier and generally of a higher quality than 10 years ago. Numerous organisations can carry out customised collections from specific patient groups.[3]

There remains, however, a strong perception that insufficient tissue is available to research, when in fact the problem is one of coordination and awareness. In the UK there are approximately 2 million animals used in experiments each year, in contrast to only a few thousand human tissue specimens collected for research (estimated from comparisons between the relative number of peer-reviewed publications in animals and human tissues). Yet the problem is not one of tissue availability – in fact there are over 600 000 surgical residual human tissues generated annually in the UK[4] and over 95% of patients are happy to donate their tissues to medical research, including commercial research. The under-utilisation of human tissues stems from logistical difficulties in collecting tissue, lack of incentive on the part of healthcare staff to carry out additional duties on top of busy workloads and the need for flexible working patterns by researchers making use of fresh tissue. Animal experiments are more convenient but this is often at the expense of biological relevance. In the UK, there are approximately 700 000 surgical procedures each year (according to NHS Hospital Statistics), but tissue is made available from a tiny fraction of this, perhaps less than 1–2%. Patients are very supportive of human tissue research, with one survey suggesting that over 95% of patients are willing to donate residual tissues.[5]

While access to fixed and frozen tissues continues to become easier, access to fresh human tissue continues to be difficult, and the costs, time and logistical problems that exist when using fresh tissue have not yet been overcome. The provision of surplus tissue for drug discovery is correctly of secondary importance to the surgeon or pathologist; therefore, there can be a conflict of interest between the pathologist and the researcher that ultimately limits the availability of fresh tissue for researchers and tissue suppliers. On a positive note, the recent increase in the number of tissue banks, although primarily focused on creating banks of frozen or fixed tissues, might, in the near future, provide a more structured system for the supply of fresh tissues. Although several commercial companies offer to supply processed, fixed or frozen human tissue, those seeking fresh human tissue for their research often need to create a bespoke tissue supply network or outsource the experimental work to a contract research organisation (CRO) with a developed tissue network.

Research teams must be flexible, to use tissue as and when it becomes available. Several studies have examined the use of various transplant solutions for the preservation of tissue function; however, little improvement has been made on standard physiological solutions. Tissue >2 mm^3 exceeds the limits of diffusion and, therefore, supplementary methods, such as continuous gassing, perfusion or cooling at 4–8 °C, are often necessary. Recent innovations, such as artificial haemoglobin and haemoglobin crosslinked to superoxide dismutase, might help to prolong tissue function but this has not yet been demonstrated. Optimisation of cryopreservation procedures for fresh tissues offers one further potential route for increasing the experimental window. Lung and vascular tissues retain many of their functional responses when thawed.

There is clearly sufficient fresh tissue and public support to allow its use as an essential and routine element of drug development; however, greater cooperation between biobanks, regulatory bodies, CROs and pharmaceutical companies can make the collection of fresh tissues for research 'the norm' rather than 'the exception'.

2.3 Common Experimental Approaches to Investigations in Isolated Fresh Human Tissues

2.3.1 Tissue Baths and Wire Myographs

Tissue baths and wire myographs are the most commonly used experimental method to study function in intact fresh tissues (Figure 2.1). Classical functional assays are strain-gauge systems that measure the contraction and relaxation of isolated muscle segments following application of a test compound.[6] In combination with imaging systems, a range of useful data (*e.g.* force, tension, pressure, volume, growth) can be collected in all types of human muscle tissue including smooth muscle (blood vessels, gut, airways, renal and reproductive tubules), cardiac muscle and skeletal muscle.

Early development of antihypertensive drugs relied heavily on assays of isolated large blood vessels usually removed during surgery or post-mortem, but the application of such techniques in drug development has until recently been fairly sporadic. With the discovery that small resistance arteries (between 150 μm and 300 μm) are primarily responsible for the control of organ-specific blood flow and overall vascular regulation of blood pressure, it has been possible to screen greater numbers of compounds for their vascular effects. Sufficient numbers of isolated functional blood vessels with measurable responses to pharmacological agents can be obtained from very small biopsies of tissue from a range of organs including the skin, kidneys, lung, heart and skeletal muscle.[7-9] Investigations in small arteries have not only contributed to our understanding of vascular biology (*e.g.* the importance of the vascular endothelium as a major modulator of vascular function,[10] the nitric oxide pathway,[11] the endothelin pathway,[12] vascular endothelial growth factor[13] and endothelium-derived hyperpolarising factor,[14] but also represent a valuable screening tool for the prediction of vascular safety concerns or efficacy. The detection of vascular effects in resistance arteries represents a potential safety concern, because even small changes in systemic blood pressure can be large enough to increase mortality and morbidity.

These same techniques can be applied to other tissue types. Respiratory tissue is significantly more delicate than blood vessels and therefore provides a major technical challenge; however, the same sensitive techniques used for blood vessels have been adapted to small airways.[15] Application of test drugs on respiratory tissue from patients suffering from chronic obstructive airway diseases and asthma have allowed bronchodilators to be assessed, providing subtle mechanistic insight in living human respiratory tissue that could not otherwise be revealed.[16] Lung parenchymal tissue also has contractile properties and the application of drug candidates to tissue strips can provide insight into drug effects.[17]

In addition, human cardiac muscle (donated from recipient hearts removed during transplant procedures, or from transplant donor hearts that are not suitable for transplant) and atrial sites are available (atrial appendages may be removed during cardiac catheterisation).[18]

(a)

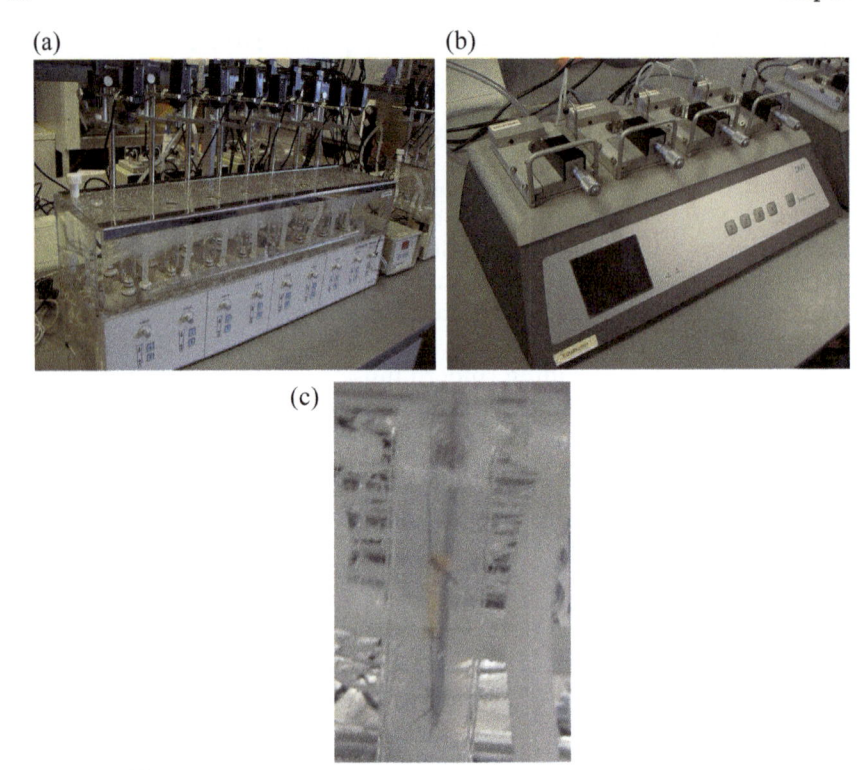

(b)

(c)

(d) **Constriction response of strips of urethra tensioned to 1.0g (longitudinal or circular)**

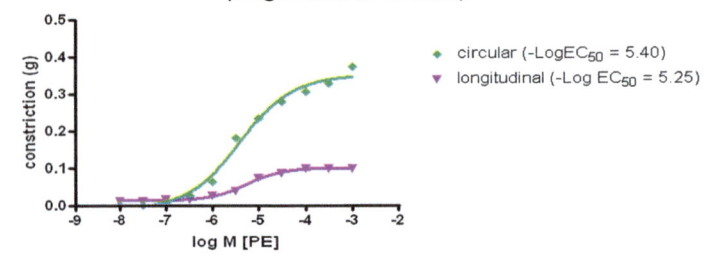

- circular (-LogEC$_{50}$ = 5.40)
- longitudinal (-Log EC$_{50}$ = 5.25)

log M [PE]

Figure 2.1 Isolated tissue specimens mounted in (a) tissue ('organ') baths and (b) wire myographs for measurement of force generation by contractile tissues. (c) shows a strip of male urethra mounted in an organ bath; contractions or relaxations of the tissue can be measured and graphed to model concentration–response relationships. (d) shows data generated from contractile responses of circular and longitudinal urethral smooth muscle in the presence of increasing concentrations of the β$_1$-adrenergic agonist phenylephrine (PE).

Electrical activity, recorded by means of tiny microelectrodes applied to stimulated tissue, can monitor the effect of applied drugs on various locations on isolated cardiac segments such as Purkinje fibres, allowing measurements of cardiac action potentials.[19] The race between the growing use of functional intact human tissues and efforts to recreate such phenotypically relevant cardiac tissue through stem cell derived cardiomyocytes will be intriguing.

Although most experiments in tissue baths and wire myographs are short-lived, lasting perhaps 8–10 hours, some researchers have sustained fresh intact cultured tissues allowing the measurement of not only contractile effects but also the release of endogenous mediators such as cytokines[20] or changes in growth or atrophy. This creates opportunities for researchers to investigate toxic effects of compounds following repeated exposures rather than 'single shot' safety pharmacology end points.

2.3.2 Perfusion Myographs

Perfusion systems aim to closely mimic *in vivo* physiological conditions and are used where dynamic interactions between the tissue, blood flow, nutrients, metabolites or gas exchange are required to properly model drug–tissue interactions. For example, perfusion myography is a technique where isolated tubular tissues such as blood vessels can be exposed to test drugs applied either to the endothelial surface or adventitial surface (Figure 2.2). Physiological pressures and flow rates can be applied to the tissue, activating mechanisms in the endothelium that are silent in static *in vitro* systems, such as endothelial shear stress and myogenic tone.[21] Sophisticated imaging analysis allows measurements not only of vessel diameter but also intracellular signalling and changes in vascular permeability. In addition, the interaction between isolated tissues and other cell types can be examined. A similar *ex vivo* perfusion technique in mouse carotid arteries has been used to investigate monocyte adhesion and rolling, a crucial step in atherosclerosis,[22] and it seems feasible that a similar approach could be applied to human fresh arteries.

2.3.3 Organoculture Systems and Precision-Cut Tissue Slices

A number of organ culture systems have been used to study local inflammatory processes in tissues such as skin,[23] gut mucosa,[24] synovium, cartilage[25] and adipose tissue.[26] Accurate assessment of inflammation using the culture method critically depends on careful handling and preparation of tissue. Because these methods retain the normal behaviour and growth of primary human tissue they provide a more relevant screening system than available artificial tissue models[27] and up to 40–50 individual 'biopsies' of tissue per donor can be cultured, given the volume of fresh tissue potentially available from surgery or transplant.

In vitro application of various challenges, *e.g.* ischaemia, lipopolysaccharide, or drug candidates, have been assessed by measuring the release of cytokines, inflammatory mediators or metabolites by radioimmunoassay following culture of tissue slices.[28] These can be compared to the responses in other important cellular components of respiratory tissues such as mast cells or pulmonary epithelial tissue.[29] *In vitro* assays that

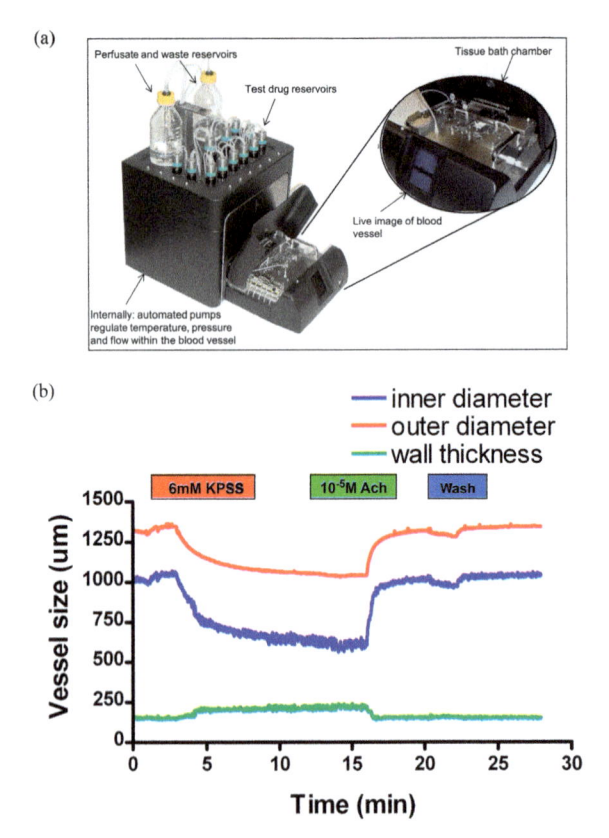

Figure 2.2 Perfusion myograph systems allow dynamic control or measurements of pressure and flow within isolated tubular tissues; in addition to simple visualisation (a) and measurements of tissue dimensions (b), such systems may also allow imaging of tissue permeability and intracellular signalling through confocal microscopy.

evaluate pulmonary toxicity have been developed using pulmonary alveolar macrophages to provide fairly simple and inexpensive screens. These measure deleterious changes in parenchymal cell populations as a marker of a toxic response and can even be combined with tracheal organ cultures to produce fibrosis after exposure to inorganic and organic fibrogens like silica or asbestos.[30]

Standardising complex human tissue assays can be challenging; one approach is to use precision-cut tissue slices of between 250 and 1000 μm thickness, which are uniform in their dimensions and minimise intra-patient variation in drug responses. Slices are typically prepared by infusion of low melting point agarose into the tissue or organ, which then solidifies upon cooling and allows slices to be prepared using equipment such as a Krumdieck tissue slicer.[31] As the temperature is increased to 37 °C, the

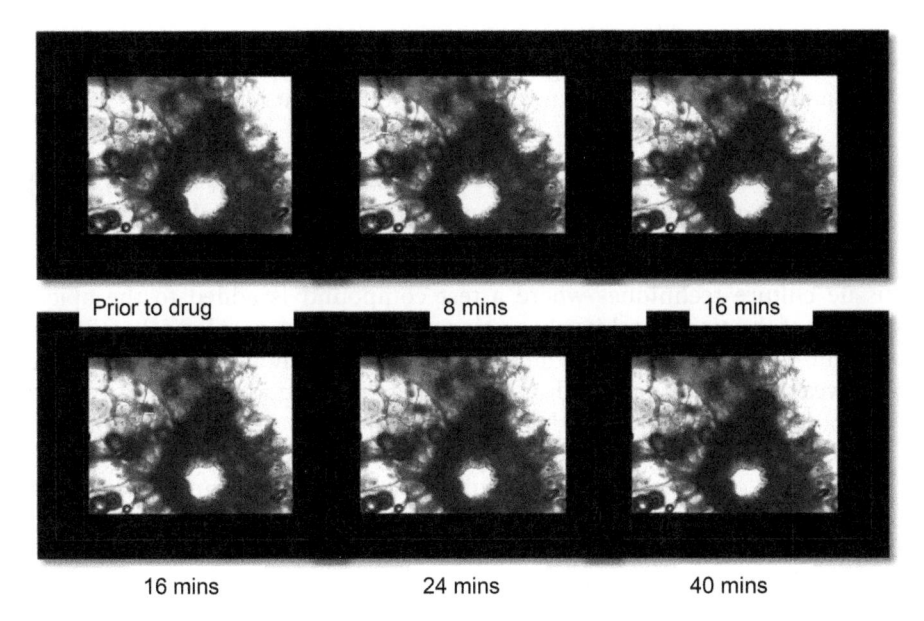

Functional Studies with Human Isolated Tissues 27

Figure 2.3 Tissue biopsies and slices can be maintained in culture for hours to days, allowing a range of functional end points to be measured. A series of images can be captured by video microscopy and image analysis allowing measurements of lumen diameter; here, a time-capture series of human isolated bronchus demonstrates contraction upon exposure to carbachol (1×10^{-5} M) at 0, 8, 16, 16, 24 and 40 minutes after drug addition (images courtesy of Graeme Macluskie, Biopta Ltd).

agarose is removed, leaving a functional tissue preparation that retains the 3D complexity of the native organ but which can be maintained in culture for days (Figure 2.3).

Probably the most difficult human tissue from which to obtain functional data is neural tissue. This is from the perspective of supply as well as the technical difficulty in preparation; however, there are some *in vitro* functional bioassays available. These use tiny thin brain slices or preparations of ganglia kept in specialised media that allow electrophysiological measures of synaptic activity to be made following pharmacological or physical stimulus.[32] The most desired assay for investigations into chronic pain within the pharmaceutical industry is the dorsal root ganglion (DRG) assay. A major obstacle is the vulnerability of neural tissue to time-dependent failure and the difficulty in sourcing DRG tissue within a short time frame. While there are a number of assays in use in academic institutions these are often not commercially useful. Moreover, assays of cultured neural tissue are difficult to validate because they only remain functional for a few hours. Finding functional human tissue assays that correlate to diseases such as Alzheimer's, Parkinson's, psychosis or depression has become a major target. It appears likely that this is an area where advances in stem cell technologies will make a significant impact, if measures to collect and transport fresh neural tissue for longer periods are not developed.

2.3.4 Membrane Transport and Ussing Chambers: Skin, Lung Mucosa, Gastrointestinal Tract and Glandular Tissues

Tissue sheets such as skin and glandular tissue can be investigated in special tissue culture techniques where a test compound is added to the apical surface of the tissue and its absorption or direct pharmacological activity is measured. Testing using skin biopsies has been common for many years and involves the application of a drug to a suspended portion of tissue; the active compound passes through the dermal region as a measurement of skin penetration (Franz cell technique).[33] Subsequent testing of each tissue type individually can also act as a useful measure of toxicity and efficacy (for dermal medication) screening. Engineered 'humanised' skin is now a reality and commercially available for this approach to testing. Whether the engineered skin possesses all of the phenotypic characteristics required is not yet clear, however early studies of human epidermal tissues appear to show morphology and phenotype that is quite similar to fresh human skin.[34] Nonetheless, fresh normal human skin is perhaps the most readily accessed of human tissue specimens, with thousands of cosmetic procedures being conducted in the UK each year, and remains the gold standard approach for studies on dermal responses and drug absorption.

An alternative to the Franz cell technique, which is useful for understanding not only permeability but also ion transport and physiological processes, is the Ussing chamber method (Figure 2.4). Voltage and current electrodes are placed on either side of the tissue allowing changes in ion transport to be measured in the presence of test drugs. One key difference between the Ussing and Franz techniques is the status of the tissue during the experiment. For Franz experiments the tissue is often no longer 'living' and may simply act as a physical barrier to drug penetration; however, in Ussing chamber experiments a section of intact mucosal tissue carefully dissected from the trachea, bladder or gastrointestinal tract remains fully functional for up to 16 hours, permitting the assessment of drug transporters, ion channels and secretory processes.

Recent publications have highlighted the value of human fresh gastrointestinal tissues in Ussing chambers both for the prediction of fraction absorbed ('Fa', the percentage of an orally delivered drug that achieves systemic absorption from the gastrointestinal tract)[35] and as a way to better understand the influence of disease status on regional permeability. Although cell-based assays used to predict drug absorption are popular it is accepted that there are differences in the expression of drug transporters (such as Pgp) and enzymes in the human small intestine, which is the site of most oral drug absorption; moreover, significant differences exist between

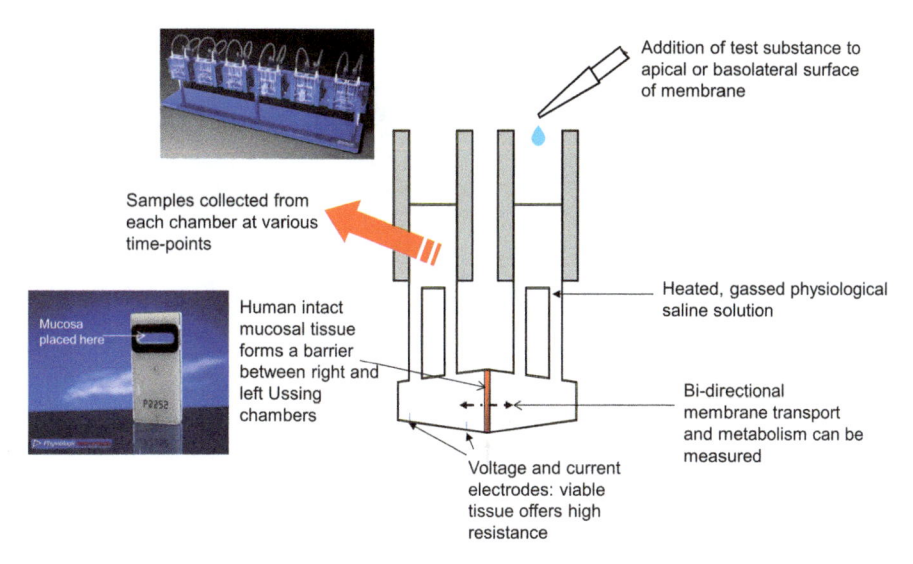

Functional Studies with Human Isolated Tissues 29

Figure 2.4 Ussing chambers are used to measure drug absorption and metabolism in intact mucosal preparations from the trachea or gastrointestinal tract. Sensitive voltage and current electrodes allow measurements of ion flux across polarised membranes.

species.[36] Gastrointestinal diseases such as ulcerative colitis and Crohn's disease may alter drug absorption, which can be modelled using diseased tissue from surgical resections.[37]

2.4 Applications of Functional Tissues in Drug Development: Safety, Efficacy and Personalised Medicines

Perhaps the most important benefit of fresh human tissue is to obtain diseased specimens from the target patient population. The main output of functional assays is to create concentration–response relationships in this target human tissue, with the goal of translating non-clinical data to relevant clinical end points that will de-risk the clinical development process (Figure 2.5).

The use of functional tissue models is not yet routine. Instead, they are often used to compare human tissue derived data with outputs from other more commonly used test systems such as cell-based assays or animal *in vitro* or *in vivo* models. Human tissue is generally considered the 'gold standard' test system against which other models may be assessed; therefore, data from cell-based assays or animal models that is in concordance with human tissue derived data increases confidence in the translation of non-clinical data to patients. In contrast, differences between human tissue derived data

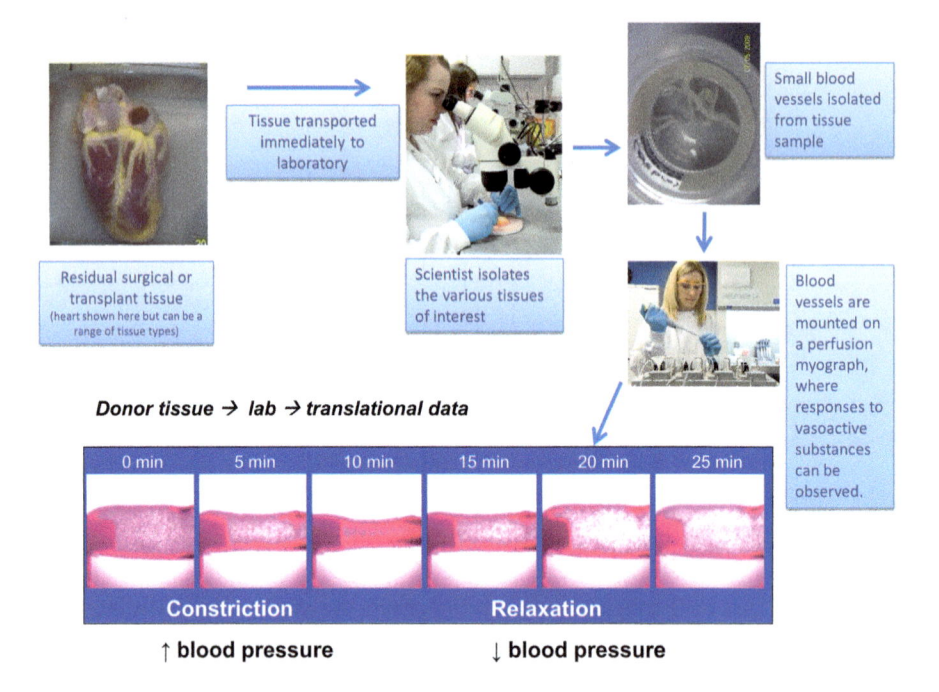

Figure 2.5 Flow chart describing the processes involved in a functional pharmacology study with the objective of translating preclinical data to relevant clinical end points such as blood pressure.

and other non-clinical methods can act as an early detection system for species differences in drug effects, or may raise questions about the predictive capacity of the cell-based assay or animal model. In either case, the generation of human data early in the drug development process increases the chances of later clinical success.

Human tissues are not only employed during early discovery and development (Figure 2.6), they may also be useful when troubleshooting unexpected clinical findings. Many drugs proceed to clinical trials with minimal human data; unanticipated species differences may lead to clinical observations in humans that were not detected in preclinical animal species.

2.4.1 Predicting Efficacy in Clinically Relevant Human Tissues

Failure to predict clinical efficacy is the single biggest cause of clinical failure, accounting for around 50–60% of phase II and III failures. Functional tissues can be used to predict clinical efficacy by understanding the behaviour of the human drug target in its natural environment.

In order to properly estimate efficacy in a non-clinical setting, various factors that can influence *in vivo* efficacy need to be replicated in the *in vitro* system, in particular for drugs that act as agonists at their intended target.

Applications of Human Tissue Research

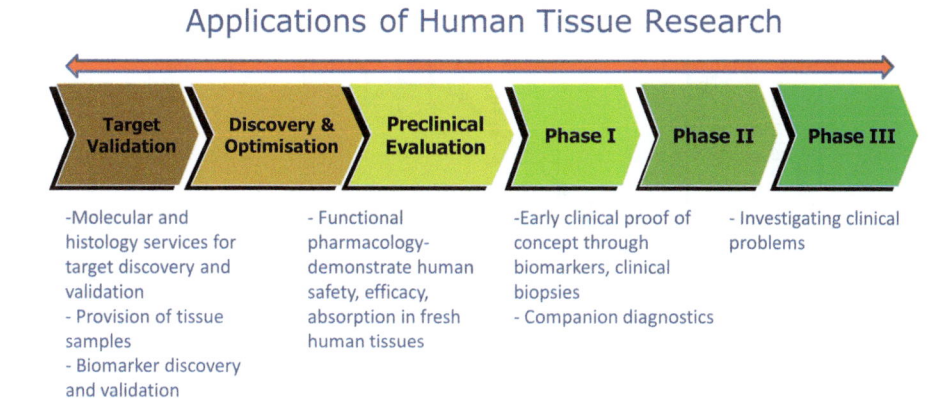

Target Validation	Discovery & Optimisation	Preclinical Evaluation	Phase I	Phase II	Phase III
-Molecular and histology services for target discovery and validation - Provision of tissue samples - Biomarker discovery and validation	- Functional pharmacology- demonstrate human safety, efficacy, absorption in fresh human tissues		-Early clinical proof of concept through biomarkers, clinical biopsies - Companion diagnostics		- Investigating clinical problems

Figure 2.6 Human functional tissue research and the drug development process.

Although many drugs have been considered selective antagonists, it is now recognised that few compounds are without at least partial agonist effects, which may vary across tissue types, not least because agonist/stimulator potency is both a compound and a tissue-dependent output. It is therefore valuable to investigate the effects of test compounds in the native tissue as well as the more commonly used high receptor expression clone systems, which can make compounds look more potent than they truly are in native tissue. For example, a natural lower expression system (native tissues) may turn full agonists to partial agonists/effective antagonists, which in turn will influence decisions on clinical effectiveness. Moreover, many drug receptors are activated preferentially along various signalling pathways by different agonists and the correct balance of signalling is tissue-dependent. Intracellular targets are also influenced by cell penetration and native tissues have a balance of transporters that can change the primary pharmacodynamics of tissue and hence the potency and time course of activity of a compound.

2.4.2 Predicting Safety and Toxicology Risks Using Functional Tissues

Other than lack of efficacy, safety and toxicology represent the greatest risks of clinical failure, with approximately 30% of drugs failing during clinical trials.[38] Human tissues are increasingly being used to investigate specific risks and concerns are being raised about the translation of animal results to humans, following the high-profile withdrawals and warnings associated with rofecoxib.[39] Particular concerns exist around the prediction of cardiovascular side effects, where small numbers of patients display significant adverse reactions which are only apparent when tens of thousands of patients receive the drug. It is questionable whether any non-clinical model exists that can predict such effects because the number of patients studied is typically quite small; however, regulators are increasingly requesting

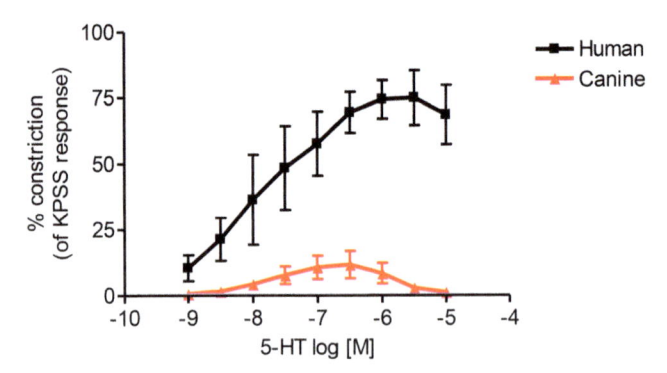

Figure 2.7 Contractile responses to 5-hydroxytryptamine ('5-HT', serotonin) in canine and human isolated subcutaneous arteries. Canine arteries are much less responsive to the effects of 5-HT and as such would underestimate the vascular effects of 5-HT agonists.

non-clinical human data to allow cross-comparisons with animal data, in particular where there are known to be significant differences between the responses of humans and the species typically used in safety pharmacology studies, such as rodents, dogs and non-human primates (Figure 2.7).

In vitro tests on human tissue or cells do have limitations in their representation of whole body drug response. A number of research groups and companies are developing 'organ-on-a-chip' methods[40] or *in vitro* recreations of organ systems to allow the simultaneous study of multiple tissue types and, more importantly, the interaction between various organs.

Human tissue assays are useful in measuring the direct toxicity of drug candidates and there are large economic and developmental benefits from obtaining more human-based information prior to clinical trials with regards to potential toxic effects. The metabolism of any chemical, influenced by gene, cell and tissue mechanisms, can be studied using human-based ADMET (ADME and toxicology) tests such as the Ussing chamber technique. Many combine chemical affinity with functional measures of metabolism to derive algorithms that closely mimic biochemical effects, at least at the molecular level. The cytochrome (CYP) family of enzymes mediates >90% of human drug metabolism and genetic expression of CYP proteins follow exposure to toxic medication. Toxicity and metabolism, predicted using human liver slices, primary hepatocytes and liver cell lines with cloned CYP enzymes have identified approved drugs (based on animal studies alone) that went on to harm humans.[41]

2.4.3 Functional Tissues and the Development of Personalised Medicines

It has been 12 years since the term 'personalised medicine' (or increasingly 'genomic medicine') was coined following the mapping of the human genome; however, with the exception of a small number of cancer therapies, the expected flow of drugs has not yet materialised. Pharmacogenomics now appears to represent only a part of the process towards achieving this goal and having moved towards proteomics on the assumption that protein expression – not gene expression – is the key factor determining drug sensitivity, it is now recognised that a true prediction of drug effects in individual patients, for the majority of diseases and chronic ailments, will require investigations that stretch from genomic level through to detailed human pharmacology at the organ/system level.[42]

It is known that inter-individual responses to drugs are variable and that a surprisingly high proportion of patients gain no benefit. For example, 40–70% of patients are classified as non-responders to bronchodilator beta-2-adrenceptor agonists in the treatment of asthma.[43]

The first example of the power of personalised medicine was the development of herceptin; human tissue research was central to the development of a personalised approach to the treatment of breast cancers that are HER-2 (human epidermal growth factor receptor) positive.[44,45] Recent efforts have moved to other similar prospects, where a clear link between a single gene mutation and the effectiveness of a drug can be reasonably predicted.

The development of a biomarker strategy has become widely accepted as a critical part of the development process, with the hope that the correct selection of biomarkers ultimately serves several different purposes including improved diagnosis and screening of patients, evaluation of risk/predisposition, assessment of prognosis, monitoring (recurrence of disease), prediction of response to treatment, and, of most relevance to early development human tissue research, as a surrogate response marker. This approach has led to a sharp increase in the number of drugs approved for cancer and HIV. Importantly, improved target selection has, under the FDA Accelerated Approval Program, brought forward the availability of 26 new chemical entities. Biomarkers have reduced the need to wait until a patient's long-term survival has been established, which is not only a commercial upside, but more importantly it is an ethically superior method by which to bring new medicines to market.

The use of biomarkers as a surrogate also has the potential to tackle the low numbers of drugs in development for orphan diseases. Miyamoto and Kakkis[46] have proposed that the use of surrogate markers could lower the barrier for accelerated approval of treatments for rare diseases, which could make such drugs commercially viable. In the vast majority of cases, access to well-characterised blood, serum, urine or sputum is needed for validation of the selected biomarker, but this can often be difficult for rare diseases and the key to future success appears to lie in networks of biobanks cooperating to standardise their collection procedures for rare samples.[47]

Surrogates need not be limited to proteins extracted from blood, urine or sputum; solid tissue samples obtained from surgery or clinical biopsies can represent an additional way to measure function in specific patient groups. Perera et al.[48] (Figure 2.8) collected gluteal biopsy samples from

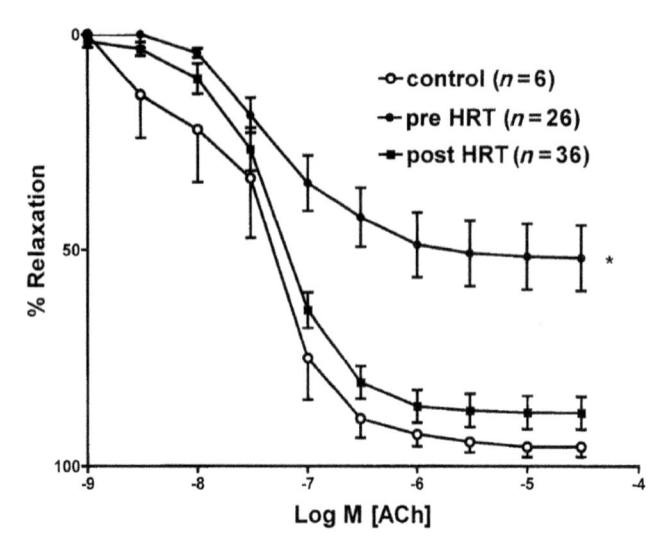

Figure 2.8 Functional tissues can be useful in longitudinal studies providing clinical biomarkers of drug activity that are not possible with biofluids. In this example, the effects of hormone replacement therapy (HRT) on the vasodilator responses of small arteries isolated from gluteal biopsies of women with and without type 2 diabetes. In healthy women, acetylcholine causes a pronounced relaxation of the arteries; however, this effect is blunted in women with type 2 diabetes. After 6 months of HRT, a second set of gluteal biopsies were taken, revealing that the vasodilator responses of women with type 2 diabetes was similar to that of healthy women.

women with and without diabetes and compared the effects of a 6 month course of hormone replacement therapy (HRT) on the responses of small arteries in both patient groups. Women with diabetes were, unsurprisingly, found to have a severely blunted baseline response (pre-HRT group) to the vasodilator acetylcholine, when compared to healthy women (control group). After 6 months, the responses of the women with diabetes were vastly improved (post-HRT group) and were comparable to the responses of healthy women. Endothelial function served as a potent biomarker of cardiovascular health and the efficacy of the HRT in improving vascular function in a specific patient group. The use of functional studies is of course limited to tissues, such as gluteal biopsies that can be safely obtained during clinical trials.

Complex data on the site-specific pharmacology of tissues is extremely valuable for safety studies, for example the expression of the 5-HT$_1$ receptor type varies greatly throughout the vascular system. The failure to recognise the site-specific effects of the migraine treatment sumatriptan, which constricts cerebral blood vessels, led to concerns once it was recognised that the drug was also capable of constricting coronary arteries.[49]

2.5 Summary

Human functional tissue research is one of the fastest growing areas of drug discovery and development and is clearly aligned with the strategic shift of the pharmaceutical industry towards biomarkers and personalised medicine. In 10–15 years' time it will probably be seen as a central pillar of all drug development programmes.

There are, however, some major differences between functional tissue research and other cornerstones of drug development. Firstly, successful provision of high-quality human tissues will be a significant factor in the impact the sector makes on development costs and timelines. Second, the dearth of human tissue-based studies in some therapeutic areas is staggering, requiring a sea change not only in pharmaceutical R&D but also academic research to better understand biological mechanisms of human cells and tissues. Thirdly, the generation of evidence of the scientific and commercial impact of functional tissue research will dramatically change the landscape, accelerating the demand for fresh, functional disease-relevant tissues and leading to specific functional tissue assays becoming part of the regulatory landscape.

Acknowledgements

Sincere thanks to Michael Finch for his help in preparing this chapter.

References

1. FDA Centre for Drug Evaluation and Research, *Gastrointestinal Drugs Advisory Committee*, 17 November 2011, p. 95 [online]. Available at www.fda.gov/downloads/AdvisoryCommittees/CommitteesMeetingMaterials/Drugs/GastrointestinalDrugsAdvisoryCommittee/UCM286759.pdf (accessed 8 August 2014).
2. S. D. Jewell, M. Srinivasan, L. M. McCart, N. Williams, W. H. Grizzle, V. LiVolsi, G. MacLennan and D. D. Sedmak, *Am. J. Clin. Pathol.*, 2002, **118**, 733.
3. J. Vaught and N. C. Lockhart, *Clin. Chim. Acta*, 2012, **413**(19–20), 1569.
4. Defined as procedures involving 'excisions or partial excisions', Hospital Episode Statistics, Admitted Patient Care 2011-12 in England. Available at: www.hscic.gov.uk (accessed 8 August 2014).
5. R. J. Bryant, R. F. Harrison, R. D. Start, A. S. Chetwood, A. M. Chesshire, M. W. Reed and S. S. Cross, *J. Clin. Pathol.*, 2008, **61**, 322.
6. J. A. Angus and C. E. Wright, *J. Pharm. Tox. Methods*, 2000, **44**, 395.
7. N. H. Buus, U. Simonsen, H. K. Pilegaard and M. J. Mulvany, *Br. J. Pharmacol.*, 2000, **129**, 184.
8. V. Johannssen, S. Maune, J. A. Werner, H. Rudert and A. Ziegler, *Rhinology*, 1997, **35**, 161.
9. M. J. Mulvany and C. Aalkjaer, *Physiol. Rev.*, 1990, **70**, 921.

10. R. F. Furchgott and J. V. Zawadzki, *Nature,* 1980, **288**, 373.
11. R. M. Palmer, A. G. Ferrige and S. Moncada, *Nature,* 1987, **327**, 524.
12. M. Yanagisawa, H. Kurihara, S. Kimura, K. Goto and T. Masaki, *J. Hypertens. Suppl.,* 1988, **6**, S188.
13. I. K. Sagar, C. N. Nagesha and J. V. Bhat, *J. Med. Microbiol.,* 1981, **14**, 243.
14. S. G. Taylor and A. H. Weston, *Trends Pharmacol. Sci.,* 1988, **9**, 272.
15. A. R. Hulsmann and J. C. de Jongste, *J. Pharmacol. Toxicol. Methods,* 1993, **30**, 117.
16. M. J. Finney, J. A. Karlsson and C. G. Persson, *Br. J. Pharmacol.,* 1985, **85**, 29.
17. C. L. Armour, J. L. Black and N. Berend, *Bull. Eur. Physiopathol. Respir.,* 1985, **21**, 545.
18. A. Duman, A. Saide Sahin, K. Esra Atalik, C. öZtin ögün, H. Basri Ulusoy, K. Durgut and S. öKesli, *J. Cardiothorac. Vasc. Anesth.,* 2003, **17**, 465.
19. S. Picard, S. Goineau and R. Rouet, *Curr. Protoc. Pharmacol.,* 2006, **11**, Unit 11.3.
20. F. Marceau, D. deBlois, E. Petitclerc, L. Levesque, G. Drapeau, R. Audet, D. Godin, J. F. Larrivée, S. Houle, T. Sabourin, J. P. Fortin, G. Morissette, L. Gera, M. T. Bawolak, G. A. Koumbadinga and J. Bouthillier, *Int. Immunopharmacol.,* 2010, **10**, 1344.
21. E. Moss, S. Lynagh, D. Smith, S. Kelly, A. McDaid and D. Bunton, *J. Pharmacol. Toxicol. Methods,* 2010, **62**, 40.
22. Z. Zhou, P. Subramanian, G. Sevilmis, B. Globke, O. Soehnlein, E. Karshovska, R. Megens, K. Heyll, J. Chun, J. S. Saulnier-Blache, M. Reinholz, M. van Zandvoort, C. Weber and A. Schober, *Cell Metab.,* 2011, **13**, 592.
23. A. R. Companjen, L. I. van der Wel, L. Wei, J. D. Laman and E. P. Prens, *Arch. Dermatol. Res.,* 2001, **293**, 184.
24. V. M. Salvati, G. Mazzarella, C. Gianfrani, M. K. Levings, R. Stefanile, B. De Giulio, G. Iaquinto, N. Giardullo, S. Auricchio, M. G. Roncarolo and R. Troncone, *Gut,* 2005, **54**, 46.
25. K. D. Rainsford, C. Ying and F. C. Smith, *J. Pharm. Pharmacol.,* 1997, **49**, 991.
26. M. Lappas, M. Permezel and G. E. Rice, *Endocrinology,* 2005, **146**, 3334.
27. D. D. Allen, R. Caviedes, A. M. Cárdenas, T. Shimahara, J. Segura-Aguilar and P. A. Caviedes, *Drug Dev. Ind. Pharm.,* 2005, **31**, 757.
28. K. Sewalk and A. Braun, *Xenobiotica,* 2013, **43**, 84.
29. S. P. Peters, E. S. Schulman, R. P. Schleimer, D. W. MacGlashan Jr, H. H. Newball and L. M. Lichtenstein, *Am. Rev. Respir. Dis.,* 1982, **126**, 1034.
30. G. L. Fisher and M. E. Placke, *Toxicology,* 1987, **47**, 71.
31. C. Martin, S. Uhlig and V. Ullrich, *Eur. Respir. J.,* 1996, **9**, 2479.
32. J. Wölfer, C. Bantel, R. Köhling, E. J. Speckmann, H. Wassmann and C. Greiner, *Eur. J. Neurosci.,* 2006, **7**, 1795.
33. L. Bartosova and J. Bajger, *Curr. Med. Chem.,* 2012, **19**, 4671.

34. C. A. Brohem, L. B. Cardeal, M. Tiago, M. S. Soengas, S. B. Barros and S. S. Maria-Engler, *Pigm. Cell Melanoma Res.,* 2011, **24**, 35.
35. Å. Sjöberg, M. Lutz, C. Tannergren, C. Wingolf, A. Borde and A. L. Ungell, *Eur. J. Pharm. Sci.,* 2013, **48**, 166.
36. G. M. Grass and P. J. Sinko, *Adv. Drug Deliver. Rev.,* 2002, **54**, 433.
37. M. Miyake, H. Toguchi, T. Nishibayashi, K. Higaki, A. Sugita, K. Koganei, N. Kamada, M. T. Kitazume, T. Hisamatsu, T. Sato, S. Okamoto, T. Kanai and T. Hibi, *J. Pharm. Sci.,* 2013, **102**, 2564.
38. I. Kola and A. Landis, *Nat. Rev. Drug Discovery,* 2004, **3**, 711.
39. P. Jüni, *Lancet,* 2004, **364**(9450), 2021.
40. M. W. Moyer, *Sci. Am.,* 2011, March, 19.
41. S. Matsumoto and Y. Yamazoe, *Br. J. Clin. Pharmacol.,* 2001, **51**, 133.
42. D. Bunton, *Eur. Pharm. Contract.,* 2011, Summer Edition, 66.
43. A. Flannery and J. McNamaraJ. Technology Strategy Board: Accelerating Stratified Medicines in the UK [online]. Available at: www.innovate10.co.uk/uploads/Stratified%20Medicine.pdf (accessed 8 August 2014).
44. E. M. Berns, J. G. Klijn, I. L. van Staveren, H. Portengen, E. Noordegraaf and J. A. Foekens, *Eur. J. Cancer,* 1992, **28**(2, 3), 697.
45. M. A. Molina, J. Codony-Servat, J. Albanell, F. Rojo, J. Arribas and J. Baselga, *Cancer Res.,* 2001, **61**, 4744.
46. B. E. Miyamoto and E. D. Kakkis, *Orphanet J. Rare Dis.,* 2011, **6**, 49.
47. H. Lochmüller and P. Schneiderat, *Adv. Exp. Med. Biol.,* 2010, **686**, 105.
48. M. Perera, J. R. Petrie, C. Hillier, M. Small, N. Sattar, J. M. Connell and M. A. Lumsden, *Hum. Reprod.,* 2002, **17**, 497.
49. B. H. C. Stricker, *Br. Med. J.,* 1992, **305**, 118.

CHAPTER 3

Translational Research in Pharmacology and Toxicology Using Precision-Cut Tissue Slices

G. M. M. GROOTHUIS[*a], A. CASINI[a], H. MEURS[b], AND P. OLINGA[c]

[a]Division of Pharmacokinetics, Toxicology & Targeting, Department of Pharmacy, University of Groningen, A. Deusinglaan 1, 9713 AV Groningen, the Netherlands; [b]Division of Molecular Pharmacology, Department of Pharmacy, University of Groningen, A. Deusinglaan 1, 9713 AV Groningen, the Netherlands; [c]Division of Pharmaceutical Technology & Biopharmacy, Department of Pharmacy, University of Groningen, A. Deusinglaan 1, 9713 AV Groningen, the Netherlands
*E-mail: g.m.m.groothuis@rug.nl

3.1 Introduction

3.1.1 Precision-Cut Tissue Slices

During drug development the efficacy, pharmacokinetics and adverse effects are tested in animal models before administration of the drug to man is allowed in phase 1 clinical studies. This paradigm is based on the assumption that experimental animals are similar to humans with respect to physiology and response to xenobiotics. However, this assumption often appears to be incorrect or only partly correct, and frequently it is not even known to what extent a particular phenomenon is comparable between a certain animal species and man. In order to reduce this uncertainty most studies are performed in at least two animal models, including one rodent and one non-rodent species. But while our knowledge is increasing, we are more and more aware of the interspecies differences and the limited translational value of many of the animal models.[1,2] Apart from these scientific drawbacks, there is an increasing resistance in society to the use of animals in research for ethical reasons. These two factors together prompted the use of human cells and tissues in drug research. For instance the use of human *in vitro* preparations like human liver microsomes, human drug metabolism enzymes and transporters expressed in transfected cells, and human hepatocytes, reduced the attrition rate of new drugs for pharmacokinetic reasons significantly, from around 40% to less than 15%, due to the early and better assessment of the pharmacokinetic characteristics of a new chemical entity using human *in vitro* methods.[1] Successful prediction of the pharmacokinetics of a new drug in man is currently achieved using such *in vitro* data combined with *in silico* modelling.[3] Also, the receptor specificity and pharmacological signalling of drugs can be assessed in human cell cultures or expressed human receptors.

However, for more complex phenomena, such as the pharmacological or the toxicological effects in a target tissue in man, such *in vitro* expression systems or even cell culture systems are often not adequate due to the lack of differentiation of the cells and absence of the interactions between the different cell types in a tissue. It is well known that all cell types in a tissue communicate with each other and that their functionality depends strongly on interactions with their extracellular matrix. Therefore, a model including all cell types in their natural environment is indispensable for studying complex, multicellular organ functions and the pharmacological and toxicological response *in vitro*. In 1980, Krumdieck developed a new instrument to produce reproducible, thin tissue preparations, which remain viable and functional during culture.[4] This technique became well known as precision-cut tissue slices (PCTS) and in recent decades PCTS have been prepared from a wide variety of animal (including rat, mouse, hamster, guinea pig, cow, pig, deer, dog, monkey, trout) and human tissues (including among others liver, intestine, lung, kidney, brain, spleen, heart, thyroid gland, prostate, lymph node and several types of tumours). In these PCTS, all cells remain in their natural environment with maintenance of the original cell–cell and cell–matrix contacts. They can be produced from healthy and diseased human tissues with sufficient reproducibility, without the use of proteolytic enzymes, and allow efficient use of the scarce human tissue. The PCTS technique is currently widely used for drug metabolism and transport studies, to investigate the mechanisms of disease and to test the toxicological and pharmacological effects of drugs in healthy and diseased human tissue. In this chapter we aim to discuss the application of human liver, intestine and lung PCTS in studies on ADME (absorption, distribution, metabolism and excretion) and toxicology, and in some selected diseases such as fibrosis in the liver and the intestine, obstructive lung diseases, viral infections and cancer. As the use of PCTS in research is steadily increasing it is impossible to give a comprehensive review of all applications of PCTS, but by highlighting some of the most important examples, with a special emphasis on the application of human PCTS, we aim to show the extensive translational potential of the use of human PCTS for pathology and drug research.

3.1.2 Availability of Different Sources of Human Organs and Tissues

Research with human tissue is subject to ethical regulations, which may be different in each country. Human tissue banks operate in some countries, such as the UK Human Tissue Bank, the International Institute for the Advancement of Medicines and the American Association for Human Tissue Users, which provide the tissue under strict medical and ethical standards. In many other countries researchers collaborate with surgeons on an individual basis, for which the approval of Medical Ethical Committees needs to be obtained.

In principle there are three categories of sources for human tissues: post-mortem tissue from patients who died before the organs were retrieved, tissue from organ donors that is unsuitable for transplantation, and tissue remaining as surgical waste from patients undergoing surgical removal of parts of organs or tissue. While post-mortem tissue is obtained sometime after the cessation of blood flow, the tissue from organ donors is obtained from brain death donors, where the blood circulation is kept intact, which may imply a potentially large difference in viability of the slices prepared from these tissues. Donor tissue is harvested under well controlled conditions with no or limited warm ischaemia as blood circulation is kept intact until *in situ* perfusion with cold preservation solution and the organs are subsequently procured optimally. However during brain death the tissue has been exposed to the proinflammatory cytokine release and activation of haemostasis induced by brain death.[5] In contrast, *post-mortem* tissue is exposed to warm ischaemia after the arrest of the blood flow and most organs like the liver, lung and intestine do not survive such warm ischaemia for a prolonged time. On the other hand, tissues from surgery patients do not suffer from responses to brain death, but usually warm ischaemia cannot be avoided due to the surgical procedures. In our institute, liver samples from patients undergoing partial hepatectomy are therefore perfused directly after removal from the body in the operating theatre, *via* veins at the cutting surface with ice-cold preservation solution (UW, Viaspan) to remove the blood and to rapidly cool the tissue. Liver tissue that is harvested after warm ischaemia of more than 1 hour loses viability and is less suitable for research.[6,7] Intestinal tissue and lung tissue is instantly submersed in ice-cold, oxygenated buffer. These differences in procedures may all have an influence on the viability and functionality of the PCTS, and it is advisable to make an extensive comparison between PCTS from the different sources with respect to viability and functionality. This comparison should be based on large numbers of individual organs as intrinsic inter-individual differences in function which may be related to age, sex, drug use, underlying disease states and feeding patterns may be large.[8,9] In addition, viability and functionality of the individual organs should be assessed by a sensitive viability test and a standard function test. If possible the tissue or the patient is tested beforehand for the presence of infectious diseases including HIV and hepatitis, not only to avoid influence of the disease state on the outcome of the experiments, but also to protect the laboratory staff. Tissue and PCTS can be preserved in tissue preservation solutions for a few hours (intestine), or up to 24 hours (for liver), without loss of viability.[10,11] Cryopreservation of PCTS has been investigated extensively but up to now the viability of the PCTS after thawing is limited to a few hours, as reviewed recently by Fahy et al.[12]

3.1.3 Techniques to Prepare and Incubate Precision-Cut Tissue Slices from Various Organs and Tissues

In 1980, C. L. Krumdieck initiated a revival of an old technique[13,14] and developed a new instrument to reproducibly prepare thin, precision-cut liver slices.[4] He realised that it was essential to preserve the viability of the tissue by providing protective conditions (such as continuous submersion in cold oxygenated buffer) and to enable the production of slices of reproducible thickness and diameter. The Krumdieck tissue slicer, which underwent regular improvements, is currently one of the most widely used instruments and can produce hundreds of slices in a semi-automatic procedure within 1 hour. Later, other instruments came onto the market, among them the Brendel/Vitron slicer, the Leica VT1200 S vibrating blade microtome with Vibrocheck and the Compresstome™ VF-300 microtome (Precisionary Instruments, San Jose, CA, USA). They all have a different design but only minor differences in quality of the PCTS were reported.[15,16] They all require the preparation of cylindrical cores of defined diameter (4–15 mm) that fit precisely in the slicer's core holder (Figure 3.1). These cores are made from solid tissues, *e.g.* liver and kidney, either with a motor-driven drill (a commercially available drill or specially designed tools like those from Alabama Research and Development Corp. or from Vitron Inc.) or by a handheld biopsy punch or the specially designed Hand Held Coring Tool (Vitron). Non-solid tissues like intestine or small tissue samples that do not allow coring the tissue can be embedded in 0.7–1.5% of low-melting agarose in a cylindrical core that fits the slicer holder. The preparation of liver and intestinal slices from human and rat tissue was recently described by de Graaf *et al.*[17] and by Fisher *et al.*[6] In order to keep the alveoli in the lung tissue open, lungs are usually filled with low-melting agarose *via* perfusion of the airways, after which the solidified tissue is cored and sliced like liver.[18,19] The thickness of the slices can be adjusted rather precisely to the required

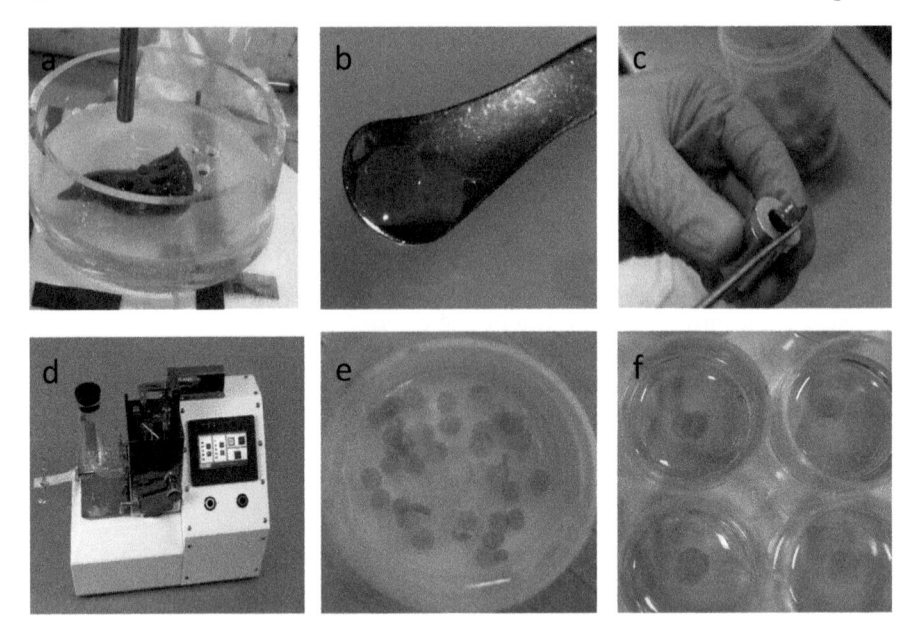

Figure 3.1 Preparation and incubation of rat or human liver slices. Liver cores are prepared from pieces of tissue using a drill and a tissue coring tool (**a**, **b**), and transferred to the cylindrical core holder of the Krumdieck slicer (**c**). Then slices are cut (**d**), collected (**e**) and transferred to well plates for incubation (**f**).

thickness, being 200–250 μm for liver, 200–400 μm for intestine and 500 μm for lung. The thickness of the slices is limited by the diffusion of substrates and oxygen to the inner cell layers, which not only affects the viability of the inner cell layers[20] but also influences the quantitative *in vitro–in vivo* extrapolation of metabolic rate.[21–23]

The slices can be stored in cold preservation solution until incubation. For liver slices it has been shown that preservation for up to 24 hours in ice-cold UW solution adequately preserves their viability as well as transport and metabolic function.[10,11,24] For intestinal slices this preservation time is limited to 3 hours, and results in better preservation in Krebs buffer than in UW (A. A. Khan and G. M. M. Groothuis, unpublished observations).

Three main groups of incubation systems are used, each with its own advantages and disadvantages. Submersion cultures are most widely used, in which slices are incubated at 37 °C in 6, 12 or 24 well plates or small Erlenmeyers in rich cell culture medium, oxygenated with 95% O_2/5% CO_2 (liver, intestine) or 95% air/5% CO_2 (lung) on a shaker (90 rpm linear shaking) for efficient medium circulation around the slices. In the so-called dynamic organ culture systems the slices are repeatedly submerged into and lifted out of the medium, with the intention of ensuring better oxygenation.[25] Both of these systems have their advantages and disadvantages: the well plates allow easy working with a large number of samples and rapid sampling, the DOC systems seem to be better equipped for long-term culturing, although good results have also been obtained in well plates. In both these incubation systems the medium composition continuously changes over time as a result of uptake of compounds and excretion of bile acids, proteins, cytokines and waste products. Therefore, the medium is usually replaced every 24 hours, which inevitably exposes the slices to a sudden environmental shock. Therefore, a third incubation method was developed, providing the slices with a constant environment by perifusion of medium.[26–29] Apart from the constant environment, such perifusion systems also allow online analysis of high concentrations of metabolites produced by the slices, online assessment of inhibition of metabolism using one slice for a range of substrates and more accurate detection of unstable metabolites.[30] Another important achievement of this perifusion system is the ability to investigate organ–organ interactions. Van Midwoud et al.[31] sequentially perfused a rat intestinal slice and a rat liver slice, and exposed them to the bile salt chenodeoxycholate, showing the involvement of FGF 15 excreted by the intestinal slice in the regulation of the enzyme CYP7A1 in the liver slice. This system now offers many opportunities to study organ–organ interactions between two or more tissues or organs. Similarly, by co-culturing slices from two organs, the toxic effect of reactive metabolites produced in the liver in a simultaneously incubated heart,[32] and the co-operation in metabolism between liver and kidney slices,[33] were shown. Liver slices have also been incubated together with blood cells to show haemolytic effects of methimazole in the blood cells, possibly mediated by liver-derived metabolites and other mediators.[34] In addition, co-cultures of liver slices and lymphocytes showed increased infiltration of lymphocytes in the liver slices upon exposure to ethanol.[35]

However the applications of co-incubation of PCTS from different human organs are hampered by the requirement that all the tissues should be available on the same day, and that the opportunity to obtain samples from several different organs from one donor is very limited. Therefore extension of the preservation possibilities, either by developing better cold preservation solutions, cryopreservation methods or long-term culturing at 37 °C, will be instrumental in the future application of multi-organ perifusion or co-incubation systems for human PCTS.

3.2 Human Precision-Cut Tissue Slices in ADME and Toxicology

Most xenobiotics, including drugs, are metabolised into more hydrophilic metabolites to enable their excretion. For these biotransformation reactions a plethora of enzymes is available in virtually every organ, but they are particularly abundant in the liver and intestine and to a lower extent in lung and kidney, and comprise both phase 1 oxidations and reduction reactions, and phase 2 conjugation reactions. It is well known that large species differences exist in the expression and substrate specificity of the enzymes involved and that human *in vitro* systems are indispensable for a proper prediction of human metabolism. The uptake and efflux transporters that are involved in the disposition (absorption, distribution and excretion) of drugs, and thereby determine the exposure of the cells in the tissues, also show high interspecies variation. Together the metabolising enzymes and transporters are important determinants for the toxicity of xenobiotics as this toxicity is often related to the formation and accumulation of reactive metabolites. In addition, many drug–drug interactions that influence the therapeutic success and the occurrence of adverse effects of drugs are based on inhibition or induction of metabolic enzymes and transporters.

Although several *in vitro* methods with human material are successfully employed to predict ADME and toxicity, PCTS is the only method that fully represents the tissue architecture and multi-cellularity. Moreover the polarised localisation of transporters on either the apical or basolateral membrane of the hepatocytes and intestinal cells remains unchanged in slices in contrast to those in freshly isolated liver cells.[36,37]

Ample literature is available showing the potential of liver, intestinal and lung slices to represent the *in vivo* metabolic functions of these organs. Several comprehensive reviews were published recently and the reader is referred to these reviews and the references therein.[18,38–41]

PCTS were successfully applied to detect species-specific metabolism. A striking example is shown in De Kanter *et al.*,[42] where the metabolism of testosterone is shown in human and rat liver, intestine, lung and kidney. It shows that not only were species-specific metabolites found, but also that the relative contribution of the different organs to the total metabolism is species-specific.

The major advantage of the slice technology is that it allows the preparation of slices from all tissues and species in much the same way without the use of proteolytic enzymes, and it is theoretically possible to study slices of several organs in the same experiment. De Graaf *et al.*[43] published data to show that good *ex vivo* to *in vivo* extrapolation of the metabolic clearance of drugs can be obtained if all organs are included, thin (*ca.* 100 µm) slices are used in case of rapidly metabolised drugs, and albumin is included in the incubation medium. Under these conditions, the metabolic clearance can be predicted with acceptable (two-fold) accuracy with organ slices.[43] Another striking advantage is the possibility of studying intra-organ heterogeneity in metabolism. In the liver, the difference between the different lobes is negligible but the acinar heterogeneity is evident and is retained in the liver slices during culturing.[44]

Also the intestine is a very heterogeneous organ with ascending and descending gradients of expression of enzymes and transporters.[45] Using precision-cut intestinal slices (PCIS) prepared from the duodenum, jejunum, ileum and colon of the rat, the gradients of activity of phase 1 and phase 2 metabolism of testosterone and 7-hydroxycoumarin were assessed,[46] as well as the induction of these enzymes by VDR, CAR, PXR, FXR and GR ligands.[47,48] Likewise, the metabolic capacity of the higher and lower airways can be assessed by preparing and incubating slices from these different regions separately.

PCTS have been successfully used to assess drug–drug interactions based on induction or inhibition of metabolism and was recently reviewed by Ioannides.[40] The induction potential of new drugs can be assessed taking into account the consequences of regional- and organ-specific expression of nuclear receptors, metabolising enzymes and transporters that determine the exposure to the inducer. Moreover the interplay between metabolic enzymes and transporters can be studied under physiologically relevant expression levels of the enzymes and proteins.

For application of human PCTS in transport studies it is important to assess whether the expression of the transporters is maintained at the proper membrane localisation. In human liver slices, the gene expression of drug transporters has been shown to be constant for up to 24 hours of incubation.[49,50] Uptake into the inner cell layers of a wide variety of compounds ranging from lipophilic and hydrophilic molecules to larger molecules such as proteins was demonstrated in rat liver slices.[51] The specific inhibition of transporter-mediated uptake could be demonstrated in rat and human liver slices for digoxin and BSP.[24,51] Also excretion of bile acids[52] and glutathione-, glucuronide- and sulfate-conjugates of many drugs like paracetamol,[53] and 7-hydroxycoumarin,[10] shows that active transports are functional. However, it remains difficult to prove whether the conjugate is excreted across the bile canalicular or the sinusoidal membrane.

Transport in PCIS has not been studied in great detail. Metabolism of drugs, known to be taken up *via* carrier-mediated transport, indirectly shows the functionality of these transporters. Similarly, the effect of bile acids on the expression of transporters and enzymes in human PCIS is an indirect proof that uptake transporters remain active.[48] The expression of several transporters involved in bile acid transport varied during culture, but in general the expression remained responsive to inducing stimuli.[54,55] Possidente *et al.*[56] were the first to investigate the inhibitory effects of drugs on excretion transporters in rat PCIS by measuring the accumulation of calcein and calcein-AM, the fluorescent substrates of MRP2 and MDR1, respectively. The obtained results showed good correlation to data collected in other *in vitro* preparations.[56] Vectorial transport from the apical to the basolateral side cannot be assessed in slices, as this needs separation of the mucosal from the serosa side like in an Ussing Chamber set-up.[57–60] These two techniques together are highly complementary to study transport and metabolism–transport interplay.

Transport studies in lung PCTS are also scarce. Saunders *et al.*[61,62] showed that in rabbit lung PCTS the uptake of the polyamines putrescine and spermidine was restricted to type II pneumocytes and alveolar macrophages, but that spermidine uptake was absent in the blood vessels and the major airways.

As the liver is the major site of metabolism of xenobiotics in the human body, most PCTS studies on xenobiotic-induced toxicity have been performed with liver slices, but lung slices have also been used for this purpose, as was reviewed recently by Morin et al.[18] In contrast, although drug-induced toxicity very often involves the gastrointestinal system, very few studies have been published on the use of PCIS for toxicity assessment.[63]

Xenobiotic-induced organ toxicity is very often the result of bioactivation by metabolism and in addition involves the interplay of several cell types in the target tissue or even the cell–matrix interactions. Moreover the toxicity may be the result of the involvement of inflammatory reactions of the tissue macrophages or the activation of endothelial cells, fibroblasts or other cell types. It is therefore increasingly recognised that organotypic models like PCTS reproduce the complexity of the processes occurring during toxicity much better than cell-based models, and all these phenomena have been reproduced in PCTS, mainly in the liver, as recently reviewed by various authors; readers are referred to these reviews for ref. 17,41,64,65. This was confirmed by transcriptomics analysis of the gene expression in liver, precision-cut liver slices, isolated hepatocytes and the HepG2 cell line, showing the greatest similarity between the intact liver and the slices.[66] In addition, *in vitro* expression of genes relevant to toxicity remained relatively unchanged during culturing of human liver slices,[67] and gene expression profiles of rat liver slices after treatment with four chemicals inducing liver toxicity *via* different mechanisms could predict the effects observed *in vivo*.[68] Moreover the interplay between the different cell types was shown in several studies on the effects of the endotoxin LPS. LPS alone does not induce toxicity in isolated hepatocytes or HepG2 cells, but induces a significant cytokine release in liver slices and the concomitant consequences for the hepatocytes such as changed expression of transporters,[50] enzymes like iNOS,[69] and synergistic toxicity with known idiosyncratic drugs.[53] The involvement of Kupffer cells was also shown for diethyldithiocarbamate[70] and venylidene[71] using PCTS. The role of Kupffer cells on the toxicity of paracetamol has not been directly proven, but the higher sensitivity of liver slices for the toxicity of paracetamol compared to hepatocytes is indicative for this.[72] The interplay between different cell types in the liver and lung is also evident in research on induction of fibrosis and matrix remodelling. This is described in more detail in Sections 3.3.1 and 3.3.2.2.

Using PCTS, large interspecies differences between human and animals were found for the toxicity of several drugs, such as in the liver for paracetamol,[72,73] its congener N-acetyl-*meta*-aminophenol,[72] coumarin,[74] voriconazole,[53] and in the lung for acroleine,[75] to name just a few.

Precision-cut lung slices are particularly interesting for studies on inhalation toxicology. To model inhalation toxicology, the dynamic organ culture methods as described above have been successfully applied to expose lung slices to aerosols and tobacco smoke. In general, precision-cut lung slices of animals can respond to toxic compounds in a comparable way as *in vivo*, showing dose-dependent toxicity of chemicals like 3-methylindole,

1-nitronaphtalene and paraquat,[76] genotoxicity of formalin and ethyl methane sulfonate,[77] covalent binding to proteins of agaritine[78] apoptosis, proinflammatory response and cytokine production by diesel exhaust,[79] endothelial dysfunction and IL1β release by cigarette smoke and nicotine,[80–82] surfactant lamellar body secretion by jet propulsion fluid,[83] inflammatory reactions and loss of viability by ozone and nitrogen dioxide,[84] and ATP and glutathione decrease by acrolein.[85] However, the number of toxicity studies in human lung slices is relatively limited. But the good *ex vivo* to *in vivo* extrapolation in animal studies and the reports on successful culturing and metabolic competence[42,86] are promising for future application of precision-cut lung slices for toxicity studies.

3.3 Application of Human Precision-Cut Tissue Slices in Disease Models

3.3.1 Fibrosis in Liver and Intestine

Fibrotic diseases, especially of the liver, cardiovascular system, intestine, kidneys and lung account for around 45% of deaths in Western societies.[87,88] Fibrosis is characterised by excessive deposition of extracellular matrix proteins, and is considered a serious complication associated with aging and/or chronic injury. Fibrosis ultimately results in structural and functional organ deterioration and loss of function. Antifibrotic therapy is one of the still unconquered fields in drug development.[87,89] The mechanisms of fibrosis are commonly studied in *in vitro* and *in vivo* animal models.[89] These animal models reflect different aspects of human liver fibrosis[90] and the use of mice with cell type-specific gene alterations has tremendously improved the understanding of the pathogenesis of liver fibrosis.[91] During the initiation, progression and regression of fibrosis, animals, mainly rodents, are exposed to considerable discomfort for extended periods of time due to specific operations, or to repeated administration of toxic substances or immune-mediating compounds.[91–93] The *in vitro* models currently utilised in fibrosis research are mainly based on cell culture models and cannot fully predict or mimic the complex cellular interactions that occur *in vivo*. In isolated primary cells, dedifferentiation rapidly occurs, partly due to the loss of the natural environment, including cues from the extracellular matrix and neighbouring or migratory cells. PCTS represent an *ex vivo* tissue culture technique that replicates most of the multicellular characteristics of whole organs *in vivo*.[94] Recently, the utilisation of PCTS for the study of organ fibrosis has been reviewed.[94] Particularly, the possibility of using (diseased) human tissue to study mechanisms of fibrosis has the potential to become a great asset of the PCTS technique. Organs share major common disease pathways, but for immune cells and especially myofibroblasts, the main extracellular matrix producing cells, there are organ- and aetiology-specific variations in cellular gene expression programmes.[95] In the liver, the platelet-derived growth factor (PDGF) and the transforming growth factor beta

(TGFβ) signalling pathways are the main pathways involved in the fibrotic process.[89]

In this section we will focus on the use of PCTS in the study of liver and intestinal fibrosis.

3.3.1.1 Precision-Cut Liver Slices in Liver Fibrosis

During chronic liver injury, mostly caused by alcohol abuse, hepatitis B or C infection, biliary diseases, non-alcoholic steatohepatitis (NASH) or chronic drug use, hepatic stellate cells, the Kupffer cells (the liver resident macrophages) and the endothelial cells are activated and act in concordance to transform the hepatic stellate cells into profibrogenic myofibroblasts. These myofibroblasts, together with the portal fibroblasts and myofibroblasts derived from bone marrow, are the collagen-producing cells in liver fibrosis.[87,89]

As fibrosis can be induced by chronic injury, most studies on liver slices concentrate on fibrotic hepatotoxins. A well-known fibrosis-inducing compound is ethanol.[96] Schaafert et al.[97] used ethanol to induce fibrogenesis in rat liver slices, and an increased α-smooth muscle actin (αSMA) and collagen 1A1 production was found. In addition, acetaldehyde, the toxic metabolite of ethanol metabolism, was formed when liver slices were incubated with ethanol[98] and Guo et al.[99] showed that acetaldehyde alone could also activate the HSC in rat liver slices. In human precision-cut liver slices, exposure of ethanol increased the chemokine receptor expression and lymphocyte recruitment into the liver tissue.[35] All these results were in line with *in vivo* results, demonstrating the added value of the slices in the early onset of alcoholic liver disease. In animal models, CCl_4 and thioacetamide (TAA) are the most commonly used hepatotoxins to induce fibrosis.[100] An increase of gene expression of early marker of HSC activation, heat-shock protein 47 (HSP47), was found in human and rat liver slices cultured in the presence of CCl_4[101,102] and TAA.[103] These results indicate that precision-cut liver slices can be used to study the early onset of fibrogenesis in chemically-induced fibrosis.

The end stage of fibrosis can be studied by utilising slices obtained from diseased animals. As it is currently not possible to culture liver slices for weeks or months, induction of end-stage fibrosis *ex vivo* has not been achieved yet. Gou et al.[104] studied the effects of established alcoholic liver fibrosis on drug metabolism, utilising fibrotic slices of livers from ethanol-fed animals; these slices remain viable for at least 6 hours. Different studies have been performed with fibrotic liver slices derived from animal models of chemically-induced fibrosis.[94] These studies focused on antifibrotic therapy,[105,106] the detection of biomarkers and the mechanism of fibrosis.[107,108] Veidal et al.[107] showed that a specific fragment of type III collagen generated by metalloproteinase-2 and metalloproteinase-9 by fibrotic liver slices is a biomarker for liver fibrosis.

The onset of biliary fibrosis in liver slices was successfully mimicked by incubating them in the presence of high concentrations of the bile acids taurodeoxycholate and taurocholate, which resulted in proliferation of biliary epithelial cells, as was also found *in vivo* during biliary fibrosis.[109] Moreover, the efficacy of antifibrotic drugs determined in fibrotic liver slices from bile duct ligated (BDL) rats correlated well with established *in vivo* fibrosis models.[94,101,110,111]

During long-term culture of normal liver slices for up to 96 hours, fibrogenic pathways are activated, which may be due to slicing.[25] Several fibrosis markers, such as gene expression of HSP47, αSMA and pro-collagen 1A1 and protein expression of αSMA and collagen 1, were increased after 48 hours of culture of (human) liver slices and further increased up to 72 hours in rat liver slices and even 96 hours in human liver slices.[25,102,112-114] The efficacy of numerous antifibrotic drugs was studied in these spontaneously activated liver slices.[112,113] Westra *et al.*[113,115] validated the methodology as to the feasibility of using it as a tool for testing *in vitro* antifibrotic drug effects. They used both long-term culture of normal rat liver slices and fibrotic liver slices from BDL rats. A major conclusion of these studies was that antifibrotic *in vivo* effects could be predicted in liver slices when measuring the gene expression of fibrosis-related markers, as exemplified with inhibitors of the PDGF pathway. However, they unexpectedly found that during the onset of fibrosis, compounds mainly acting on the TGFβ pathway showed only limited efficacy, indicating a predominance of the PDGF pathway in the early onset of fibrosis in liver slices.[113] Furthermore, liver slices from advanced fibrotic livers after bile duct ligation showed a good correlation with the *in vivo* model described in the literature with respect to the antifibrotic effect of drugs.[115] Thus Westra *et al.*[113,115] established that liver slices could be successfully used as a functional *ex vivo* model for incipient and established liver fibrosis, and to assess the antifibrotic effects of inhibitors of the PDGF or TGFβ pathways. In addition, the onset of fibrosis can be induced *ex vivo* by the culture of liver slices from healthy tissues in the presence of profibrotic stimuli.[94,116]

In conclusion, rat and human precision-cut liver slices are successfully used to study the onset of alcohol, chemical and biliary induced fibrosis. Moreover, long-term incubation of precision-cut liver slices induces fibrosis in these liver slices. Furthermore, fibrotic liver slices from diseased (human) tissue represent a good model for the end stage of fibrosis. The data from the described studies confirm that the fibrotic changes in the liver slices replicate the *in vivo* situation, can reveal species differences in the mechanisms involved in liver fibrosis and are an excellent tool to study the efficacy of potential antifibrotic drugs during different stages of liver fibrosis.

3.3.1.2   *Precision-Cut Intestinal Slices in Intestinal Fibrosis*

Intestinal fibrosis occurs in patients after radiation therapy or as a reaction of intestinal tissue to chronic inflammation such as in Crohn's disease (CD), after radiation therapy, or post-transplantation.

There is a lack of adequate *in vivo* and *in vitro* models for intestinal fibrosis. The PCIS could be an ideal multicellular model to study intestinal fibrosis. Pham *et al.*[117] recently established that PCIS could be used as a model to study the early onset of fibrosis in the intestine. PCIS from rat, mouse and human remain viable for up to 24, 48 and 48 hours, respectively. During culture of rat and mouse PCIS, the gene expression of HSP47 and the extracellular matrix protein fibronectin increased, indicating that spontaneous activation of fibrosis markers due to incubation was also found in PCIS. In human PCIS, however, while incubation periods of up to 48 hours also resulted in the up-regulation of gene expression of HSP47, no such up-regulation of fibronectin was seen, instead there was an increase in gene expression of synaptophysin, a marker of stellate cells. These results indicate that PCIS offer the opportunity to test the effect of antifibrotic drugs and study the mechanism of fibrosis in a human model of early intestinal fibrogenesis.

In conclusion, PCIS are a valuable tool to study the mechanism of incipient and established intestinal fibrosis.

3.3.2 Obstructive Lung Diseases

Asthma and chronic obstructive pulmonary disease (COPD) are chronic obstructive lung diseases with increasing prevalence worldwide and a high burden to patients and society.

Asthma patients experience attacks of breathlessness due to periods of intermittent airflow obstruction, often associated with exposure to airborne allergens.[118] Characteristic features of this disease are allergen-induced early and late asthmatic reactions, airway inflammation (particularly eosinophilia), airway hyper-responsiveness (AHR) to a variety of physical, chemical and pharmacological stimuli, and airway remodelling, characterised by increased airway smooth muscle mass, sub-epithelial fibrosis and mucus cell hyperplasia.[118-121] Although for many years asthma has been considered primarily as a large airways disease, there is increasing evidence that peripheral airways are also significantly involved.[122]

COPD is characterised by an abnormal inflammatory response of the lung to noxious particles and gases, mainly cigarette smoke, associated with a progressive and largely irreversible airflow limitation.[123] COPD includes both emphysema and chronic bronchitis and these conditions often coexist. Chronic inflammation in COPD is predominantly caused by infiltration of macrophages, neutrophils and lymphocytes in the lung, which can trigger structural alterations and narrowing of particularly the small airways, as well as emphysema, characterised by parenchymal and alveolar destruction.[123,124] The loss of lung function may result from airway remodelling, characterised by peribronchiolar fibrosis, increased airway smooth muscle mass and mucus cell hyperplasia, as well as loss of elastic recoil by parenchymal damage.[120,123,124]

Many patients with asthma and most patients with COPD are not sufficiently controlled with current drug therapy.[125,126] The development of novel therapeutic approaches to treat these diseases requires better understanding of the (patho)physiology of the lung. In this regard recent studies have indicated that precision-cut lung slices may be highly instrumental in investigating the impact of cellular responses and pharmacological treatment on airway mechanics, inflammation and remodelling under *in vitro* or *ex vivo* conditions.[19,127-129] This is particularly important for the investigation of peripheral airway function, which is not easily addressed by other methods.[122]

3.3.2.1 Precision-Cut Lung Slices as a Model to Study Airway Physiology and Pharmacology

Aberrant airway mechanics, determined by airway smooth muscle function and its interaction with the environmental tissue, are key to airways obstruction in asthma and COPD. Although several approaches may be used to assess airway mechanics by whole lung function measurements *in vivo* and *ex vivo*, they do not give mechanistic insight into the functional behaviour of cells and tissue at the microscopic level. Alternatively, the use of isolated airway smooth muscle strips and cells is hindered by the lack of structural organisation and functional communication with other cells and extracellular matrix components within the tissue, whereas intact airway segments lack the surrounding parenchymal attachments. Due to the preservation of morphological and functional characteristics at the macroscopic and microscopic level, precision-cut lung slices provide a unique additional approach to investigate integrated molecular, cellular and mechanical responses of the lung.

The use of lung slices to study intrapulmonary airway mechanics was introduced by Dandurand *et al.*,[130] using hand-cut transverse slices prepared from low-melting agarose-filled rat lungs. After removal of agarose, methacholine-induced bronchoconstriction in the slices was visualised with an inverted microscope equipped with a video camera, using digital image processing to allow quantitative analysis of individual airway closure. Since variable responses were measured in relatively thick slices of variable thickness (0.5–1 mm), the method was refined by the group of Martin *et al.*,[128] by preparing 250 µm thick lung slices using a mechanical slicer.

The use of precision-cut lung slices in airway physiology and pharmacology has several advantages. Due to the intact structural organisation in lung slices, the mechanical interdependence between airway smooth muscle and its environment is maintained and translates to auxotonic smooth muscle contraction (shortening against an increasing load, due to tethering forces by the parenchyma and stiffness of the airway wall) as *in vivo*, as opposed to isometric or isotonic contractions that are commonly used in organ bath experiments, requiring an artificial static preload. Moreover, airway responses in precision-cut lung slices can be measured under dynamic stretch conditions, simulating the effects of breathing.[131] Importantly, it is possible to study intrapulmonary airways of different airway sizes with different physiological characteristics within the same slice or in sequential slices along the bronchial tree. Due to the microscopic visualisation it is possible to perform these studies down to the smallest peripheral airways. This is a major breakthrough for the study of small airway function in health and disease; these airways cannot be studied in the organ bath, because contraction of these airways relies on the use of parenchymal lung strips that contain contractile elements other than airway smooth muscle.[132] In addition to airway dynamics, the regulation of vascular tone can be studied in the same slices, which may be useful for pharmacological studies[133] and to investigate mechanisms of pulmonary hypertension as in COPD.[134] As for other organs, many slices can usually be obtained from the same tissue, reducing experimental error, allowing the simultaneous assessment of various experimental conditions and internal controls, and limiting the use of experimental animals. Precision-cut lung slices are viable for at least 3 days, enabling the study of long-term effects of mechanical and immunological stimuli,[19,127] as well as transfection studies.[135]

The use of lung slices has now been widely adopted for many species, including rats, mice, guinea pigs, horses, sheep, non-human primates and humans.[129,136] In these species, differential airway and vascular responses to numerous contractile and relaxant stimuli have been investigated, including cholinergic agonists, histamine, serotonin, leukotriene D_4, thromboxane, endothelin-1, β-adrenoceptor agonists and phosphodiesterase inhibitors.[129,133] In some of these studies, it was demonstrated that small peripheral airways of the rat are more sensitive to methacholine, serotonin and thromboxane, but not to endothelin-1, than relatively large airways.[128,137,138] In addition, neurogenic contractile responses have recently been studied by electrical field stimulation (EFS), indicating that nerve endings remain intact in precision-cut lung slices.[136,139] In the rat, EFS responses were almost fully mediated by cholinergic nerves and most prominent in the larger airways, which is probably due to decreasing cholinergic innervation towards the periphery.[139] In distal airways of guinea pigs and humans, but not of rats, excitatory non-adrenergic non-cholinergic responses have also been detected.[136] Importantly, both pharmacological and neurogenic responses vary between the different species. In the most common experimental animals, *i.e.* mice and rats, these responses differ considerably from those in humans, whereas good correlations are generally found for guinea pig and human responses.[133,136] This corresponds well to other experimental models, and may at least partially be explained by differences in (the sensitivity to) endogenous mediators, innervation of the airways and airway anatomy.[121]

Real-time high magnification confocal microscopy and two-photon laser scanning microscopy have been used for extensive investigation of the role of Ca^{2+}-signalling in airway and vascular smooth muscle contraction in precision-cut lung slices, which was pioneered by the group of Sanderson.[129] In lung slices from rats, mice and humans it was demonstrated that contractile agonists propagate Ca^{2+} oscillations in smooth muscle cells, the frequency of which increases with the agonist concentration and correlates with contraction. The cellular mechanisms underlying the agonist-induced Ca^{2+} oscillations have been extensively investigated and are primarily dependent on Ca^{2+} release from internal stores *via* the phosphatidylinositol signalling pathway. Moreover, using precision-cut lung slices with clamped intracellular Ca^{2+} concentrations in the smooth muscle cells, it was demonstrated that, in addition to Ca^{2+} oscillation, contractile agonists may also induce enhanced Ca^{2+} sensitivity to mediate contraction, which involves activation of Rho-kinase and protein kinase C causing inhibition of myosin phosphatase. Conversely, β₂-adrenoceptor agonist-induced relaxation of airway smooth muscle primarily relies on inhibition of agonist-induced Ca^{2+} oscillations and reducing Ca^{2+} sensitivity (reviewed by Sanderson[129]). Of potential clinical importance, in human precision-cut lung slices it was demonstrated that prolonged treatment with both short- and long-acting β₂-agonists induced β₂-adrenoceptor tolerance in peripheral small airways, which involved different molecular mechanisms and was differentially reversed by glucocorticosteroids.[140,141]

3.3.2.2 Precision-Cut Lung Slices as a Model to Study Airway Pathophysiology in Obstructive Lung Diseases

The use of precision-cut lung slices as a model to study pathophysiological processes in asthma and COPD is just emerging. Allergen-induced airways obstruction is a common feature of asthma that has been investigated in airways of different sizes *in vitro*, using animal and human precision-cut lung slices. In both passively sensitised rat and human lung slices, allergen challenge caused an immediate allergic airway constriction that increased in magnitude with decreasing airway size, although different mediators were involved in the two species.[138,142] By contrast, mediators involved in allergic airway constriction induced in passively sensitised lung slices from guinea pigs were very similar to those in human lung slices.[133] The studies indicate an important role of the small airways in asthmatic airways obstruction, which may fit well with some recent clinical observations in asthma patients.[122]

Attempts to measure allergen-induced AHR in precision-cut lung slices from ovalbumin-sensitised mice after acute or chronic allergen challenge were unsuccessful,[129,143] despite AHR *in vivo* and airway remodelling in one study.[143] Remarkably, we have recently shown that ovalbumin-induced AHR may be observed in precision-cut lung slices of chronically challenged guinea pigs, which could be attenuated by inhibition of arginase activity *in vivo* (Maarsingh *et al.*, unpublished), in line with previous *in vivo* and *ex vivo* findings.[121]

In patients with mild asthma it has recently been established that bronchoconstriction induced by repeated methacholine challenge can induce features of airway remodelling.[144] As a potential mechanism, we recently demonstrated in guinea pig precision-cut lung slices that prolonged airway constriction by various contractile stimuli *in vitro* promotes the release of TGF-β, inducing increased expression of contractile proteins both in large and in small airways.[19] The study indicates that lung slices are a useful *in vitro* model to study mechanisms involved in airway remodelling. Precision-cut lung slices from asthmatic patients have thus far not been studied.

Finally, only few studies have been done in animal models of COPD. In one study it was shown that lung slices from rats chronically exposed to tobacco smoke exerted small airway hyper-responsiveness to carbachol, which was associated with increased airway smooth muscle mass and increased levels of IL-13 that could be related to the observed AHR and remodelling.[127] Reasoning that elastase is involved in the development of emphysema and may thus cause increased airway narrowing to acetylcholine by reduced parenchymal tethering forces, cholinergic responsiveness was investigated in precision-cut lung slices from mice intranasally exposed to elastase *in vivo*. However, no changes were found in the acetylcholine concentration–response relationship, despite a loss of parenchymal attachments. Remarkably, overnight *in vitro* exposure of lung slices from control animals to elastase caused increased magnitudes and velocities of airway narrowing, with impaired relaxation.[145] Recently, using a guinea pig model of LPS-induced COPD, we demonstrated that the maximal effect of particularly small airway constriction in response to methacholine was significantly enhanced in precision-cut lung slices of the LPS-treated animals, which was correlated with the degree of emphysema. This indicates that reduced parenchymal tethering forces may indeed be involved in exaggerated airway narrowing in this COPD model. Interestingly, a similar increase in methacholine-induced constriction was found in small airways from patients with COPD (Maarsingh *et al.*, unpublished).

3.3.3 Cancer Research

Currently, cancer is one of the major causes of death worldwide. In this context, it appears that current therapies are not sufficiently effective and novel treatments need to be established.

Development of new drugs for cancer diseases consists of a variety of single steps, starting from the discovery of potential pharmacological effects in cell cultures *in vitro*, leading to the assessment of both anti-cancer properties and toxicity in animal models, and finally to the demonstration of efficacy and safety in humans. Standard two-dimensional (2D) cell culture studies are widely used to delineate the biological, chemical and molecular cues of living cancer cells, but as reductionist models they have limitations. The development of increasingly complex 3D *in vitro* models, which aims to recapitulate the tumour microenvironment, in terms of cell types and acellular constituents, has in part addressed the problem of mimicking cell–cell and cell–matrix interactions occurring during tumour growth and invasion.[146] Nevertheless, as previously mentioned, PCTS have a number of advantages over other *in vitro* and *in vivo* systems, especially in relation to the study of drug-dependent effects. However, up to now, these promising *ex vivo* models have seen only scarce application in the field of anti-cancer therapeutics. Below, we provide an overview of the studies available in the literature based on this technology, which makes use of freshly isolated tissue slices from tumours, therefore maintaining the heterogeneity and complex phenotype of the tumour, including its vasculature, as well as of healthy organs to assess drug toxicity and selectivity.

3.3.3.1 Precision-Cut Tissue Slices for Testing of Chemotherapeutics in Cancer and to Investigate Tumour Progression

As it has been shown in the previous chapters, the PCTS system is uniquely suited to examine molecular responses to toxicant exposures and compare species differences. As an example, PCTS from human hepatocellular carcinoma have been used to analyse the effects of a cyclooxygenase-2 (COX-2) inhibitor, meloxicam, on cancer cell proliferation.[147] The compound was shown to cause apoptosis in cancer tissues, while having limited toxic effects in non-tumorous liver parenchyma. Notably, this analysis confirmed the results obtained in cancer cells *in vitro*, as well in animal models *in vivo*.

Similarly, *ex vivo* COX-2 inhibition with assessment of apoptosis was performed using PCTS in human liver metastases of colorectal carcinoma.[148] Slices of these liver metastases were treated with different concentrations of meloxicam, and apoptosis was increased after COX-2 inhibition, in accordance with several *in vitro* data using cancer cells.

Sonnenberg *et al.*[149] used tissue slices to demonstrate for the first time that not only tumour cells but also carcinoma-associated fibroblasts are targeted by cytotoxic chemotherapy (using cisplatin or taxol). Effects of taxol[150] and of the PI3K inhibitor LY294002[151] were also investigated using the PCTS technique.

Nephrotoxic side-effects induced by the anti-cancer drug cisplatin were also investigated in human and rat kidney slices and characterised morphologically, as well as in terms of gene expression and functional changes, evidencing mechanisms of apoptosis induction.[152] Overall, the acute nephrosis of tubular epithelium induced by cisplatin *in vivo* was reproduced in both human and rat kidney slices, while the glomerulus appeared resistant even at high drug concentration (80 µM).

Very recently, inflammatory cytokine responses resulting from treatment with the cytotoxic drug NSC 710305 (Phortress), a metabolically activated prodrug that causes DNA adduct formation, was evaluated in rat lung PCTS.[153] The rationale for this study was the observed pulmonary toxicity in animal models, which could not be reliably monitored by clinical pulmonary function tests.

PCTS from liver metastasis were also used to study the effects of activating the so-called extrinsic apoptosis pathway by the CD95 antibody receptor and tumour necrosis factor (TNFα) directly in tumour tissue.[154] Thus, tumours derived from MMTV-*neuN* mice, which develop mammary adenocarcinomas, were sliced and used to study the effects of the cytokine INF-α and the drug doxorubicin on tumour tissues using biochemical and histological methods.[155] Interestingly, in this study PCTS have also been used to prepare explant cultures of tumour epithelial cells (ECs), because the ultra-thin slices have high surface area, allowing outgrowth of tumour ECs at a rate superior to that observed with tumour pieces obtained with conventional cutting procedures.

In addition, the toxicity of TNFα itself, known to act as a potent antineoplastic agent *via* receptor-mediated cell death, was evaluated in human liver slices undergoing a fructose-mediated ATP depletion.[156] The obtained results showed that these non-malignant human liver specimens were protected transiently against TNF-induced death by systemic pretreatment with fructose used in non-toxic/physiological concentrations. Conversely, samples of human tissues including hepatocellular carcinoma, colorectal carcinoma and pancreatic carcinoma, were not protected by energy-dependent cell death pathways induced by TNFα upon fructose pretreatment.

Interestingly, PCTS have been used to study therapeutic efficacy and toxicity of conditionally replicative adenovirus,[157] and have been shown promising to test virus-mediated gene transfer. In gene therapy approaches, a therapeutic gene for mutation compensation, immunopotentiation, or prodrug activation is transferred. At present, the most promising gene delivery vehicle is the recombinant adenoviral vector. Instead, in the case of virotherapy, tumour cell killing is achieved by oncolysis – virus replication cell killing.

Within this frame, some studies described the use of PCTS to assess the best tissue-specific promoter (TSP) to restrict transgene expression or viral replication necessary to increase specificity towards tumour tissues, and to reduce adverse effects in non-target tissues such as the liver. Curiel and co-workers were the first to report on the PCTS technique applied in human breast cancer to systematically test a series of promoters whose genes are overexpressed in a variety of cancers, but minimally expressed in normal host tissues.[158] The best promoter out of this study was found to be the stromal cell derived factor CXCR4, showing the highest luciferase activity in the breast cancer tissue slices in comparison to non-tumorigenic liver slices.

Similarly, Rots *et al.*[159] used PCTS to test the specificity and efficiency of a conditionally replicated adenovirus, based on the epithelial glycoprotein-2 (EPG-2) promoter, to eradicate EGP-2 positive tumour tissues from mice with respect to healthy human liver samples. Since the issue of selectivity of conditionally replicated adenoviruses towards cancer cells with respect to healthy ones is crucial in this field, *ex vivo* slices from healthy human liver were also used to validate targeted virus selectivity.[160]

Liver slices were also validated as a predictive system to test the oncolytic effectiveness of different measles vaccine viruses in primary or secondary tumours of human liver.[161] Employing mixed tumorous/non-tumorous liver slices from patients exhibiting colorectal liver metastasis for infection by non-targeted MeV-GFP vectors, it was possible to detect tumour areas with prevalent expression of the MeV encoded GFP marker gene.[162] Furthermore, confocal microscopy demonstrated virus penetration throughout tumour tissues into deep cell layers. Notably, and in discrepancy to rat liver tissue slices infected with recombinant non-replicating adenovirus exhibiting an infection pattern restricted only to the outer layer of the liver slices,[163] this study showed that MeV can actually penetrate throughout the entire tissue slice, mainly caused by cellular fusions.[162]

Interestingly, the PCTS system has also been applied to study the chemo-preventive effects of certain phytochemicals including xanthohumol[164] and sulforaphane.[165,166] The latter compound, for example, is believed to elicit its chemo-preventive activity by suppressing the interaction of the reactive intermediates of carcinogenic chemicals (*e.g.* polycyclic aromatic hydrocarbons) with DNA. The pivotal mechanism through which isothiocyanates of this type antagonise the carcinogenicity of chemicals seems to rely on the up-regulation of detoxification enzymes such as NAD(P)H:quinone oxidoreductase (NQO1) and glutathione S-transferases (GSTs). Thus, sulforaphane was incubated with human liver slices, in comparison with rat liver ones, and the expressions of the above-mentioned enzymes were determined.[167] The obtained results showed that sulforaphane elevates GST expression in isoform-specific manner in both rat and human liver. However, NQO1, either at the activity or protein level, is inducible by these compounds only in rat, and very poorly, or not at all, in human liver, therefore highlighting important species differences.

Overall, the promising results described above on the various applications of PCTS in anti-cancer drug development, from the search for concentration-dependent effects of compounds in tumours or healthy organs to model different stages of cancer diseases, should constitute the basis to invest in this type of *ex vivo* technology as a powerful tool for research in preclinical trials. Nowadays, this would be particularly important in the study of the effects of targeted therapies, including small molecule inhibitors and adenoviral vectors. Moreover this method may enable a molecular profiling of all tumour compartments using laser micro-dissection techniques. Subsequently, it can be used to analyse the molecular response of each tissue component to both cytotoxic drugs and signal transduction inhibitors *via* genomic or proteomic approaches.

3.3.4 Viral Infection Research

Limited access to *in vivo* and *in vitro* models has hampered studies on the mechanisms of hepatitis C virus (HCV) infection. Recently it was shown that HCV infection could be studied in human liver slices.[168] For the first time, the ability of primary HCV isolates to undergo *de novo* viral replication in human liver slices was demonstrated. This opens up the possibility of using liver slices to study virus infections that can lead to liver fibrosis. Furthermore, human liver slices is a new model to study the biology of HCV *ex vivo* and represents a powerful tool for studying the viral life cycle, dynamics of virus spread in native tissue and to evaluate the efficacy of new antiviral drugs. Furthermore, Carranza-Rosales *et al.*[169,170] showed that liver slices are an appropriate *ex vivo* parasitic disease model, to study the pathogenesis of hepatic amoebiasis.

3.4 Concluding Remarks

PCTS have been applied successfully by many researchers because they represent an organ mini-model that closely resembles the organ from which it is prepared, with all cell types present in their original tissue-matrix configuration. They are particularly useful for studies on drug transport, metabolism and toxicity, and as such, deserve a prominent place in the toolbox of the pharmaceutical industry. They reproduce adequately the organ's physiology and pathology and enable the study of human-specific disease mechanisms and allow the testing of new drugs. Finally they have potent significance for studying tumour responses in the complex environment of cancer tissue.

References

1. I. Kola and J. Landis, *Nat. Rev. Drug Discov.*, 2004, **3**, 711.
2. J. A. Kramer, J. E. Sagartz and D. L. Morris, *Nat. Rev. Drug Discov.*, 2007, **6**, 636.
3. C. Emoto, N. Murayama, A. Rostami-Hodjegan and H. Yamazaki, *Curr. Drug Metab.*, 2010, **11**, 678.
4. C. L. Krumdieck, *Xenobiotica*, 2013, **43**, 2.
5. T. Lisman, H. G. D. Leuvenink, R. J. Porte and R. J. Ploeg, *J. Thromb. Haemost.*, 2011, **9**, 1959.
6. R. L. Fisher and A. E. Vickers, *Xenobiotica*, 2013, **43**, 8.
7. A. E. Vickers and R. L. Fisher, *Xenobiotica*, 2013, **43**, 29.
8. P. Olinga, M. Merema, I. H. Hof, K. P. de Jong, M. J. H. Slooff, D. K. F. Meijer and G. M. M. Groothuis, *Drug Metab. Dispos.*, 1998, **26**, 5.
9. J. T. Heinonen, R. Fisher, K. Brendel and D. L. Eaton, *Toxicol. Appl. Pharm.*, 1996, **136**, 1.
10. P. Olinga, M. T. Merema, I. H. Hof, M. H. De Jager, K. P. De Jong, M. J. Slooff, D. K. Meijer and G. M. Groothuis, *Xenobiotica*, 1998, **28**, 346.
11. J. Plazar, I. Hrejac, P. Pirih, M. Filipic and G. M. M. Groothuis, *Toxicol. In Vitro*, 2007, **21**, 1134.
12. G. M. Fahy, N. Guan, I. A. de Graaf, Y. Tan, L. Griffin and G. M. Groothuis, *Xenobiotica*, 2013, **43**, 113.
13. H. A. Krebs, *Hoppe-Seyl Z.*, 1933, **217**, 190.
14. O. Warburg, *Biochem. Z.*, 1923, **142**, 317.
15. R. J. Price, S. E. Ball, A. B. Renwick, P. T. Barton, J. A. Beamand and B. G. Lake, *Xenobiotica*, 1998, **28**, 361.
16. M. Zimmermann, J. Lampe, S. Lange, I. Smirnow, A. Königsrainer, C. Hann-von-Weyhern, F. Fend, M. Gregor, M. Bitzer and U. M. Lauer, *Cytotechnology*, 2009, **61**, 145.
17. I. A. de Graaf, P. Olinga, M. H. de Jager, M. T. Merema, R. de Kanter, E. G. van de Kerkhof and G. M. Groothuis, *Nat. Protoc.*, 2010, **5**, 1540.
18. J. P. Morin, J. M. Baste, A. Gay, C. Crochemore, C. Corbiere and C. Monteil, *Xenobiotica*, 2013, **43**, 63.

19. T. A. Oenema, H. Maarsingh, M. Smit, G. M. Groothuis, H. Meurs and R. Gosens, *PLoS One,* 2013, **8**, e65580.
20. R. L. Fisher, S. J. Hasal, J. T. Sanuik, A. J. Gandolfi and K. Brendel, *Toxicol. Methods,* 1995, **5**, 115.
21. I. A. M. de Graaf, R. de Kanter, M. H. de Jager, R. Camacho, E. Langenkamp, E. G. van de Kerkhof and G. M. M. Groothuis, *Drug Metab. Dispos.,* 2006, **34**, 591.
22. P. D. Worboys, A. Bradbury and J. B. Houston, *Drug Metab. Dispos.,* 1997, **25**, 460.
23. R. J. Price, A. B. Renwick, P. T. Barton, J. B. Houston and B. G. Lake, *ATLA, Altern. Lab. Anim.,* 1998, **26**, 541.
24. M. Terdoslavich, I. A. de Graaf, J. H. Proost, A. Cocolo, S. Passamonti and G. M. Groothuis, *Drug Metab. Lett.,* 2012, **6**, 165.
25. A. E. Vickers, M. Saulnier, E. Cruz, M. T. Merema, K. Rose, P. Bentley and P. Olinga, *Toxicol. Sci.,* 2004, **82**, 534.
26. P. M. van Midwoud, M. T. Merema, N. Verweij, G. M. M. Groothuis and E. Verpoorte, *Biotechnol. Bioeng.,* 2011, **108**, 1404.
27. P. M. van Midwoud, M. T. Merema, E. Verpoorte and G. M. M. Groothuis, *JALA, J. Lab. Autom.,* 2011, **16**, 468.
28. K. Schumacher, Y. M. Khong, S. Chang, J. Ni, W. X. Sun and H. Yu, *Tissue Eng.,* 2007, **13**, 197.
29. F. Goethals, D. Deboyser, V. Lefebvre, I. Decoster and M. Roberfroid, *Toxicol. In Vitro,* 1990, **4**, 435.
30. P. M. van Midwoud, J. Janssen, M. T. Merema, I. A. de Graaf, G. M. Groothuis and E. Verpoorte, *Anal. Chem.,* 2011, **83**, 84.
31. P. M. van Midwoud, M. T. Merema, E. Verpoorte and G. M. Groothuis, *Lab Chip,* 2010, **10**, 2778.
32. A. R. Parrish, R. T. Dorr, A. J. Gandolfi and K. Brendel, *Toxicol. In Vitro,* 1994, **8**, 1233.
33. O. Kretz, C. Guenat, R. Beilstein and G. Gross, *J. Pharmacol. Toxicol.,* 2002, **48**, 119.
34. A. E. Vickers, J. R. Sinclair, R. L. Fisher, S. R. Morris and W. Way, *Toxicol. Appl. Pharmacol.,* 2010, **244**, 354.
35. S. Karim, E. Liaskou, S. Hadley, J. Youster, J. Faint, D. H. Adams and P. F. Lalor, *Toxicol. Sci.,* 2013, **132**, 131.
36. D. A. Bow, J. L. Perry, D. S. Miller, J. B. Pritchard and K. L. Brouwer, *Drug Metab. Dispos.,* 2008, **36**, 198.
37. G. M. M. Groothuis, C. E. Hulstaert, D. Kalicharan and M. J. Hardonk, *Eur. J. Cell Biol.,* 1981, **26**, 43.
38. I. A. Graaf, G. M. Groothuis and P. Olinga, *Expert Opin. Drug Metab. Toxicol.,* 2007, **3**, 879.
39. G. M. Groothuis and I. A. de Graaf, *Curr. Drug Metab.,* 2013, **14**, 112.
40. C. Ioannides, *Xenobiotica,* 2013, **43**, 15.
41. B. G. Lake and R. J. Price, *Xenobiotica,* 2013, **43**, 41.
42. R. De Kanter, M. H. De Jager, A. L. Draaisma, J. U. Jurva, P. Olinga, D. K. F. Meijer and G. M. M. Groothuis, *Xenobiotica,* 2002, **32**, 349.

43. I. A. de Graaf, R. de Kanter, M. H. de Jager, R. Camacho, E. Langenkamp, E. G. van de Kerkhof and G. M. Groothuis, *Drug Metab. Dispos.*, 2006, **34**, 591.
44. A. Lupp, M. Danz and D. Muller, *Toxicology*, 2005, **206**, 427.
45. E. G. van de Kerkhof, I. A. de Graaf and G. M. Groothuis, *Curr. Drug Metab.*, 2007, **8**, 658.
46. E. G. van de Kerkhof, I. A. M. de Graaf, M. H. de Jager, D. K. F. Meijer and G. M. M. Groothuis, *Drug Metab. Dispos.*, 2005, **33**, 1613.
47. E. G. van de Kerkhof, I. A. M. de Graaf, M. H. de Jager and G. M. M. Groothuis, *Drug Metab. Dispos.*, 2007, **35**, 898.
48. A. A. Khan, E. C. Y. Chow, A. M. M. A. van Loenen-Weemaes, R. J. Porte, K. S. Pang and G. M. M. Groothuis, *Eur. J. Pharm. Sci.*, 2009, **37**, 115.
49. D. Jung, M. G. L. Elferink, F. Stellaard and G. M. M. Groothuis, *Liver Int.*, 2007, **27**, 137.
50. M. G. L. Elferink, P. Olinga, A. L. Draaisma, M. T. Merema, K. N. Faber, M. J. H. Slooff, D. K. F. Meijer and G. M. M. Groothuis, *Am. J. Physiol.*, 2004, **287**, G1008.
51. P. Olinga, I. H. Hof, M. T. Merema, M. Smit, M. H. de Jager, P. J. Swart, M. J. Slooff, D. K. Meijer and G. M. Groothuis, *J. Pharmacol. Toxicol.*, 2001, **45**, 55.
52. K. Barth, R. Bläsche and M. Kasper, *Histochem. Cell Biol.*, 2006, **126**, 563.
53. M. Hadi, Y. X. Chen, V. Starokozhko, M. T. Merema and G. M. M. Groothuis, *Chem. Res. Toxicol.*, 2012, **25**, 1938.
54. A. A. Khan, E. C. Chow, R. J. Porte, K. S. Pang and G. M. Groothuis, *Biopharm. Drug Dispos.*, 2009, **30**, 241.
55. A. A. Khan, E. C. Y. Chow, R. J. Porte, K. S. Pang and G. M. M. Groothuis, *Toxicol. In Vitro*, 2011, **25**, 80.
56. M. Possidente, S. Dragoni, G. Franco, M. Gori, E. Bertelli, E. Teodori, M. Frosini and M. Valoti, *Eur. J. Pharm. Biopharm.*, 2011, **79**, 343.
57. H. Lennernas, S. Nylander and A. L. Ungell, *Pharm. Res.*, 1997, **14**, 667.
58. B. I. Polentarutti, A. L. Peterson, A. K. Sjoberg, E. K. I. Anderberg, L. M. Utter and A. L. B. Ungell, *Pharm. Res.*, 1999, **16**, 446.
59. E. G. van de Kerkhof, A. L. B. Ungell, A. K. Sjoberg, M. H. de Jager, C. Hilgendorf, I. A. M. de Graaf and G. M. M. Groothuis, *Drug Metab. Dispos.*, 2006, **34**, 1893.
60. S. Siissalo, H. de Waard, M. H. de Jager, R. Hayeshi, H. W. Frijlink, W. L. J. Hinrichs, H. Dinter-Heidorn, A. van Dam, J. H. Proost, G. M. M. Groothuis and I. A. M. de Graaf, *Drug Metab. Dispos.*, 2013, **41**, 1557.
61. N. A. Saunders, P. J. Rigby, K. F. Ilett and R. F. Minchin, *Lab. Invest.*, 1988, **59**, 380.
62. N. A. Saunders, J. K. Mcgeachie, K. F. Ilett and R. F. Minchin, *Am. J. Physiol.*, 1989, **257**, C579.
63. X. Niu, I. A. de Graaf and G. M. Groothuis, *Xenobiotica*, 2013, **43**, 73.
64. A. E. Vickers and R. L. Fisher, *Xenobiotica*, 2013, **43**, 29.

65. R. de Kanter, M. Monshouwer, D. K. F. Meijer and G. M. M. Groothuis, *Curr. Drug Metab.*, 2002, **3**, 39.
66. F. Boess, M. Kamber, S. Romer, R. Gasser, D. Muller, S. Albertini and L. Suter, *Toxicol. Sci.*, 2003, **73**, 386.
67. M. G. L. Elferink, P. Olinga, E. M. van Leeuwen, S. Bauerschmidt, J. Polman, W. G. Schoonen, S. H. Heisterkamp and G. M. M. Groothuis, *Toxicol. Appl. Pharm.*, 2011, **253**, 57.
68. M. G. L. Elferink, P. Olinga, A. L. Draaisma, M. T. Merema, S. Bauerschmidt, J. Polman, W. G. Schoonen and G. M. M. Groothuis, *Toxicol. Appl. Pharm.*, 2008, **229**, 300.
69. P. Olinga, M. T. Merema, M. H. de Jager, F. Derks, B. N. Melgert, H. Moshage, M. J. H. Slooff, D. K. F. Meijer, K. Poelstra and G. M. M. Groothuis, *J. Hepatol.*, 2001, **35**, 187.
70. H. Ishiyama, K. Ogino, T. Hobara and T., *Eur. J. Pharm. Environ.*, 1995, **292**, 135.
71. J. B. Wijeweera, A. J. Gandolfi, D. A. Badger, I. G. Sipes and K. Brendel, *Fundam. Appl. Toxicol.*, 1996, **34**, 73.
72. M. Hadi, S. Dragovic, R. van Swelm, B. Herpers, B. van de Water, F. G. M. Russel, J. N. M. Commandeur and G. M. M. Groothuis, *Arch. Toxicol.*, 2013, **87**, 155.
73. M. G. Miller, J. Beyer, G. L. Hall, L. A. Degraffenried and P. E. Adams, *Toxicol. Appl. Pharm.*, 1993, **122**, 108.
74. B. G. Lake, *Food Chem. Toxicol.*, 1999, **37**, 423.
75. R. L. Fisher, M. S. Smith, S. J. Hasal, K. S. Hasal, A. J. Gandolfi and K. Brendel, *Hum. Exp. Toxicol.*, 1994, **13**, 466.
76. R. J. Price, A. B. Renwick, P. T. Wield, J. A. Beamand and B. G. Lake, *Arch. Toxicol.*, 1995, **69**, 405.
77. S. Switalla, J. Knebel, D. Ritter, C. Dasenbrock, N. Krug, A. Braun and K. Sewald, *Toxicol. In Vitro*, 2013, **27**, 798.
78. R. J. Price, D. G. Walters, C. Hoff, H. Mistry, A. B. Renwick, P. T. Wield, J. A. Beamand and B. G. Lake, *Food Chem. Toxicol.*, 1996, **34**, 603.
79. E. Le Prieur, E. Vaz, A. Bion, F. Dionnet and J. P. Morin, *Arch. Toxicol.*, 2000, **74**, 460.
80. E. Streck, R. A. Jorres, R. M. Huber and A. Bergner, *J. Invest. Allerg. Clin.*, 2010, **20**, 324.
81. J. L. Wright and A. Churg, *J. Appl. Physiol.*, 2008, **104**, 1462.
82. J. C. J. Lin, J. P. Roy, J. Verreault, S. Talbot, F. Cote, R. Couture and A. Morin, *Toxicology*, 2012, **293**, 125.
83. A. M. Hays, R. C. Lantz and M. L. Witten, *Toxicol. Pathol.*, 2003, **31**, 200.
84. S. Switalla, J. Knebel, D. Ritter, N. Krug, A. Braun and K. Sewald, *Toxicol. Lett.*, 2010, **196**, 117.
85. C. Monteil, E. Le Prieur, S. Buisson, J. P. Morin, M. Guerbet and J. M. Jouany, *Toxicology*, 1999, **133**, 129.
86. R. Nave, R. Fisher and N. McCracken, *Respir. Res.*, 2007, **8**, 65.
87. S. L. Friedman, D. Sheppard, J. S. Duffield and S. Violette, *Sci. Transl. Med.*, 2013, **5**, 167sr1.

88. T. A. Wynn, *J. Pathol.*, 2008, **214**, 199.
89. D. Schuppan and Y. O. Kim, *J. Clin. Invest.*, 2013, **123**, 1887.
90. Y. Popov and D. Schuppan, *Hepatology*, 2009, **50**, 1294.
91. I. Mederacke, *Z. Gastroenterol.*, 2013, **51**, 55.
92. A. A. Eddy, J. M. Lopez-Guisa, D. M. Okamura and I. Yamaguchi, *Pediatr. Nephrol.*, 2012, **27**, 1233.
93. F. Rieder, S. Kessler, M. Sans and C. Fiocchi, *Am. J. Physiol. Gastrointest. Liver Physiol.*, 2012, **303**, G786.
94. I. M. Westra, B. T. Pham, G. M. Groothuis and P. Olinga, *Xenobiotica*, 2013, **43**, 98.
95. J. L. Rinn, C. Bondre, H. B. Gladstone, P. O. Brown and H. Y. Chang, *PLoS Genet.*, 2006, **2**, e119.
96. J. Levitsky and M. E. Mailliard, *Semin. Liver Dis.*, 2004, **24**, 233.
97. C. S. Schaffert, M. J. Duryee, R. G. Bennett, A. L. DeVeney, D. J. Tuma, P. Olinga, K. C. Easterling, G. M. Thiele and L. W. Klassen, *Am. J. Physiol.*, 2010, **299**, G661.
98. L. W. Klassen, G. M. Thiele, M. J. Duryee, C. S. Schaffert, A. L. DeVeney, C. D. Hunter, P. Olinga and D. J. Tuma, *Biochem. Pharmacol.*, 2008, **76**, 426.
99. Y. Guo, X. Q. Wu, C. Zhang, Z. X. Liao, Y. Wu and H. Wang, *J. Med. Food*, 2012, **15**, 557.
100. H. Hayashi and T. Sakai, *Am. J. Physiol.*, 2011, **300**, G729.
101. M. van de Bovenkamp, G. M. Groothuis, D. K. Meijer and P. Olinga, *Toxicol. In Vitro*, 2008, **22**, 771.
102. M. van de Bovenkamp, G. M. M. Groothuis, D. K. F. Meijer, M. J. H. Slooff and P. Olinga, *Chem. Biol. Interact.*, 2006, **162**, 62.
103. C. Guyot, C. Combe, H. Clouzeau-Girard, V. Moronvalle-Halley and A. Desmouliere, *Virchows Arch.*, 2007, **450**, 503.
104. Y. Guo, H. Wang and C. Zhang, *Clin. Exp. Pharmacol. Physiol.*, 2007, **34**, 406.
105. T. Nakamura, H. Akiyoshi, I. Saito and K. Sato, *J. Hepatol.*, 1999, **30**, 101.
106. K. W. Wang, B. L. Lin, J. J. Brems and R. L. Gamelli, *Apoptosis*, 2013, **18**, 566.
107. S. S. Veidal, M. J. Nielsen, D. J. Leeming and M. A. Karsdal, *BMC Res. Notes*, 2012, **5**, 686.
108. C. Guyot, C. Combe, C. Balabaud, P. Bioulac-Sage and A. Desmouliere, *J. Hepatol.*, 2007, **46**, 142.
109. H. Clouzeau-Girard, C. Guyot, C. Combe, V. Moronvalle-Halley, C. Housset, T. Lamireau, J. Rosenbaum and A. Desmouliere, *Lab. Invest.*, 2006, **86**, 275.
110. L. Beljaars, P. Olinga, G. Molema, P. de Bleser, A. Geerts, G. M. Groothuis, D. K. Meijer and K. Poelstra, *Liver*, 2001, **21**, 320.
111. W. I. Hagens, P. Olinga, D. K. Meijer, G. M. Groothuis, L. Beljaars and K. Poelstra, *Liver Int.*, 2006, **26**, 232.
112. Z. Manojlovic, J. Blackmon and B. Stefanovic, *PLoS One*, 2013, **8**, e65897.
113. I. M. Westra, G. M. Groothuis and P. Olinga, *Gastroenterology*, 2011, **140**, S981.

114. I. M. Westra, D. Oosterhuis, G. M. Groothuis and P. Olinga, *Hepatology*, 2011, **54**, 754A.
115. I. M. Westra, D. Oosterhuis, G. M. Groothuis and P. Olinga, *PLoS One*, 2014, **9**, e95462.
116. P. Olinga and D. Schuppan, *J. Hepatol.*, 2013, **58**, 1252.
117. B. T. Pham, W. T. van Haaften, J. Nieken, I. A. M. De Graaf, D. Oosterhuis and P. Olinga, *Gastroenterology*, 2014, **146**, S791.
118. J. Bousquet, P. K. Jeffery, W. W. Busse, M. Johnson and A. M. Vignola, *Am. J. Respir. Crit. Care Med.*, 2000, **161**, 1720.
119. D. S. Postma and H. A. Kerstjens, *Am. J. Respir. Crit. Care Med.*, 1998, **158**, S187.
120. D. S. Postma and W. Timens, *Proc. Am. Thorac. Soc.*, 2006, **3**, 434.
121. H. Meurs, R. Gosens and J. Zaagsma, *Eur. Respir. J.*, 2008, **32**, 487.
122. M. van den Berge, N. H. ten Hacken, J. Cohen, W. R. Douma and D. S. Postma, *Chest*, 2011, **139**, 412.
123. J. C. Hogg and W. Timens, *Ann. Rev. Pathol.*, 2009, **4**, 435.
124. J. C. Hogg, F. Chu, S. Utokaparch, R. Woods, W. M. Elliott, L. Buzatu, R. M. Cherniack, R. M. Rogers, F. C. Sciurba, H. O. Coxson and P. D. Pare, *New Engl. J. Med.*, 2004, **350**, 2645.
125. P. J. Barnes, *Chest*, 2008, **134**, 1278.
126. P. J. Barnes, *J. Allerg. Clin. Immunol.*, 2012, **129**, 48.
127. P. R. Cooper, C. T. Poll, P. J. Barnes and R. G. Sturton, *Am. J. Respir. Cell Mol. Biol.*, 2010, **43**, 220.
128. C. Martin, S. Uhlig and V. Ullrich, *Eur. Respir. J.*, 1996, **9**, 2479.
129. M. J. Sanderson, *Pulm. Pharmacol. Ther.*, 2011, **24**, 452.
130. R. J. Dandurand, C. G. Wang, N. C. Phillips and D. H. Eidelman, *J. Appl. Physiol.*, 1993, **75**, 364.
131. T. L. Lavoie, R. Krishnan, H. R. Siegel, E. D. Maston, J. J. Fredberg, J. Solway and M. L. Dowell, *Am. J. Respir. Crit. Care Med.*, 2012, **186**, 225.
132. J. N. Evans and K. B. Adler, *Exp. Lung Res.*, 1981, **2**, 187.
133. A. R. Ressmeyer, A. K. Larsson, E. Vollmer, S. E. Dahlen, S. Uhlig and C. Martin, *Eur. Respir. J.*, 2006, **28**, 603.
134. J. L. Wright and A. Churg, *J. Appl. Physiol.*, 2008, **104**, 1462.
135. Z. Yang, N. Balenga, P. R. Cooper, G. Damera, R. Edwards, C. E. Brightling, R. A. Panettieri, Jr and K. M. Druey, *Am. J. Respir. Cell Mol. Biol.*, 2012, **46**, 823.
136. M. Schleputz, A. D. Rieg, S. Seehase, J. Spillner, A. Perez-Bouza, T. Braunschweig, T. Schroeder, M. Bernau, V. Lambermont, C. Schlumbohm, K. Sewald, R. Autschbach, A. Braun, B. W. Kramer, S. Uhlig and C. Martin, *PLoS One*, 2012, 7, e47344.
137. C. Martin, V. Ullrich and S. Uhlig, *Eur. Respir. J.*, 2000, **16**, 316.
138. A. Wohlsen, S. Uhlig and C. Martin, *Am. J. Respir. Crit. Care Med.*, 2001, **163**, 1462.
139. M. Schleputz, S. Uhlig and C. Martin, *J. Appl. Physiol.*, 2011, **110**, 545.
140. P. R. Cooper, R. C. Kurten, J. Zhang, D. J. Nicholls, I. A. Dainty and R. A. Panettieri, *Br. J. Pharmacol.*, 2011, **163**, 521.

141. P. R. Cooper and R. A. Panettieri Jr, *J. Allerg. Clin. Immunol.*, 2008, **122**, 734.
142. A. Wohlsen, C. Martin, E. Vollmer, D. Branscheid, H. Magnussen, W. M. Becker, U. Lepp and S. Uhlig, *Eur. Respir. J.*, 2003, **21**, 1024.
143. A. D. Chew, J. A. Hirota, R. Ellis, J. Wattie, M. D. Inman and L. J. Janssen, *Eur. Respir. J.*, 2008, **31**, 532.
144. C. L. Grainge, L. C. Lau, J. A. Ward, V. Dulay, G. Lahiff, S. Wilson, S. Holgate, D. E. Davies and P. H. Howarth, *New Engl. J. Med.*, 2011, **364**, 2006.
145. M. A. Khan, S. Kianpour, M. R. Stampfli and L. J. Janssen, *Eur. Respir. J.*, 2007, **30**, 691.
146. E. Cukierman, R. Pankov, D. R. Stevens and K. M. Yamada, *Science*, 2001, **294**, 1708.
147. M. A. Kern, A. M. Haugg, E. Eiteneuer, E. Konze, U. Drebber, H. P. Dienes, K. Breuhahn, P. Schirmacher and H. U. Kasper, *Liver Int.*, 2006, **26**, 604.
148. H. U. Kasper, E. Konze, H. P. Dienes, D. L. Stippel, P. Schirmacher and M. Kern, *Anticancer Res.*, 2010, **30**, 2017.
149. M. Sonnenberg, H. van der Kuip, S. Haubeis, P. Fritz, W. Schroth, G. Friedel, W. Simon, T. E. Murdter and W. E. Aulitzky, *BMC Cancer*, 2008, **8**, 364.
150. H. van der Kuip, T. E. Murdter, M. Sonnenberg, M. McClellan, S. Gutzeit, A. Gerteis, W. Simon, P. Fritz and W. E. Aulitzky, *BMC Cancer*, 2006, **6**, 86.
151. V. Vaira, G. Fedele, S. Pyne, E. Fasoli, G. Zadra, D. Bailey, E. Snyder, A. Faversani, G. Coggi, R. Flavin, S. Bosari and M. Loda, *Proc. Natl. Acad. Sci. U.S.A.*, 2010, **107**, 8352.
152. A. E. M. Vickers, K. Rose, R. Fisher, M. Saulnier, P. Sahota and P. Bentley, *Toxicol. Pathol.*, 2004, **32**, 577.
153. H. P. Behrsing, M. J. Furniss, M. Davis, J. E. Tomaszewski and R. E. Parchment, *Toxicol. Sci.*, 2013, **131**, 470.
154. H. U. Kasper, E. Konze, M. Kern and D. L. Stippel, *In Vivo*, 2010, **24**, 653.
155. N. Parajuli and W. Doppler, *In Vitro Cell. Dev. Biol.: Anim.*, 2009, **45**, 442.
156. T. Weiland, K. Klein, M. Zimmermann, T. Speicher, S. Venturelli, A. Berger, H. Bantel, A. Konigsrainer, M. Schenk, T. S. Weiss, A. Wendel, M. Schwab, M. Bitzer and U. M. Lauer, *PLoS One*, 2012, 7, e52496.
157. T. O. Kirby, A. Rivera, D. Rein, M. Wang, I. Ulasov, M. Breidenbach, M. Kataram, J. L. Contreras, C. Krumdieck, M. Yamamoto, M. G. Rots, H. J. Haisma, R. D. Alvarez, P. J. Mahasreshti and D. T. Curiel, *Clin. Cancer Res.*, 2004, **10**, 8697.
158. M. A. Stoff-Khalili, A. Stoff, A. A. Rivera, N. S. Banerjee, M. Everts, S. Young, G. P. Siegal, D. F. Richter, M. H. Wang, P. Dall, J. M. Mathis, Z. B. Zhu and D. T. Curiel, *Breast Cancer Res.*, 2005, 7, R1141.
159. W. M. Gommans, P. M. J. McLaughlin, J. A. Schalk, G. M. M. Groothuis, H. J. Haisma and M. G. Rots, *J. Controlled Release*, 2007, **117**, 1.

160. J. E. Carette, H. C. Graat, F. H. Schagen, D. C. Mastenbroek, M. G. Rots, H. J. Haisma, G. M. Groothuis, G. R. Schaap, J. Bras, G. J. Kaspers, P. I. M. Wuisman, W. R. Gerritsen and V. W. van Beusechem, *Virology*, 2007, **361**, 56.
161. M. Zimmermann, T. Weiland, M. Bitzer and U. M. Lauer, in *Human Cell Culture Protocol*, eds R. Ragai and D. Robin, Springer, Berlin, 2012, vol. 806, p. 121.
162. M. Zimmermann, S. Armeanu, I. Smirnow, S. Kupka, S. Wagner, M. Wehrmann, M. G. Rots, G. M. M. Groothuis, T. S. Weiss, A. Konigsrainer, M. Gregor, M. Bitzer and U. M. Lauer, *Int. J. Oncol.*, 2009, **34**, 1247.
163. M. G. Rots, M. G. L. Elferink, W. M. Gommans, D. Oosterhuis, J. A. C. Schalk, D. T. Curiel, P. Olinga, H. J. Haisma and G. M. M. Groothuis, *J. Gene Med.*, 2006, **8**, 35.
164. J. Plazar, M. Filipic and G. M. M. Groothuis, *Toxicol. In Vitro*, 2008, **22**, 318.
165. A. F. A. Razis, R. Iori and C. Ioannides, *Int. J. Cancer*, 2011, **128**, 2775.
166. N. Hanlon, C. L. Poynton, N. Coldham, M. J. Sauer and C. Ioannides, *Mol. Nutr. Food Res.*, 2009, **53**, 836.
167. N. Hanlon, N. Coldham, M. J. Sauer and C. Ioannides, *Chem. Biol. Interact.*, 2009, **177**, 115.
168. S. Lagaye, H. Shen, B. Saunier, M. Nascimbeni, J. Gaston, P. Bourdoncle, L. Hannoun, P. P. Massault, A. V. Pichard, V. Mallet and S. Pol, *Hepatology (Baltimore)*, 2012, **56**, 861.
169. P. Carranza-Rosales, M. G. Santiago-Mauricio, N. E. Guzmán-Delgado, J. Vargas-Villarreal, G. Lozano-Garza, E. Viveros-Valdez, R. Ortiz-López, J. Morán-Martínez and A. J. Gandolfi, *Exp. Parasitol.*, 2012, **132**, 424.
170. P. Carranza-Rosales, M. G. Santiago-Mauricio, N. E. Guzmán-Delgado, J. Vargas-Villarreal, G. Lozano-Garza, J. Ventura-Juárez, I. Balderas-Rentería, J. Morán-Martínez and A. J. Gandolfi, *Exp. Parasitol.*, 2010, **126**, 117.

CHAPTER 4

Modelling the Human Respiratory System: Approaches for in Vitro Safety Testing and Drug Discovery
Human-Derived Lung Models; The Future of Toxicology Safety Assessment

ZOË PRYTHERCH* AND KELLY BÉRUBÉ

School of Biosciences, Cardiff University, The Sir Martin Evan Building, Museum Avenue, Cardiff CF10 3AX, Wales, UK
*E-mail: prytherchzc@cardiff.ac.uk

4.1 Introduction
4.1.1 Overview of the Respiratory System

Functionally, the human respiratory system can be divided into three main regions: (i) the extra-thoracic region – nose/mouth and larynx; (ii) the tracheobronchial region – trachea, bronchi and bronchioles; and (iii) the pulmonary region – respiratory bronchioles and all alveolar regions involved with gas exchange.[1] As a general rule, the epithelium lining the airways from the nasal cavity to the terminal bronchioles consists of ciliated, goblet and basal cells. However, gradual changes in populations of cell types occur as the airways move from conducting to respiratory, *e.g.* the replacement of goblet with Clara cells, which begins in the bronchioles. The pseudo-stratified columnar epithelium eventually presents as a simple, cuboidal, non-ciliated epithelium, with the alveoli consisting of type I and II pneumocytes.[2,3] This almost constant change in cell composition and morphology makes studying the lung a unique challenge and the choice of a representative model is paramount to obtaining relevant results.

4.1.2 Inhalation Toxicology

The natural bodily function of 'breathing' exposes the human respiratory system to a variety of ambient detritus in the form of particulates and gases from a plethora of anthropogenic (*e.g.* nanoparticles and chemicals), biogenic (*e.g.* aero-allergens and microbes), geogenic (*e.g.* volcanic ash and minerals) and technogenic (*e.g.* coal fly ash and smelter particles) sources.[4] The respiratory epithelium provides one of the first lines of defence in response to the constant assault on the body by airborne detritus. Consequently, it represents an ideal region to investigate the toxicological consequences to inhaled substances. In addition to inhalation hazards, a wide variety of airway diseases, such as chronic obstructive pulmonary disorder (COPD), asthma and cancer inflict regio-specific damage to the resident epithelia, resulting in subsequent tissue remodelling. The structural and functional integrity of the airway epithelial barrier is therefore maintained through active repair mechanisms directed by tissue-specific progenitor/stem cells (*e.g.* basal cells of the bronchial epithelium).[5] To better understand the impact of inhaled substances, disease mechanisms, as well as to accurately test general chemicals, cosmetics, pharmaceuticals and nanoparticles in the human respiratory system, *in vivo*-like pulmonary models are required.[6]

Historically, animals such as mice,[7,8] rats,[9,10] rabbits[11,12] and guinea pigs[13,14] have been used as the principal testing systems for inhalation toxicology and pulmonary disease modelling. The ability of animal models to provide a systemic effect in response to an inhalant has long served as the critical argument for using *in vivo* over *in vitro* models. Nevertheless, animal models possess significant disparities when compared to the human *in vivo* situation, including dissimilar airway dichotomies,[15] types and composition of cells,[16,17] different biotransforming enzymes[18] and physiological variations in breathing patterns and metabolic rates.[19] Consequently, a degree of adaptation and extrapolation to the actual human *in vivo* response is required. In addition, an increasing reluctance by the general public to see animals used in research, along with new government legislation, is directing ever-increasing 3 Rs research practices (*i.e.* replacement, reduction and refinement of animals in research; EU (European Union) Directive, 2003;[20] REACH (Registration, Evaluation, Authorisation and Restriction of Chemicals regulation) 2006;[21] FRAME (*Fund for the Replacement of Animals in Medical Experiments*), 2014;[22] NC3R (National Centre for the Replacement, Refinement and Reduction of Animals in Research 2014).[23] For these and many other reasons, the pursuit of human cell- and tissue-based *in vitro* models are at the forefront of toxicity screening assays for safety assessments (SA) and chemical/drug development, at both the academic and industrial levels.[6]

4.1.3 *Status Quo* in Safety Assessment

Exposure to chemicals is a source of great concern to the public and so adequate and relevant toxicological SA need to be undertaken. In response to this, there has been increased legislative pressure to perform SA on old and new (current backlog) of chemical substances in use within the EU. The EU REACH legislation requires the registration and evaluation of all (new and existing) chemicals and promotes the use of alternative (to animals) tests.[21] In the USA the National Research Council (NRC) produced a report in 2007 entitled *Toxicity Testing in the 21st Century: A Vision and a Strategy*.[23] This report was created to guide the future of toxicity testing, directing chemical characterisation, with testing to become less expensive and laborious, using fewer animals and requiring the development of more robust tests for assessing health effects of environmental agents.[23]

However, despite numerous well-established methods in the scientific literature, there are few methods for 'integrated (or intelligent) testing strategies' available to make these assessments.[25] SA of chemicals across the medical, dental and veterinary sciences, as well as the industrial chemical and consumer product sectors, have long adopted the traditional paradigm of relying on data from laboratory animal tests to undertake hazard labelling and SA. This long-established safety strategy involves high-dose treatments in animals with default methods for extrapolating observed results to low-level exposures in human populations. Such 'whole animal' methods are labour-intensive, require many animals, inter-species extrapolations, are not cost-effective and more often than not, are misleading with respect to human safety. This stagnation in the advancement of non-animal models has taken place despite new and existing technologies in *in silico* (not covered in this review)[26] and *in vitro*[6,27] techniques that could provide more accurate human 'dose-response' data. All these caveats have necessitated the need for predictive *in vitro* models for end points of toxicity that provide the 'mode of action' to be determined.

4.1.4 *In Vitro* Toxicology

In vitro toxicology can be defined as the testing of a substance(s) on cells cultured outside of the body, in an attempt to determine whether the substance(s) is harmful to those cells and potentially to predict human *in vivo* toxicity. This field can also include *ex vivo* toxicology testing, which involves using material extracted from the body, such as whole organs (*e.g.* the liver), or tissue explants (*e.g.* the trachea) and maintaining them outside the body for the duration of the experiment. Both have their pros and cons, which will be discussed in more detail later in this chapter. In general (and currently), these testing methods are employed primarily to identify potentially hazardous chemicals and/or to confirm the lack of certain toxic properties in the early stages of the development of new substances (*e.g.* therapeutics, agricultural chemicals, *etc.*). An ideal *in vitro* toxicity screen needs to encompass not only the target organ or cell type, but also broader categories of toxicity such as reactive metabolites. From the pharmaceutical industry's viewpoint in particular, the fundamental attraction of *in vitro* toxicology is the low cost, high-throughput format provided. These factors offer the ability to rapidly rank candidate compounds prior to further investigation into more costly test systems. However, the type of cell-based safety screens, applications and strategies may differ between companies.

4.1.5 Trends and Technology Uptake

The development of non-animal platforms has progressed rapidly over the last 20 years (Figure 4.1). It is increasingly recognised by all that *in vitro*, *ex vivo* and *in silico* methods have an important role in hazard identification and SA. Critical to this development has been the fact that a number of alternative methods have been scientifically validated and accepted by competent regulatory bodies, and can be used for regulatory toxicology purposes.[28] The acceptance of the alternative methods as valuable tools of modern toxicology has been recognised and encouraged by regulators, including the Organisation for Economic Co-operation and Development (OECD), United States Food and Drug Administration (FDA) and United States Environmental Protection Agency (EPA). The continued uptake by both academic and industrial scientists of *in vitro* tools in itself displays the potential benefit from such techniques.

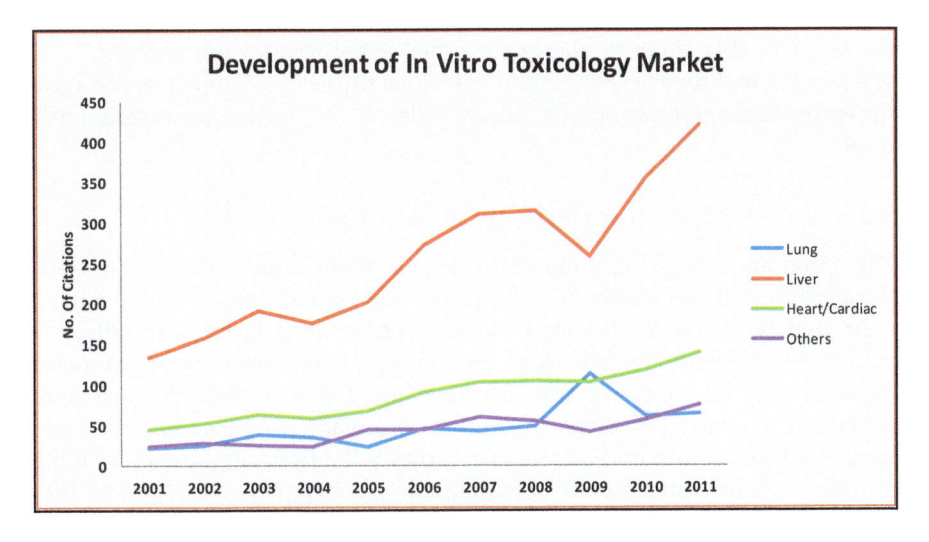

Figure 4.1 *In Vitro* Toxicology Research Development, 2001 to 2011 (Evolution Analysis of ScienceDirect Database and Pharma Research Symposia).

4.2 Current Human-Based Models of the Respiratory System: An Overview

As previously stated, animals have been the models of choice for respiratory scientists but many researchers and studies have questioned the reliability of animals to predict the human response.[29,30] For this reason this chapter is going to focus on non-animal models and the potential they hold for the future of safety testing and drug discovery. Below we will discuss different cell and tissue models, as well as different culture conditions.

4.2.1 Cells and Tissues

The main focus of human-based respiratory models are cell culture platforms, which can be divided into three main groups: carcinoma-derived, virus-transformed and primary, each with their various advantages and disadvantages detailed below. Alternatively, there are *ex vivo* models, which can either be the whole lung or a section of the respiratory system.

4.2.1.1 Cell Lines

Cell lines are immortalised cells of a particular origin, and they can be broadly categorised as either carcinoma-derived or virus-transformed. In non-cancer based respiratory research, virus-transformed cell lines are preferred over carcinoma-derived cell lines, mainly due to the ability of virally-transformed cells to retain more differentiated characteristics of the cell of origin.[31] (For good reviews on the use of cell lines in cancer research please see Gazdar *et al.*[32,33]) A complete list of all cell lines used in respiratory research and their advantages and disadvantages is beyond the scope of this chapter; please see other references for more detail.[31,34,35] We will instead focus on the main cell lines utilised, *i.e.* those which have the most desirable/promising characteristics, or those most frequently used.

4.2.1.1.1 Carcinoma-Derived Cells 1: Calu-3

The predominant carcinoma-derived cell lines used are the Calu-3s (bronchial) and the A549s (alveolar). The Calu-3 cell line is derived from a bronchial adenocarcinoma; it forms confluent polarised monolayers, exhibits *trans*-epithelial electrical resistance (TEER; with inconsistencies between laboratories[4]), contains numerous P450 enzymes,[36] possesses functional transport systems and efflux pumps,[37] produces and secretes mucus glycoproteins and contains microvilli.[38] Limitations of the Calu-3 model include non-pseudo-stratification, deficiencies in some biotransforming enzymes, lack of real cilia, deficiency in the tight junction protein ZO-1,[39] as well as some uncharacteristic phenotypes due to their carcinogenic origin.[34,38]

4.2.1.1.2 Carcinoma-Derived Cells 2: A549

The A549 cell line was established from a human adenocarcinoma and has the characteristics of alveolar type II (ATII) pulmonary epithelial cells.[34,40,41] A549 cell cultures form a confluent monolayer and contain lamellar bodies.[41] They express some ATII cell P450s, such as cytochrome 1A1 (CYP1A1), CYP1B1 and CYP2B6,[42] however their phase I biotransformation capacity does not adequately represent that of the human lung. Paradoxically, the A549's phase II activity does seem to resemble that of the human lung.[18] Other major limitations of the A549 cell lines include poor barrier potential (low TEER),[41] lack of functional tight junctions,[43,44] scarcity of phosphatidylglycerol, fibroblast-like phospholipid profile,[45] absence of SP-A expression,[46] deficiency of domes (presence indicates active ion transport)[40] and appearance of uncharacteristic phenotypes due to carcinogenic origin of cell type.[34,47] An additional problem with using the A549 cell line for drug transport or metabolism studies is the fact that they only represent ATII pneumocytes, which, although they account for 60% of alveolar cells, only make up around 5% of the total alveolar surface area.[48] Therefore, the A549 cell line is not a suitable system to study drug absorption, but may be useful in determining type II pneumocyte biology.

4.2.1.1.3 Immortalised Cells 1: BEAS-2B

Viral genes or vectors have commonly been used to transform or 'immortalise' cells for use in cell culture. The main virus-transformed respiratory cell lines are the BEAS-2B and the 16HBE14o-. The BEAS-2B cell line was derived from normal human epithelial cells immortalised using the adenovirus 12-simian virus 40 hybrid virus.[49] This cell line retains its ability to undergo squamous differentiation in response to FBS (foetal bovine serum) or TGF-β (tumour growth factor-beta)[18] and their use has been popular in studies analysing cytokine regulation,[50,51] glucocorticoid receptors[52] and response to challenges, such as tobacco smoke[53,54] and particulate matter.[55,56] However, characterisation of this cell line has revealed that they have deficiencies in forming adequate TEER,[57] display low expression of junctional proteins such as ZO-1 and E-cadherin[39] and subsequently have difficulty forming tight junctions.[34] Although BEAS-2B cells form primary cilia (solitary), they do not form motile cilia characteristic of ciliated airway epithelial cells.[58] They also lack basal expression levels of numerous mucin proteins or the ability to adequately up-regulate those mucin proteins they possess.[35,39,57] In addition to these basic deficiencies, they also lack numerous CYP enzymes.[59,60]

4.2.1.1.4 Immortalised Cells 2: 16HBE14o-

The 16HBE14o- cells are human bronchial epithelia cells transformed by SV40,[61] and were originally developed to study the cystic fibrosis conductance regulator (CFTR).[62] This cell culture system is known to form polarised, multilayered cultures, possessing microvilli and cilia, containing tight junctions and expressing many drug transporters and drug-related proteins.[34,63,64] The major disadvantages associated with this cell line are deficiency in producing the protective mucus covering as well as the possibility that important transcription factors of relevant genes may be altered or not expressed.[34]

As we know, transformed epithelial cells have an extended life, which is one of their main advantages. However, they may undergo 'crisis' during early stages, which can terminate their proliferative ability. The slowing of cell growth, cell senescence, cellular vacuolisation and general deterioration of cultures are characteristic when undergoing crisis.[62]

4.2.1.2 Primary and Stem Cells

Here primary cells refer to cells removed from humans either during a surgical procedure or post-mortem, and we will also include adult or tissue-specific stem cells under this classification. Stem cells here will refer to embryonic stem cells or induced pluripotent stem (iPS) cells (somatic cells reverted to an embryonic stem cell-like state).

4.2.1.2.1 Primary Cells

Once primary cells have been removed from the body they must undergo a series of enzymatic digestion and filtration treatments to separate them from contaminating cells.[65] Human nasal and bronchial epithelium,[5,66–68] as well as alveolar[65] primary cell cultures, have been utilised as investigative models. Both the nasal and bronchial epithelial cells have been demonstrated to form confluent cultures, providing relatively accurate, *in vivo*-like morphological and physiological properties.

Normal Nasal Epithelial (NHNE) Cells have been removed and disassociated by enzymatic digestion and successfully cultured to recreate the human nasal epithelium *in vitro*.[68,69] NHNE cultures have been shown to differentiate, secrete mucin, develop apical cilia,[68,70] achieve robust barrier properties (TEER > 1000 Ω cm^2),[70] contain aquaporins, ion channels and transporters.[69,71]

Normal Human Bronchial Epithelial (NHBE) Cells removed by enzymatic digestion, have been cultured *in vitro* to produce a highly *in vivo*-like, pseudo-stratified muco-ciliary epithelium (Figure 4.2).[5,72] Mature NHBE cultures have been shown to possess numerous differentiated cell types, such as basal, intermediate, ciliated, goblet, Clara and serous phenotypes.[5] The cultures attain high TEERs (>2000 Ω cm^2), through formation of tight junctional (TJ) protein complexes featuring the zona occludens (*i.e.* ZO-1; anchors TJ strand proteins to the actin-based cytoskeleton) and the claudins (*i.e.* establish *para*-cellular barrier that controls the flow of molecules in the intercellular space between the cells of an epithelium[5]), as well as secreting mucin and cytokines.[5,31]

Human Airway Epithelial Cells (HAEpC) have also been utilised in order to recreate the alveolar epithelium, but to a much lesser extent than nasal or tracheobronchial primary cells. ATII pneumocytes isolated from human lung

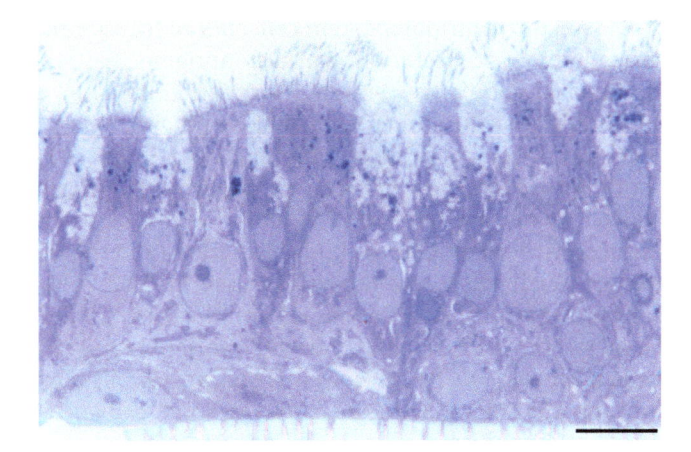

Figure 4.2 Image of a fully-differentiated NHBE model (Toluidine blue stained semi-thin section). The epithelium is pseudo-stratified and contains, basal, intermediate, goblet, ciliated and Clara cells. Scale bar = 10 μm.

tissue can be cultured to form a confluent monolayer with high TEER values (>2000 Ω cm^2).[65] These monolayers of HAEpC mainly consist of flattened type I-like cells (>5 days ALI culture). These type I-like cells are morphologically similar to alveolar type I (ATI) pneumocytes, express ZO-1 and occludin, possess desmosomes, a type I lectin profile, as well as plasma membrane invaginations possessing the typical features of caveolae.[65,73] Upon day 8 in cell culture, the HAEpC cells appear to develop another distinct cell type, a 'rounder' and more prominent cell type that morphologically resembles a ATII pneumocyte. This type II-like cell also exhibits residual multilamellar bodies, characteristic of type II pneumocytes.[35,73]

The major limitation of primary cell cultures is the fact that they can only undergo a finite number of passages and cell divisions before they begin to lose their viable characteristics.[35,74] Another possible limitation with primary cell cultures is donor variation,[34] however this does depend on what aspect of the cultures one investigates.[75] The main problem seems to lie in the metabolic profile of donors and obviously the fact that certain CYPs may be 'switched on', for instance in smoking donors.[34] However, this donor variability may be advantageous in drug development, whereby differences between the diseased and normal populations may enable specific targeting of drugs or even to ensure that most people will respond the same to a certain drug (*i.e.* a well-known issue when new compounds are released onto the market).[76]

4.2.1.2.2 Stem Cells

Embryonic stem (ES) cells are undifferentiated pluripotent cells isolated from the inner cell mass of blastocysts.[77] Induced pluripotent stem (iPS) cells are created from the reprogramming of differentiated human somatic cells (such as fibroblasts) into pluripotent stem cells. iPS cells resemble ES cells in terms of morphology, proliferation, surface antigens, gene expression, epigenetic status and telomerase activity.[78] Human ES and iPS cells can be induced to differentiate into numerous somatic cells.[77,78]

Researchers were able to initially differentiate ES into ATII cells *via* embryonic body formation, but this was not particularly efficient in generating ATII cells.[79] Wang and co-workers (2007) have subsequently derived ATII cells from ES, without embryonic body formation (and therefore sooner) and with greater efficiency.[80] Morphologically the ES-ATII cells resembled ATII cells, containing typical lamellar bodies and they also produced SP-A, -B and -C.[80] The generation of specialised full-differentiated lung epithelia from human ES cells has proved difficult. However, human ES cells have been used to generate a lung epithelial-like tissue, showing some cell-specific markers for ATI, ATII, Clara and ciliated cells.[81] Similar results have been found with a human ES cell line RUES2, with cells expressing markers for ATI, ATII, goblet, ciliated, Clara and epithelial basal cells.[82]

Takahashi and co-workers (2007) were one of the first groups to reprogramme human somatic cells (fibroblasts) into pluripotent stem (iPS) cells.[78] Once generated, iPS cells can be differentiated either into endoderm cells, the precursors for lung and liver tissues,[83,84] and can subsequently be differentiated further into lung epithelial cells.[85] The differentiation of ES or iPS cells into a bronchial epithelium has recently been achieved.[86] However, cultures were not homogeneous and also contained SP-C expressing (function specific to ATII cells) cells, although co-localisation of both bronchial and alveolar cells does not occur.[86] Bronchial epithelial cells expressing a CFTR mutation have been derived *in vitro* from iPS cells.[87] Nasal epithelium has also been used as the somatic 'starter' cell used to generate iPS cells.[88] As nasal brushings are easily and relatively painless to obtain, the nasal epithelium provides an easy and accessible source for the creation of iPS cells.

One of the major disadvantages of human ES cells is that their use involves the destruction of human embryos, which in itself carries obvious ethical issues. In addition to this, the use of ES cells involves the risk of teratoma formation.[89] As iPS cells are acquired from the reprogramming of somatic stem cells they offer a real advantage over ES, both ethically and in terms of ease of access, and ultimately they will allow autologous transplantation of stem cells to repair damaged tissues.[90] Although the generation of specific *in vivo* accurate epithelium has yet to be generated from ES or iPS cells, researchers are getting progressively closer, and it is likely to be achieved in the near future.

4.2.1.3 Ex vivo

The use of *ex vivo* models of the lung have proven useful as they are able to maintain a great deal of the *in vivo* complexity and can aid in bridging the gap between *in vitro* and *in vivo*. There are many different *ex vivo* options, all with their various advantages and disadvantages, and animal *ex vivo* models are much more widely used than human models, mainly due to ease of access and the more developed culture protocols. However, the main disadvantage (at least at the moment) to all *ex vivo* models is their extremely limited lifespan.

4.2.1.3.1 Precision Cut Lung Slices

Precision cut lung slices (PCLS) are one of the most widely used *ex vivo* models to have overcome the reliability issues associated with standard tissue slices.[91] The use of human PCLS became feasible by initially perfusing lungs with an agar solution (although this may have negative effects),[92] and allowing it to gel prior to tissue cores being precision cut into 500 µm slices.[91,93] Human PCLS have been utilised for many different studies, including immune[94] and allergic[92] responses as well as a test platform for drug efficacy.[93,95] The main advantages of PCLS are that the tissue slices maintain *in vivo* structure, in terms of cellular composition and spatial conformation.[96] This enables the alveolar region and conducting airways to be observed in themselves, but also their synergistic effects as well as that from the surrounding connective tissue and associated lung parenchyma. As previously stated, one of the major limitations of PCLS is their limited lifespan, with general consensus limiting this to 1–3 days in culture, as well as the fact that any treatment is given to the whole PCLS and therefore 'inhalation' routes may not be mimicked.[91-94]

4.2.1.3.2 Tissue Explants

Tissue explants are another *ex vivo* option, allowing the removal of a specific part of the lung (*e.g.* alveolar, respiratory bronchiole, trachea) to be examined *in vitro*. In some ways they are very similar to PCLS as you are cultivating a small *ex vivo* section of the lung, and indeed possess many of the same benefits and limitations. For instance the tissue explants maintain the structure, cellular composition and spatial conformation of that section *in vivo*. To that end they have been used to study bronchoactive drugs,[97] pharmacological agents,[98] procarcinogen metabolism[99] and infection studies.[100] Tissue explants have also been used as a method to promote outgrowth of cells from the explant to a surrounding substrate.[101] However, these experiments have not been very successful in maintaining differentiated cultures away from the explant.[101] Their main limitations also centre around their lack of *in vivo* preservation in culture, with most reports stating 3–7 days as their viable limit, with de-differentiation, morphological changes and necrotic centres developing after this time.[99,102]

4.2.1.3.3 Isolated Perfused Lungs

The use of human isolated perfused lungs (hIPL) is much less common than animal IPL. This is mainly due to the difficulty of obtaining human lungs, the complexity of the method, the limited lifespan of hIPL and the fact that one lung or lobe could produce several PCLS (*i.e.* lower replicates per lung). Being an intact lung or lobe, the architecture, cell composition, spatial relationships and a certain amount of the vascularity of the *in vivo* lung are maintained. These are very major advantages in modelling the human lung. Human IPLs have been used in studies on vasoconstriction,[103] drug treatments[104] and creating disease models.[105] However, of all the *ex vivo* models, all of which suffer from limited culture viability, IPL is limited to a 4 hour exposure window.[106,107] This is a serious disadvantage of this *ex vivo* model and limits the types of studies that can be undertaken and potentially the validity of such studies.

4.2.2 General Culture Conditions

When culturing cells, many things must be taken into consideration to ensure the most *in vivo*-like morphology and behaviour of the cells is obtained, and these parameters change depending on which cell type one aims to culture. Some common considerations for primary cell culture are passage number and seeding/cell density.[74] Generally, primary cell cultures do not produce properly differentiated cultures until after approximately passage 3 or 4.[68,107] Cell/seeding density is also an important parameter in producing differentiated cultures,[65] whereby insufficient densities caused cells to undergo squamous differentiation.[108]

During the 1980s, major advances were made with respect to culture conditions and media, allowing differentiation of cells and formation of tissue-like constructs. Airway epithelial cells were found to be particularly sensitive to serum.[109,110] This led to the development of serum-free media[72] supplemented with selected growth factors, hormones and antibiotics. The main factors added during this time were hydrocortisone, insulin, epidermal growth factor, transferrin, cholera toxin, bovine hypothalamus extract, triiodothyronine and retinoic acid (RA).[72,109] Hill and colleagues (1998) demonstrated the importance of sufficient RA within the media of mucociliary epithelia, with RA-sufficient cultures displaying a columnar, mucociliary epithelium.[67] RA-deficient cultures were shown to undergo metaplastic squamous differentiation.[67] In terms of ES cell differentiation the particular medium used is essential in determining cell differentiation and fate. Wnt and BPM4, along with RA, are known to be essential for the development of hPSCs into lung progenitors, with the presence of RA potentially favouring proximal airway development.[81] The differences in culture conditions not only affect cell morphology and differentiation, but they can also affect the TEER and therefore the barrier potential of cultures, an important parameter of any pharmacological model.[111] Therefore, the composition of the medium for each cell type needs to be carefully defined.

Further to this, the culture substrate needs to be considered. Grobstein (1953) initially proposed that extracellular matrix (ECM) had an effect on cellular differentiation.[112] This was originally achieved by using denuded tracheal grafts to culture *in vivo*-like characteristic cells.[113,114] However, denuded tracheal grafts are difficult to come by and use. As a result, more accessible, artificially derived substrates such as fibronectin-collagen coated surfaces[65] and commercially available ECM-based compositions such as

Matrigel®[80,115] can be utilised, all of which enhanced the culture's *in vivo*-like morphology. More recently, researchers have been looking beyond the 2D culture substrate, to the effect 3D cell scaffolds have on the differentiation and *in vivo*-like characteristics of cells. 3D cell scaffolds are most commonly used by those looking to recreate a whole organ, or section of the organ (*e.g.* trachea), for transplantation purposes. However, the use of scaffolds is becoming increasingly popular in general cell culture techniques to improve the original platform. Common materials used include (i) hydrogels – crosslinking natural base materials such as collagen or hyaluronic acid with a high water content, which may contain ECM components such as laminin[116] and (ii) polystyrene – porous scaffolds, *e.g.* Alvetex®.[117] Therefore, factors such as seeding density, passage number, media composition and substrate conditions need to be carefully defined for each cell type used in order to gain the most from alternative cell models.

4.2.2.1 Submerged versus ALI

Traditional cell culture is undertaken with cells submerged in liquid, which is logical considering the *in vivo* cell situation for most of the cells in the body. However, this is not the case with lung cells. Basic lung physiology would tell you that the apical side of pulmonary cells are exposed (to some degree) to the air and therefore lung cells should be cultured at an air–liquid interface (ALI). However, some would argue that the protective mucus or surfactant layers present *in vivo* protect the cells from being truly at an ALI, thus should not be cultured as such. Wu and colleagues (1986) were one of the first to suggest that culturing at an ALI may enable better morphology of airway epithelial cells.[72] This was achieved by the development of a 'special chamber' on which a nitrocellulose membrane was glued and crosslinked with gelatine; the collagen gel substratum was formed on top of the membrane. A spacer was placed in a tissue culture dish and the chamber rested on the spacer, allowing the culture to be fed basally.[118] This culture system demonstrated a polarity (*i.e.* apical and basal regions) in the differentiation of the cultured cells. Polarity confers development of cilia and the secretion of mucus granules on the apical surface. Furthermore, the columnar appearance of apical cells could be demonstrated.[72] This allowed the development of ALI cultures originally with guinea pig tracheal epithelial cells.[118] ALI cultures were rapidly adopted for use with primate,[119] human,[120,121] rat,[122] bovine[123] and canine[124] tracheal epithelial cells.

Under ALI conditions, the levels of aerobic respiration are higher than under submerged conditions, and this is thought to be the cause of their increased differentiation when cultured at an ALI.[31] Indeed pulmonary cells (specifically primary cells) produce epithelium with more *in vivo*-like morphology and greater differentiation of cell types.[5,40,72,121] Pulmonary stem cells also appear to benefit from increased differentiation when cultured at an ALI.[125] However, even though culturing at an ALI appears to have the greatest impact on respiratory epithelial differentiation for primary cells, this is not always the case with cell lines. The A549 cell line is thought to have a more physiological response to nanoparticles when cultured at an ALI,[126] whilst there are conflicting reports on the benefits of ALI culture for Calu-3 cells.[111] For instance TEER values (barrier properties) are generally higher under submerged conditions,[111,127] but differentiation (including production of mucus) is increased under ALI conditions.[35] The bronchial cell line 16HBE14o- has been consistently found to be less in vivo-like when cultured at an ALI. Morphology,[128] the presence of cell junction proteins[35] and TEER[35,111] are all less desirable under ALI conditions for the 16HBE14o⁻ cell line. Taking into consideration that ALI conditions mimic in vivo conditions and that most (especially primary) cell cultures are more in vivo-like under ALI conditions; the validity of the 16HBE14o⁻ cell line as a model of the bronchial epithelium should be questioned.

4.2.2.2 Static versus Continuous Flow

Traditionally, cell culture has been undertaken using static conditions. Static culture, however, does not mimic the flow of nutrients, hormones and waste that occurs in vascularised tissues such as the lung. There are a variety of flow-based cell feeding systems available, including pump-less and pump-based systems. Pump-less technologies include those developed by Kim and Cho[129] or commercially available platforms such as the FloWell™ 2W Plate.[130] Pump-based technologies are usually small bioreactors sequentially connected by a flow of media, such as the Quasi-Vivo®[131,132] or microfluidic based chips, like the lung-on-a-chip microdevice.[133,134] Continuous flow cultures consistently produce cultures of more desirable morphology, phenotypes and general in vivo-like properties.[129,135–137] Therefore, if the aim is to produce a model that mimics the in vivo situation as closely as possible, then flow-based cultures are the way to go. Nevertheless, the cost and technical difficulty of such techniques need to be weighed against the benefits provided before deciding on static or continuous flow cultures. Recent work at the Wyss Institute have generated one of the most desirable alveolar-based models, lung-on-a-chip.[133,134] This methodology not only uses flow culture conditions for the media (which can also circulate inflammatory cells), but air flow on the apical side of the cells (which can contain bacteria), endothelial and A549 co-culture (see more on co-culture below), as well as two vacuum chambers, which mimics the mechanical stress of breathing.[133,134] The organ-on-a-chip platforms that have been developed are currently at the forefront of highly in vivo-like tissue engineering. These platforms will be more beneficial if cell lines can be avoided.

4.2.2.3 Co-culture

Once a well-differentiated epithelium has been established, the next logical step is to build on the complexity of the model. Co-culturing may require minor or major modifications to the culture conditions mentioned above, and depending on the cell type(s) used, can be challenging. In a simplistic way, co-culturing can be viewed as building a model using building blocks, where each cell type is a building block added to the existing model, making it as simple or as complex as required.

The use of co-culturing to create a more *in vivo*-like situation can take three forms: (i) the addition of other cell types within or closely related to the organ in question; (ii) the inclusion of immune cells to provide a 'systemic' immune response; or (iii) the use of cells from other organs to generate a multi-organ based approach. The first scenario could logically involve the co-culture of airway cells (bronchial or alveolar) with fibroblasts (to create a full-thickness epithelial–mesenchymal model),[138,139] endothelial cells (to incorporate epithelial–endothelial interactions and assess transport in/out of the systemic circulation)[140,141] or all three cell types.[142] Modelling the immune response to inhaled pathogens, particles or drugs is becoming increasingly popular in order to bridge the gap between animal and cell culture models. The addition of immune cells to an airway culture can take numerous forms. For instance, an immune cell type(s) may be added to the airway epithelium alone or a more complex co-culture model of the epithelium. In addition, these immune cell type(s) may be: (i) placed on the apical surface of the differentiated epithelium (*e.g.* macrophages);[143] (ii) cultured under the epithelium, either directly (*e.g.* dendritic cells)[143] or on top/surrounded by fibroblasts;[144] or (iii) within the basolateral media. This is particularly effective under flow conditions, with endothelial cells also involved in the co-culture.[133,134]

The final type of co-culture is that of cells from different organs in order to create a multi-organ response. This may be achieved with either the respiratory epithelium alone, organ-related co-culture and/or immune cell involvement with the cells from a different organ. Multi-organ based co-cultures are attempting to recreate the potential downstream fate of inhalable compounds. For instance, the majority of potential therapeutics fail due to cardio- and/or hepatotoxicity.[145] Therefore, if these therapeutics are delivered *via* the lungs and enter the systemic circulation downstream, toxicity testing is a real concern. Multi-organ co-cultures can either exist in static (less common) or in flow-based cultures. In static-based co-cultures the 'organs' are separated by a membrane, as is seen in the Metabo-Lung model (unpublished observations; *WO2012/123712*), whereby the bronchial epithelium is cultured on the apical part of the cell culture insert and the hepatocytes are attached to the base of the well.[6,25] Flow-based cultures however allow each 'organ' model to be cultured in separate bioreactors that are connected by a directional media flow.[132] It is therefore also possible with the flow-based approach to have more than two 'organ' cultures connected sequentially.[132] Such models provide great promise in predicting *in vivo* outcomes, if each individual 'organ' is *in vivo*-like.

4.3 Discussion

The field of human toxicology and SA is currently shifting away from failing animal models in all fields of human toxicology (*e.g.* pharmaceutical, agrochemical, cosmetics, chemical) to more *in vivo*-relevant human cell and tissue models. There have been many drivers for this shift, which have now tipped the balance in favour of *in vitro* and *in silico* toxicology platforms. One of these drivers has been the fact that the pharmaceutical industry has struggled with the high attrition rate of new compounds at Phase II and III of clinical trials.[30] This is mainly due to inadequate or inappropriate screening prior to these phases.[146,147] The flip side to this is that inadequate animal models may not only fail to predict toxicity in humans but they may even be screening out compounds that would not be toxic to humans, meaning potential therapeutics may have been lost to society. A change in legislation within the EU[20] has resulted in the cosmetics industry having to change their SA strategies to non-animal based tests. This legislative change within the EU has resulted in several other countries following suit (*e.g.* Turkey, India), with a global ban on cosmetics testing on animals surely to follow. This legislative change will have a two-fold impact on the remainder of the toxicology/SA fields. Firstly, resources are being poured into the research, development and validation of alternative test systems, which could then be utilised by other sectors (*e.g.* agrochemical SA). Secondly, the example that will be set by the cosmetics industry that a wholly non-animal based SA is possible will enable other legislative changes and acceptance of such methods across the board.

The validity of human-based *in vitro* and *in silico* methods has been recognised in the NRC report *Toxicity Testing in the 21st Century: A Vision and a Strategy.*[24] This report has called for a fundamental shift in the testing of chemicals for human health effects and SA. Seven years on and the NRC proposals approach has seen a decrease in the current reliance on animal studies and a move towards *in vitro* methods, typically using human cells in a high-throughput context. This new framework is expected to have several advantages over current practice, including a more relevant scientific foundation for human health risk assessments, lower costs and shorter duration of testing and assessment, broader coverage of chemicals, health effects, personalised medicine (given the new tools' higher throughput, lower cost and shorter duration), and cumulative reductions in animal usage as the new models and approaches gain scientific and regulatory acceptance.[24]

This shift from animal based toxicology would not be possible without the improved accessibility of human cells/tissues, improved culturing methods and equipment. This chapter has detailed the advancements made in the field of *in vitro* models of the human respiratory system. The co-culture of primary or stem cells under flow conditions currently represents the future of *in vitro* toxicology. However, the choice of *in vitro* tools needs to be carefully assessed and validated, and will probably form part of a panel of *in vitro* toxicology tests, otherwise *in vitro* models may be no better than current animal models in predicting human toxicity outcomes.

References

1. D. E. Gardner, *Toxicology of the Lung*, 4th edn, Taylor & Francis, London, UK, 2006.
2. S. I. Fox, *Human Physiology*, 8th edn, McGraw-Hill, New York, USA, 2004.
3. P. S. Hasleton and A. Curry, *Anatomy of the Lung, Spencer's Pathology of the Lung*, 5th edn, McGraw-Hill, New York, USA, 1996.
4. K. BéruBé, M. Aufderheide, D. Breheny, R. Clothier, R. Combes, R. Duffin, B. Forbes, M. Gaça, A. Gray, I. Hall, M. Kelly, M. Lethem, M. Liebsch, L. Merolla, J. P. Morin, J. Seagrave, M. A. Swartz, T. D. Tetley and M. Umachandran, *ATLA, Altern. Lab. Anim.*, 2009, **37**, 89.
5. Z. Prytherch, T. Hughes, C. Job, H. Marshall, V. Oreffo, M. Foster and K. BéruBé, *Macromol. Biosci.*, 2011, **11**, 1467.
6. K. A. BéruBé, *ATLA, Altern. Lab. Anim.*, 2013, **41**, 229.
7. M. W. Himmelstein, M. J. Turner, B. Asgharian and J. A. Bond, *Carcinogenesis*, 1994, **15**, 1479.
8. M. Mall, B. R. Grubb, J. R. Harkema, W. K. O'Neal and R. C. Boucher, *Nat. Med.*, 2004, **10**, 487.
9. C. H. Chu, D. D. Liu, Y. H. Hsu, K. C. Lee and H. I. Chen, *Pulm. Pharmacol. Ther.*, 2007, **20**, 503.
10. A. Tronde, B. Nordén, H. Marchner, A.-K. Wendel, H. Lennernäs and U. H. Bengtsson, *J. Pharm. Sci.*, 2003, **92**, 1216.
11. J. R. Brentjens, D. W. O'Connell, I. B. Pawlowski, K. C. Hsu and G. A. Andres, *J. Exp. Med.*, 1974, **140**, 105.
12. S. Lin, J. Walker, L. Xu, D. Gozal and J. Yu, *Exp. Physiol.*, 2007, **92**, 749.
13. E. S. Fiala, O. S. Sohn, C. X. Wang, E. Seibert, J. Tsurutani, P. A. Dennis, K. El-Bayoumy, R. S. Sodum, D. Desai, J. Reinhardt and C. Aliaga, *Carcinogenesis*, 2005, **26**, 605.
14. T. J. Toward and K. J. Broadley, *J. Pharmacol. Exp. Ther.*, 2002, **302**, 814.
15. A. R. Brody, *Environ. Health Perspect.*, 1984, **56**, 149.
16. R. G. Breeze and E. B. Wheeldon, *Am. Rev. Respir. Dis.*, 1977, **116**, 705.
17. D. G. Thomassen, P. Nettesheim, *Biology, Toxicology, and Carcinogenesis of Respiratory Epithelium*, Hemisphere Publishing Corporation, London, UK, 1990.
18. J. V. Castell, M. T. Donato and M. J. Gomez-Lechon, *Exp. Toxicol. Pathol.*, 2005, **57**, 189.
19. W. J. Mautz, *Environ. Res.*, 2003, **92**, 14.
20. EU Directive 76/768/EEC, Directive 2003/15/EC of the European Parliament and of the Council of 27 February 2003, *Off. J. Eur. Union*, 2003, **L66**, 26–35.
21. Regulation (EC) No 1907/2006 of the European Parliament and of the Council of 18 December 2006 concerning the Registration, Evaluation, Authorisation and Restriction of Chemicals (REACH), establishing a European Chemicals Agency.
22. FRAME, 2014, http://www.frame.org.uk. Date last accessed: Jan 2014.
23. NC3R, 2014, http://www.nc3rs.org.uk. Date last accessed: Jan 2014.

24. National Research Council (NRC), *Toxicity Testing for Assessment of Environmental Agents*, Washington, DC, National Academies Press, 2006.
25. K. A. BéruBé, *Biotechnol. Online Mag.*, 2011, **22**, 10.
26. R. D. Combes, *In Silico Methods for Toxicity Prediction in New Technologies for Toxicity Testing*. Springer, USA, 2012.
27. K. A. BéruBé, *ATLA, Altern. Lab. Anim.*, 2011, **39**, 121.
28. AltTox, 2014: http://www.alttox.org/ttrc/validation-ra/validated-ra-methods.html. Date last accessed: May 2014.
29. T. Hartung and M. Leist, *ALTEX*, 2008, **25**, 91.
30. I. Kola and J. Landis, Can the Pharmaceutical Industry Reduce Attrition Rates?, *Nat. Rev Drug Discovery*, 2004, **3**, 711.
31. D. C. Gruenert and W. E. Finkbeiner, *Am. J. Physiol.*, 1995, **268**, L347.
32. A. F. Gazdar, B. Gao and J. D. Minna, *Lung Cancer*, 2010, **68**, 309.
33. A. F. Gazdar, L. Girard, W. W. Lockwood, W. L. Lam and J. D. Minna, *J. Nat. Cancer Inst.*, 2010, **102**, 1310.
34. B. Forbes, *Pharm. Sci. Technol. Today.*, 2000, **3**, 18.
35. A. Steimer, E. Haltner and C.-M. Lehr, *J. Aerosol. Med.*, 2005, **18**, 137.
36. K. A. Foster, C. G. Oster, M. M. Mayer, M. L. Avery and K. L. Audus, *Int. J. Pharm.*, 2000, **208**, 1.
37. B. I. Florea, M. L. Cassara, H. E. Junginger and G. Borchard, *J. Controlled Release*, 2003, **87**, 131.
38. C. I. Grainger, L. L. Greenwell, D. J. Lockley, G. P. Martin and B. Forbes, *Pharm. Res.*, 2006, **23**, 1482.
39. C. E. Stewart, E. E. Torr, N. H. M. Jamili, C. Bosquillon and I. Sayers, Evaluation of Differentiated Human Bronchial Epithelial Cell Culture Systems for Asthma Research, *J. Allerg.*, 2012, **2012**, 943982, doi: 10.1155/2012/943982.
40. N. R. Mathias, F. Yamashita and V. H. L. Lee, *Adv. Drug Deliv. Rev.*, 1996, **22**, 215.
41. K. A. Foster, *Exp. Cell Res.*, 1998, **243**, 359–366.
42. J. Hukkanen, A. Lassila, K. Päivärinta, S. Valanne, S. Sarpo, J. Hakkola, O. Pelkonen and H. Raunio, *Am. J. Respir. Cell Mol. Biol.*, 2000, **22**, 360.
43. Z. Wang and Q. Zhang, *Int. J. Pharm.*, 2004, **269**, 451.
44. H. L. Winton, H. Wan, M. B. Cannell, D. C. Gruenert, P. J. Thompson, D. R. Garrod, G. A. Stewart and C. Robinson, *Clin. Exp. Allerg.*, 1998, **28**, 1273.
45. R. J. Mason and M. C. Williams, *Biochim. Biophys. Acta*, 1980, **617**, 197.
46. R. J. Korst, B. Bewig and R. G. Crystal, *Hum. Gene Ther.*, 1995, **6**, 277.
47. L. M. Simon, E. D. Robin and J. Theodore, *J. Cell. Physiol.*, 1981, **108**, 393.
48. J. D. Crapo, B. E. Barry, P. Gehr, M. Bachofen and E. R. Weibel, *Am. Rev. Respir. Dis.*, 1982, **126**, 332.
49. R. R. Reddel, Y. Ke, B. I. Gerwin, M. G. McMenamin, J. F. Lechner, R. T. Su, D. E. Brash, J.-B. Park, J. S. Rhim and C. C. Harris, *Cancer Res.*, 1988, **48**, 1904.

50. J. Mullol, J. N. Baraniuk, C. Logun, T. Benfield, C. Picado and J. H. Shelhamer, *Neuropeptides,* 1996, **30**, 551.
51. J. M. Veranth, E. G. Kaser, M. M. Veranth, M. Koch and G. S. Yost, *Part. Fibre Toxicol.,* 2007, **4**, 2.
52. N. S. Holden, T. George, C. F. Rider, A. Chandrasekhar, S. Shah, M. Kaur, M. Johnson, D. P. Siderovski, R. Leigh, M. A. Giembycz and R. Newton, *J. Pharmacol. Exp. Ther.,* 2014, **348**, 12.
53. M. J. Scian, M. J. Oldham, D. B. Kane, J. S. Edmiston and W. J. McKinney, *Inhalation Toxicol.,* 2009, **21**, 234.
54. W. Sun, R. Wu and J. A. Last, *Toxicology,* 1995, **100**, 163.
55. M. Gualtieri, J. Øvrevik, J. A. Holme, M. Grazia Perronea, E. Bolzacchini, P. E. Schwarze and M. Camatini, *Toxicol. in Vitro,* 2010, **24**, 29.
56. A. Penn, G. Murphy, S. Barker, W. Henk and L. Penn, *Environ. Health Perspect.,* 2005, **113**, 956.
57. T. L. Noah, J. R. Yankaskas, J. L. Carson, T. M. Gambling, L. H. Cazares, K. P. McKinnon and R. B. Devlin, *In Vitro Cell. Dev. Biol.: Anim.,* 1995, **31**, 738.
58. R. Jain, J. Pan, J. A. Driscoll, J. W. Wisner, T. Huang, S. P. Gunsten, Y. You and S. L. Brody, *Am. J. Respir. Cell Mol. Biol.,* 2010, **43**, 731.
59. J. V. Castell and M. J. Gomez-Lechon, In Vitro Methods in Pharmaceutical Research, ch. 18, *Drug Metabolism and Carcinogen Activation Studies with Human Genetically Engineered Cells.* K. Macé, E. A. Offord, A. M. A. Pfeifer, 1996. pp. 433–456.
60. C. Garcia-Cantona, E. Minet, A. Anadonb and C. Meredith, *Toxicol. in Vitro,* 2013, **27**, 1719.
61. C. Westmoreland, T. Walker, J. Matthews and J. Murdock, *Toxicol. in Vitro,* 1999, **13**, 761.
62. D. C. Gruenert, C. B. Basbaum, M. J. Welsh, M. Li, W. E. Finkbeiner and J. A. Nadel, *Proc. Natl. Acad. Sci. U. S. A.,* 1988, **85**, 5951.
63. C. Ehrhardt, C. Kneuer, M. Laue, U. F. Schaefer, K.-J. Kim and C.-M. Lehr, *Pharm. Res.,* 2003, **20**, 545.
64. H. Wan, H. L. Winton, C. Soeller, G. A. Stewart, P. J. Thompson, D. C. Gruenert, M. B. Cannell, D. R. Garrod and C. Robinson, *Eur. Respir. J.,* 2000, **15**, 1058.
65. K. J. Elbert, U. F. Schäfer, H.-J. Schäfers, K.-J. Kim, V. H. L. Lee and C.-M. Lehr, *Pharm. Res.,* 1999, **16**, 601.
66. T. Gray, P. Nettesheim, C. Loftin, J.-S. Koo, J. Bonner, S. Peddada and R. Langenbach, *Mol. Pharmacol.,* 2004, **66**, 337.
67. E. M. Hill, T. Eling and P. Nettesheim, *Toxicol. Lett.,* 1998, **96**, 239.
68. H.-J. Roh, E.-K. Goh, S.-G. Wang, K.-M. Chon, J.-H. Yoon and Y.-S. Kim, *J. Rhinol.,* 1999, **6**, 107.
69. R. Agu, C. MacDonald, E. Cowley, D. Shao, K. Renton, D. B. Clarke and E. Massoud, *Int. J. Pharm.,* 2011, **406**, 49.
70. M. K. Lee, J. W. Yoo, H. Lin, Y. S. Kim, D. D. Kim, Y. M. Choi, S. K. Park, C. H. Lee and H. J. Roh, *Drug Delivery,* 2005, **12**, 305.

71. E. S. Jun, Y. S. Kim, M. A. Yoo, H. J. Roh and J. S. Jung, *Life Sci.*, 2001, **68**, 827.
72. R. Wu, G. H. Sato and M. J. Whitcutt, *Fundam. Appl. Toxicol.*, 1986, **6**, 580.
73. S. Fuchs, A. J. Hollins, M. Laue, U. F. Schaefer, K. Roemer, M. Gumbleton and C. M. Lehr, *Cell Tissue Res.*, 2003, **311**, 31.
74. R. I. Freshney, *Culture of Animal Cells: A Manual of Basic Techniques*. 5th edn Wiley, New Jersey, USA, 2005.
75. H. Maunders, S. Patwardhan, J. Phillips, A. Clack and A. Richter, *Am. J. Physiol.*, 2007, **292**, L1248.
76. U. A. Meyer, *Lancet*, 2000, **356**, 1667.
77. A. M. Wobus, *Mol. Aspects Med.*, 2001, **22**, 149.
78. K. Takahashi, K. Tanabe, M. Ohnuki, M. Narita, T. Ichisaka, K. Tomoda and S. Yamanaka, *Cell*, 2007, **131**, 861.
79. A. Samadikuchaksaraei, S. Cohen, K. Isaac, H. J. Rippon, J. M. Polak, R. C. Bielby and A. E. Bishop, *Tissue Eng.*, 2006, **12**, 867.
80. D. Wang, D. L. Haviland, A. R. Burns, E. Zsigmond and R. A. Wetsel, *Proc. Natl. Acad. Sci. U. S. A.*, 2007, **104**, 4449.
81. L. van Haute, G. De Block, I. Liebaers, K. Sermon and M. De Rycke, *Respir. Res.*, 2009, **10**, 105.
82. S. X. L. Huang, M. N. Islam, J. O'Neill, Z. Hu, Y.-G. Yang, Y.-W. Chen, M. Mumau, M. D. Green, G. Vunjak-Novakovic, J. Bhattacharya and H.-W. Snoeck, *Nat. Biotechnol.*, 2014, **32**, 84.
83. A. Somers, J. C. Jean, C. A. Sommer, A. Omari, C. C. Ford, J. A. Mills, L. Ying, A. G. Sommer, J. M. Jean, B. W. Smith, R. Lafyatis, M. F. Demierre, D. J. Weiss, D. L. French, P. Gadue, G. J. Murphy, G. Mostoslavsky and D. N. Kotton, *Stem Cells*, 2010, **28**, 1728.
84. M. D. Green, A. Chen, M. C. Nostro, S. L. d'Souza, C. Schaniel, I. R. Lemischka, V. Gouon-Evans, G. Keller and H. W. Snoeck, *Nat. Biotechnol.*, 2011, **29**, 267.
85. A. P. Wong and J. Rossant, *Curr. Pathobiol. Rep.*, 2013, **1**, 137.
86. B. A. McIntyre, C. Alev, R. Mechael, K. R. Salci, J. B. Lee, A. Fiebig-Comyn, B. Guezguez, Y. Wu, G. Sheng and M. Bhatia, *Stem Cells Transl. Med.*, 2014, **3**, 7.
87. A. P. Wong, C. E. Bear, S. Chin, P. Pasceri, T. O. Thompson, L.-J. Huan and Rossant, *J. Nat. Biotechnol.*, 2012, **30**, 876.
88. M. Ono, Y. Hamada, Y. Horiuchi, M. Matsuo-Takasaki, Y. Imoto, K. Satomi, T. Arinami, M. Hasegawa, T. Fujioka, Y. Nakamura and E. Noguchi, *PLoS One*, 2012, 7(8), e42855.
89. A. S. Lee, C. Tang, F. Cao, X. Xie, K. van der Bogt, A. Hwang, A. J. Connolly, R. C. Robbins and J. C. Wu, *Cell Cycle*, 2009, **8**, 2608.
90. I. H. Park, N. Arora, H. Huo, N. Maherali, T. Ahfeldt, A. Shimamura, M. W. Lensch, C. Cowan, K. Hochedlinger and G. Q. Daley, *Cell*, 2008, **134**, 877.
91. A. R. Parrish, A. J. Gandolfi and K. Brendel, *Life Sci.*, 1995, **57**, 1887.
92. A. Wohlsen, C. Martin, E. Vollmer, D. Branscheid, H. Magnussen, W. M. Becker, U. Lepp and S. Uhlig, *Eur. Respir. J.*, 2003, **21**, 1024.

93. R. L. Fisher, M. S. Smith, S. J. Hasal, K. S. Hasal, A. J. Gandolfi and K. Brendel, *Hum. Exp. Toxicol.*, 1994, **13**, 466.
94. S. Switalla, L. Lauenstein, F. Prenzler, S. Knothe, C. Förster, H. G. Fieguth, O. Pfennig, F. Schaumann, C. Martin, C. A. Guzman, T. Ebensen, M. Müller, J. M. Hohlfeld, N. Krug, A. Braun and K. Sewald, *Toxicol. Appl. Pharmacol.*, 2010, **246**, 107.
95. R. G. Sturton, A. Trifilieff, A. G. Nicholson and P. J. Barnes, *J. Pharmacol. Exp. Ther.*, 2008, **324**, 270.
96. J. Adamson, L. E. Haswell, G. Phillips and M. D. Gaça, *In Vitro Models of Chronic Obstructive Pulmonary Disease (COPD), in Bronchitis*, ed. Dr. Ignacio MartÃn-Loeches, 2011.
97. D. F. Hawkins and H. O. Schild, *Br. J. Pharmacol. Chemother.*, 1951, **6**, 682.
98. R. G. Goldie, J. F. Bertram, J. M. Papadimitriou and J. W. Paterson, *Trends Pharmacol. Sci.*, 1984, **5**, 7.
99. G. D. Stoner, C. C. Harris, H. Autrup, B. F. Trump, E. W. Kingsbury and G. A. Myers, *Lab. Invest.*, 1978, **38**, 685.
100. J. Jäger, S. Marwitz, J. Tiefenau, J. Rasch, O. Shevchuk, C. Kugler, T. Goldmann and M. Steinert, *Infect. Immun.*, 2014, **82**, 275.
101. J. F. Lechner, A. Haugen, H. Autrup, I. A. McClendon, B. F. Trump and C. C. Harris, *Cancer Res.*, 1981, **41**, 2294.
102. L. A. Barrett, E. M. McDowell, A. L. Frank, C. C. Harris and B. F. Trump, *Cancer Res.*, 1976, **36**, 1003.
103. R. T. Bennett, R. D. Jones, A. H. Morice, C. F. Smith and M. E. Cowen, *Thorax*, 2004, **59**, 401.
104. T. E. Mürdter, G. Friedel, J. T. Backman, M. McClellan, M. Schick, M. Gerken, K. Bosslet, P. Fritz, H. Toomes, H. K. Kroemer and B. Sperker, *J. Pharmacol. Exp. Ther.*, 2002, **301**, 223.
105. J. W. Lee, X. Fang, N. Gupta, V. Serikov and M. A. Matthay, *Proc. Natl. Acad. Sci. U. S. A.*, 2009, **106**, 16357.
106. A. Linder, G. Friedel, P. Fritz, K. T. Kivistö, M. McClellan and H. Toomes, *Thorac. Cardiovasc. Surg.*, 1996, **44**, 140.
107. J. H. Widdicombe, *J Anat.*, 2002, **201**, 313.
108. Y. Ke, B. I. Gerwin, S. E. Ruskie, A. M. Pfeifer, C. C. Harris and J. F. Lechner, *Am. J. Pathol.*, 1990, **137**, 833.
109. J. F. Lechner, I. A. McClendon, M. A. LaVeck, A. M. Shamsuddin and C. C. Harris, *Cancer Res.*, 1983, **43**, 5915.
110. R. Wu and D. Smith, Continuous multiplication of rabbit tracheal epithelial cells in a defined, hormone-supplemented medium, *In Vitro*, 1982, **18**, 800.
111. B. Forbes and C. Ehrhardt, *Eur. J. Pharm. Biopharm.*, 2005, **60**, 193.
112. C. Grobstein, *J. Exp. Zool.*, 1953, **124**, 383.
113. J. F. Engelhardt, E. D. Allen and J. M. Wilson, *Proc. Natl. Acad. Sci. U. S. A.*, 1991, **88**, 11192.
114. M. Terzaghi, P. Nettesheim and M. L. Williams, *Cancer Res.*, 1978, **38**, 4546.

115. Corning Matrigel® Matrix, 2014: http://www.corning.com/lifesciences/surfaces/en/matrigel.aspx?gclid=CjgKEAjwnfGbBRDlxoHrl6uikyESJAD-nzCFTFmAAZZxG8bg5JffFgJ-U_xsMbrP9N4bBGcNHTHrmfD_BwE. Date last accessed: May 2014.
116. D. J. Maltman and S. A. Przyborski, *Biochem. Soc. Trans.,* 2010, **38**, 1072.
117. Alvatex® Reinnervate, 2014: http://reinnervate.com/alvetex-3d-cell-culture-fit-purpose/. Date last accessed: May 2014.
118. K. B. Adler and Y. Li, *Am. J. Respir. Cell Mol. Biol.,* 2001, **25**, 397.
119. W. R. Martin, C. Brown, Y. J. Zhang and R. Wu, *J. Cell. Physiol.,* 1991, **147**, 138.
120. T. E. Gray, K. Guzman, C. W. Davis, L. H. Abdullah and P. Nettesheim, *Am. J. Respir. Cell Mol. Biol.,* 1996, **14**, 104.
121. M. Yamaya, W. E. Finkbeiner, S. Y. Chun and J. H. Widdicombe, *Am. J. Physiol.,* 1992, **6**, L713.
122. L. Kaartinen, P. Nettesheim, K. B. Adler and S. H. Randell, *In Vitro Cell. Dev. Biol.: Anim.,* 1993, **29**, 481.
123. M. Kondo, W. E. Finkbeiner and J. H. Widdicombe, *In Vitro Cell. Dev. Biol.: Anim.,* 1993, **29**, 19.
124. H. Mochizuki, A. Morikawa, K. Tokuyama, T. Kuroume and A. C. Chao, *Eur. J. Pharmacol.,* 1994, **252**, 183.
125. V. Sueblinvong and D. J. Weiss, *Transl. Res.,* 2010, **56**, 188.
126. A.-G. Lenz, E. Karg, E. Brendel, H. Hinze-Heyn, K. L. Maier, O. Eickelberg, T. Stoeger and O. Schmid, *BioMed Res. Int.,* 2013, **2013**, 652632, doi: 10.1155/2013/652632.
127. M. Sakagami, *Adv. Drug Delivery Rev.,* 2006, **58**, 1030.
128. C. Ehrhardt, B. Forbes, K.-J. Kim. In Vitro Models of the Tracheo-Bronchial Epithelium. ed. C. Hrhardt, K.-J. Kim,in *Drug Absorption Studies.* US: Springer; 2008.
129. T. Kim and Y. H. Cho, A pumpless cell culture chip with the constant medium perfusion-rate maintained by balanced droplet dispensing, *Lab Chip,* 2011, **11**, 1825.
130. Corning 2014: http://csmedia2.corning.com/LifeSciences//media/pdf/Corning_FloWell_2W_Plate_User_Guide.pdf. Date last accessed: May 2014.
131. Kirkstall 2014: http://kirkstall.org/index.php/quasi-vivo-system/. Date last accessed May 2014.
132. T. Sbrana and A. Ahluwalia in *Engineering Quasi-Vivo® in vitro Organ Models, in New Technologies for Toxicity Testing,* ed. M. Balls, R. D. Combes and N. Bhogal, Springer, 2012.
133. D. Huh, G. A. Hamilton and D. E. Ingber, *Trends Cell Biol.,* 2011, **21**, 745.
134. D. Huh, B. D. Matthews, A. Mammoto, M. Montoya-Zavala, H. Y. Hsin and D. E. Ingber, *Science,* 2010, **328**, 1662.
135. L. Bjerre, C. E. Bünger, M. Kassem and T. Mygind, *Biomaterials,* 2008, **29**, 2616.
136. C. E. Ghezzi, P. A. Risse, B. Marelli, N. Muja, J. E. Barralet, J. G. Martin and S. N. Nazhat, *Biomaterials,* 2013, **34**, 1954.

137. T. T. Goodman, C. P. Ng and S. H. Pun, *Bioconjugate Chem.*, 2008, **19**, 1951.
138. M. Griffin, R. Bhandari, G. Hamilton, Y. C. Chan and J. T. Powell, *J. Cell Sci.*, 1993, **105**, 423.
139. D. S. Lang, R. A. Jörres, M. Mücke, W. Siegfried and H. Magnussen, *Toxicol. Lett.*, 1998, **96–97**, 13.
140. F. Chowdhury, W. J. Howat, G. J. Phillips and P. M. Lackie, *Exp. Lung. Res.*, 2010, **36**, 1.
141. M. I. Hermanns, R. E. Unger, K. Kehe, K. Peters and C. J. Kirkpatrick, *Lab. Invest.*, 2004, **84**, 736.
142. H. G. Thompson, D. T. Truong, C. K. Griffith and S. C. George, A three-dimensional in vitro model of angiogenesis in the airway mucosa, *Pulm. Pharmacol. Ther.*, 2007, **20**, 141.
143. B. M. Rothen-Rutishauser, S. G. Kiama and P. Gehr, *Am. J. Respir. Cell Mol. Biol.*, 2005, **32**, 281.
144. A. T. Nguyen Hoang, P. Chen, J. Juarez, P. Sachamitr, B. Billing, L. Bosnjak, B. Dahlén, M. Coles and M. Svensson, *Am. J. Physiol. Lung. Cell Mol. Physiol.*, 2012, **302**, L226.
145. D. Schuster, C. Laggner and T. Langer, *Curr. Pharm. Des.*, 2005, **11**, 3545.
146. S. Matsuda, M. Hisama, H. Shibayama, N. Itou and M. Iwaki, *Pharm. Soc. Japan*, 2009, **129**, 1113.
147. J. E. May, J. Xu, H. R. Morse, N. D. Avent and C. Donaldson, *Br. J. Biomed. Sci.*, 2009, **66**, 160.

CHAPTER 5

Complex Primary Human Cell Systems for Drug Discovery

ELLEN L. BERG* AND ALISON O'MAHONY

BioSeek, A Division of DiscoveRx, 310 Utah Avenue, Suite 100, South San Francisco, CA 94080, USA
*E-mail: eberg@bioseekinc.com

5.1 Introduction

5.1.1 Challenges of Target-Based Drug Discovery

Despite reports documenting that phenotypic assays have, historically, been the most effective means of discovering new drugs,[1] the vast majority of pharmaceutical drug discovery programs remains target based. A large proportion of target-centric pipelines focus on kinases, and with over 500 members, kinases remain a compelling target class with a number of inhibitors already having achieved regulatory approval in oncology. These include drugs that inhibit BCR-Abl (bosutinib, dasatinib, imatinib and nilotinib), EGFR (afatinib, erlotinib, gefitinib and lapatinib), ErbB1/B2 (erlotinib and lapatinib), VEGFR (axitinib, cabozantinib, pazopanib, regorafenib, sorafenib, sunitinib and vandetanib), ALK/Met (crizotinib), BRAF (debrafenib and vemurafenib), MEK (trametinib) and JAK (ruxolitinib). Only recently has a kinase inhibitor been approved for a non-oncology indication. Tofacitinib (CP-690,550), a JAK inhibitor, was approved in November 2012 by the US FDA for the treatment of rheumatoid arthritis.

The success of target-based programs depends on sufficient validation of the target and selectivity of the compound. Most target validation efforts employ relatively simple, often transformed cells and animal models, with many relying on gene knock-out or overexpression approaches. Transformed cell lines have evolved markedly altered signalling networks and proliferation properties relative to normal, non-tissue culture adapted cells. Moreover, disease mechanisms in mouse and human have significant differences, making the translation from animal models to human patients unreliable at best. In addition, traditional target validation studies using knock-down techniques are problematic for kinases as these proteins have both enzymatic and structural roles. For this class, knock-down in combination with overexpression of catalytically inactive forms of the protein are used to study the potential effects of inhibition of the active site. Compound selectivity can also be challenging, particularly for target classes that have highly homologous members or where the ATP-binding loop may be highly conserved across multiple kinases. Indeed most of the approved inhibitors inhibit multiple kinases, particularly in a concentration-dependent manner. Indeed higher clinical exposure rates that differ from patient to patient can lead to adverse effects related to secondary targets. Notably, the recently withdrawn BCR-Abl inhibitor, ponatinib, inhibits a broad range of kinases including VEGFR, PDGFR, FGFR, Eph, SRC family kinases, Kit, RET, Tie2 and Flt3, in addition to BCR-Abl, and was associated with the incidence of blood clots.

While biochemical selectivity panels have been very effective for characterising kinase inhibitor selectivities, they have proven to be of limited use for certain kinase families. As a case in point, the JAK family members are not only highly homologous, they also operate in pairs or multimers in functional cascades, thus determining selectivity alone will not be sufficient to fully characterise compounds. Here we describe the application of a broad panel of primary human cell-based tissue and disease models toward the characterisation of two kinase inhibitors that have been in development for inflammatory disease indications, the JAK3/1 kinase inhibitor tofacitinib and the SYK kinase inhibitor fostamatinib.

5.1.2 Kinase Inhibitors – An Example Target Class

Protein kinases are a large family (>500 members) of cytoplasmic and cell surface protein enzymes that mediate signalling networks in almost all cellular functions and responses. Kinases typically contain both regulatory domains that couple to the pathway intermediates and catalytic ATP-binding domains that transfer a phosphate group onto target amino acid residues (serine or tyrosine) within specified protein substrates. Kinases, together with phosphatases, that remove phosphate groups, function to transmit signals to evoke a particular cellular response typically *via* gene transcription and translation programs, although some kinase targets may serve a transient, functional response that may or may not result in a phenotypic change (*e.g.* gated channel proteins). Most kinases can phosphorylate multiple targets, and most targets can be phosphorylated at multiple sites, often by different kinases. These features lead to complex patterns of phosphorylation and dephosphorylation, the cadence and timing of which can produce complex signalling networks that can display sophisticated oscillatory behaviours that are very difficult to predict.

The large number of kinases, the similarity of their ATP-binding loops, and their often promiscuous target selectivity renders the development of highly selective compounds technically challenging. In addition, the functions of many kinases remain poorly understood and the impact of their inhibition difficult to detect and interpret. Together, these obstacles combine to make the identification, optimisations and clinical development of compounds with optimal selectivity and efficacy difficult. A number of kinase inhibitors have been approved by regulatory agencies for treatment of cancer and more recently in the USA for treatment of rheumatoid arthritis.

5.1.3 Complex Primary Human Cell Systems for Translational Biology

5.1.3.1 *BioMAP® Systems*

We have previously described standardised panels of primary human cell and co-culture models of tissue and disease biology, BioMAP systems, for the characterisation of drug leads.[2–6] The Diversity PLUS panel of 12 BioMAP systems is shown in Table 5.1. Each system consists of one or more primary human cell types and is stimulated by the addition of one or more activating factors. The selection of cell types and stimuli is designed to recapitulate tissue states and cell signalling networks that are relevant to in *vivo* disease settings. In each system, the levels of a panel of translational biomarker readouts (primarily proteins including cell surface and secreted proteins, mediators, *etc.*) are measured after the addition of test agents to cells prior to stimulation. The resulting changes in biomarker levels are then measured after 1–4 days in culture. There have been numerous publications on the effects of well-characterised reference compounds in these systems.[2–4,6–10] Kunkel *et al.*[2] described the value of testing compounds in multiple assay systems and particularly co-culture assays. Follow-up studies have described the utility of this platform for kinase inhibitors, natural product mechanisms of action, and for defining toxicity mechanisms. A panel of eight BioMAP systems have been used by the US EPA as part of the ToxCast program, for characterising environmental chemicals, toxicity mechanisms, and for the development of predictive signatures.[4,8]

We have recently demonstrated how a database of reference BioMAP profiles generated for a number of well-characterised, target-selective compounds was used to develop predictive models for 28 mechanism classes using support vector machines. This approach provides a useful

Table 5.1 List of 12 BioMAP systems in the Diversity PLUS panel.

System	Symbol	Primary human cell types	Disease/tissue relevance	Read-out parameters	System description
3C		Venular endothelial cells	Cardiovascular disease, chronic inflammation	MCP-1, VCAM-1, TM, TF, ICAM-1, E-selectin, uPAR, IL-8, MIG, HLA-DR, proliferation, SRB	The 3C system models vascular inflammation of the Th1 type, an environment that promotes monocyte and T cell adhesion and recruitment and is anti-angiogenic. This system is relevant for chronic inflammatory diseases, vascular inflammation and restenosis.
4H		Venular endothelial cells	Asthma, allergy, autoimmunity	MCP-1, eotaxin-3, VCAM-1, P-selectin, uPAR, SRB, VEGFRII	The 4H system models vascular inflammation of the Th2 type, an environment that promotes mast cell, basophil, eosinophil, T and B cell recruitment and is pro-angiogenic. This system is relevant for diseases where Th2-type inflammatory conditions play a role such as allergy, asthma and ulcerative colitis.
LPS		Peripheral blood mononuclear cells + venular endothelial cells	Cardiovascular disease, chronic inflammation	MCP-1, VCAM-1, TM, TF, CD40, E-selectin, CD69, IL-8, IL1α, M-CSF, sPGE2, SRB, sTNFα	The LPS system models chronic inflammation of the Th1 type and monocyte activation responses. This system is relevant to inflammatory conditions where monocytes play a key role including atherosclerosis, restenosis, rheumatoid arthritis, and other chronic inflammatory conditions, as well as metabolic diseases.

(continued)

Table 5.1 (continued)

System	Symbol	Primary human cell types	Disease/tissue relevance	Read-out parameters	System description
SAg		Peripheral blood mononuclear cells + venular endothelial cells	Autoimmune disease, chronic inflammation	MCP-1, CD38, CD40, E-selectin, CD69, IL-8, MIG, PBMC cytotoxicity, proliferation, SRB	The SAg system models chronic inflammation of the Th1 type and T cell effector responses to TCR signalling with co-stimulation. This system is relevant to inflammatory conditions where T cells play a key role including organ transplantation, rheumatoid arthritis, psoriasis, Crohn's disease and multiple sclerosis.
BT		B Cells + peripheral blood mononuclear cells	Asthma, allergy, oncology, autoimmunity	B Cell proliferation, PBMC cytotoxicity, secreted IgG, sIL17A, sIL17F, sIL-2, sIL-6, sTNFα	The BT system models T cell dependent B cell activation and class switching as would occur in a germinal centre. This system is relevant for diseases and conditions where B cell activation and antibody production are relevant. These include autoimmune disease, oncology, asthma and allergy.
BF4T		Bronchial epithelial cells + dermal fibroblasts	Asthma, allergy, fibrosis, lung inflammation	MCP-1, eotaxin-3, VCAM-1, ICAM-1, CD90, IL-8, IL1α, keratin 8/18, MMP-1, MMP-3, MMP-9, PAI-1, SRB, tPA, uPA	The BF4T system models lung inflammation of the Th2 type, an environment that promotes the recruitment of eosinophils, mast cells and basophils as well as effector memory T cells. This system is relevant for allergy and asthma, pulmonary fibrosis, as well as COPD exacerbations.

BE3C	Bronchial epithelial cells	Lung inflammation, COPD	ICAM-1, uPAR, IP-10, I-TAC, IL-8, MIG, EGFR, HLA-DR, IL1α, keratin 8/18, MMP-1, MMP-9, PAI-1, SRB, tPA, uPA	The BE3C system models lung inflammation of the Th1 type, an environment that promotes monocyte and T cell adhesion and recruitment. This system is relevant for sarcoidosis and pulmonary responses to respiratory infections.
CASM3C	Coronary artery smooth muscle cells	Cardiovascular inflammation, restenosis	MCP-1, VCAM-1, TM, TF, uPAR, IL-8, MIG, HLA-DR, IL-6, LDLR, M-CSF, PAI-1, proliferation, SAA, SRB	The CASM3C system models vascular inflammation of the Th1 type, an environment that promotes monocyte and T cell recruitment. This system is relevant for chronic inflammatory diseases, vascular inflammation and restenosis.
HDF3CGF	Dermal fibroblasts	Fibrosis, chronic inflammation	MCP-1, VCAM-1, ICAM-1, collagen I, collagen III, IP-10, I-TAC, IL-8, MIG, EGFR, M-CSF, MMP-1, PAI-1, prolif_72 h, SRB, TIMP-1, TIMP-2	The HDF3CGF system models wound healing and matrix/tissue remodelling in the context of Th1-type inflammation. This system is relevant for various diseases including fibrosis, rheumatoid arthritis, psoriasis as well as stromal biology in tumours.
KF3CT	Keratinocytes + dermal fibroblasts	Psoriasis, dermatitis, skin biology	MCP-1, ICAM-1, IP-10, IL-8, MIG, IL1α, MMP-9, PAI-1, SRB, TIMP-2, uPA	The KF3CT system models cutaneous inflammation of the Th1 type, an environment that promotes monocyte and T cell adhesion and recruitment. This system is relevant for cutaneous responses to tissue damage caused by mechanical, chemical or infectious agents as well as certain states of psoriasis and dermatitis.

(continued)

Table 5.1 (continued)

System	Symbol	Primary human cell types	Disease/tissue relevance	Read-out parameters	System description
MyoF		Lung fibroblasts	Fibrosis, chronic inflammation, wound healing, matrix remodelling	α-SM actin, bFGF, VCAM-1, collagen I, collagen III, collagen IV, IL-8, decorin, MMP-1, PAI-1, SRB, TIMP-1	The MyoF system models the development of pulmonary myofibroblasts, and are relevant to respiratory disease settings as well as other chronic inflammatory settings where fibrosis occurs such as rheumatoid arthritis.
/Mphg		Venular endothelial cells + macrophages	Cardiovascular inflammation, restenosis, chronic inflammation	MCP-1, MIP-1α, VCAM-1, CD40, E-selectin, CD69, IL-8, IL1α, M-CSF, sIL10, SRB, SRB-Mphg	The /Mphg system models chronic inflammation of the Th1 type and macrophage activation responses. This system is relevant to inflammatory conditions where monocytes play a key role including atherosclerosis, restenosis, rheumatoid arthritis, and other chronic inflammatory conditions.

method for assigning mechanism class to compounds and bioactive agents, and can be applied for triaging hits from phenotypic drug discovery screens.

Here we describe the application of these systems for characterisation of the recently approved JAK kinase inhibitor, tofacitinib (CP-690,550), and fostamatinib (R788), a SYK kinase inhibitor, both of which have been tested in numerous clinical trials for the treatment of rheumatoid arthritis. We demonstrate how BioMAP profile data is consistent with the well-documented biology of the JAK and SYK kinases, while also correlating with clinical effects of these compounds. We posit that compound characterisation using a broad panel of primary human cell-based assays can help guide selection of improved second-generation compounds for these target classes.

5.2 Case Studies – Anti-Inflammatory Kinase Inhibitors

5.2.1 JAK Kinase Inhibitors

5.2.1.1 Background – JAK Kinases

Janus kinases (JAK) are a small family of receptor-associated tyrosine kinases, comprised of JAK1, JAK2, JAK3 and Tyk2. JAK activation of downstream signalling intermediates provides a rapid signalling pathway for cytokines acting through both type I and type II receptors. Present in all cells, JAK signalling has been implicated in cell proliferation, survival, development and differentiation, but are critically required in immune and haematopoietic cell function.[11] JAK kinases transduce their effects through a family of transcriptional proteins termed signal transducers and activators of transcription (STATs). Cytokine or growth factor binding results in the autophosphorylation and/or transphosphorylation of select tyrosines in the JAK kinases. As is the case with most kinases, this is an activating/signal transducing event, and these phosphotyrosines serve as a docking site for STAT proteins, which in turn are phosphorylated by the JAKs. The phosphorylated STAT proteins then dimerise and translocate to the nucleus where they bind DNA promoters and regulate gene transcription. Phosphorylation of select tyrosines in the JAKs can also be a negative regulatory event. Other negative regulating events include association with tyrosine phosphatases, SOCS and CIS proteins.[12]

Unlike other ligand-activated receptors, signalling through JAK-associated receptors proceeds through a heterodimeric or multimeric JAK kinase complex coupled to the receptor. This pairing facilitates *trans*-, auto- and substrate phosphorylation, all of which appear to be required for a full cellular response. While gene knock-out or siRNA knock-down studies have revealed critical catalytic roles for each JAK isoform in various signalling paradigms, they do not recapitulate fully the structural roles of these proteins and, more importantly, do not effectively predict outcomes from selective small molecule inhibitors that interfere with enzymatic functions.[13] A number of interesting studies document the functional and structural partnership between pairs/complexes of JAK kinases. With most gene knock-out studies, selection pressure will drive the redundant phenotype, whereas knock-in of enzymatically-inactive kinase genes (KD) preserves the structural hierarchy but not the phosphorylation activity. Selective small molecule inhibitors would, in theory, be closer to the knock-out/knock-in models and potentially limit the degree of redundancy. A good example of this is the well-documented role of JAK1 in IFN-γ signalling; JAK1 is clearly required for a full IFN-γ response yet, in cells lacking endogenous JAK1, expression of a catalytically inactive JAK1 can sustain substantial IFN-γ inducible gene expression, consistent with a structural as well as an enzymatic role for JAK1. In contrast, IFN-γ responses remain attenuated, despite low-level activation of STAT1 DNA binding activity, in JAK2−/− cells reconstituted with kinase-negative JAK2.[14] These data suggest that there exists a functional hierarchy in pairs of JAKs shared among different types of receptors. In contrast, JAK3 is the only family member that associates with cytokine receptors *via* the common γ (γc) chain, exclusively used by the haematopoietic receptors for IL-2, IL-4, IL-7, IL-9, IL-15 and IL-21. Human X-linked SCID with mutations in the γc have a phenotype highly similar to the autosomal recessive T- B + SCID phenotype associated with JAK3 mutations,[15,16] indicating that JAK1, 2 or Tyk2 do not functionally compensate for disrupted JAK3 in haematopoietic cytokine signalling.

Furthermore, JAK3 is primarily expressed in haematopoietic cells and, through association with the common γc chain, is selectively coupled to cytokine and growth factor receptors in these cells. Evidence from transgenic or knock-out mice as well as clinical reports support the notion that JAK3 or Tyk2 mediate innate and adaptive immune responses. Homozygous mutations in JAK3 or in the coupling to the γc receptor chain in humans gives rise to a range of B and T cell disorders commonly known as severe combined immune deficiency or SCID. This disorder results in severe recurring and opportunistic infections from early childhood, with haematopoietic stem-cell transplantation being the primary therapeutic strategy to date.[15,16] In addition to its critical role in the development and function of lymphocytes, JAK3 is pivotal to the development of a TH2 response involving IL-4 binding, STAT6 activation and GATA3 induction. Individuals with a deficiency in either JAK1 or JAK2 have not been reported, and indeed, JAK1-2 knock-out mice exhibit perinatal lethality.[17,18] In contrast, the presence of activating mutations in JAKs is associated with malignant transformations in lymphoid/myeloid cells in humans. Most notable is the gain-of-function mutations in JAK2 (V617F) that underlie a subset of disorders collectively referred to as myeloproliferative diseases including polycythaemia vera.[19] For these reasons JAK3 has been pursued as a therapeutic target for autoimmune conditions where inhibition of an overactive immune system may prove beneficial.

Complex Primary Human Cell Systems for Drug Discovery 97

Tofacitinib
CP-690,550

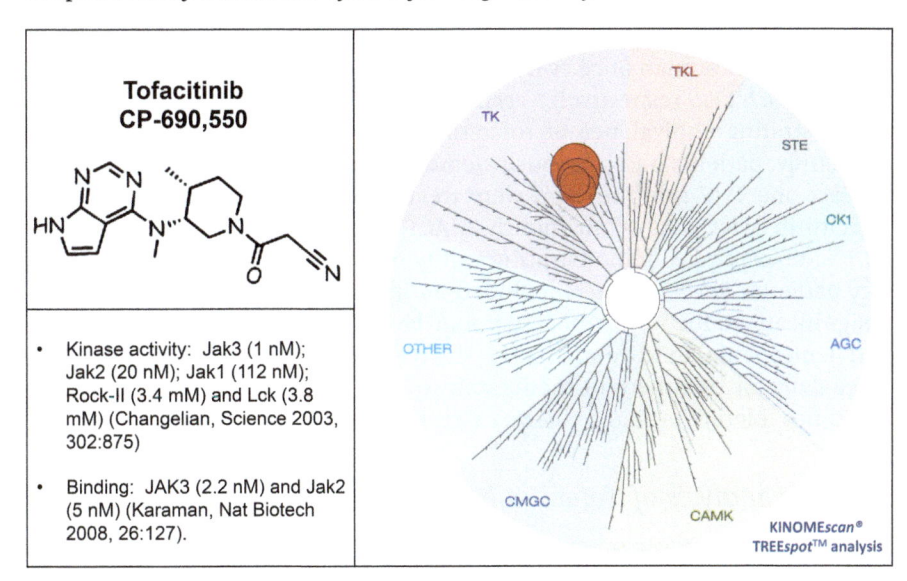

- Kinase activity: Jak3 (1 nM); Jak2 (20 nM); Jak1 (112 nM); Rock-II (3.4 mM) and Lck (3.8 mM) (Changelian, Science 2003, 302:875).

- Binding: JAK3 (2.2 nM) and Jak2 (5 nM) (Karaman, Nat Biotech 2008, 26:127).

KINOMEscan® TREEspot™ analysis

Figure 5.1 Tofacitinib structure and kinase selectivity. Kinase binding data from Davis et al.[22]

5.2.1.2 JAK3 and Tofacitinib

The development of JAK3 inhibitors as immune-selective inhibitors was first suggested by the observation that JAK3 expression appears to be highest in haematopoietic cells and that JAK3 deficits primarily affect lymphoid cells. The first-in-class JAK3 inhibitor to achieve regulatory approval in an inflammatory indication is Pfizer's (PFE) tofactinib (CP-690,550), approved by the US FDA for rheumatoid arthritis (RA) in November 2012.[20,21] Tofacitinib is also being evaluated for other inflammatory diseases including asthma, ulcerative colitis, Crohn's disease, psoriasis and prevention of transplant rejection. Although tofacitinib was first reported to be a JAK3-selective kinase inhibitor, subsequent studies have shown that while it is highly selective for JAK-family kinases, the selectivity among the other JAK family members is relatively modest (Figure 5.1).[22]

Results from the first phase II study of tofacitinib revealed strong efficacy in RA (70–81% patients reported higher ACR20 scores compared to 29% in the placebo group and 13–28% *versus* 3% achieved an ACR70 score).[23] Of note, these patients had previously failed to respond to MTX or anti-TNF biologics. Despite the role for JAK3 deficiency in lymphopenia, no major changes in CD4+ and CD8+ lymphocytes were detected. Similar positive outcomes were reported in clinical trials for the treatment of psoriasis and renal transplantation. The phase III program for tofacitinib included five clinical studies. Two pivotal trials studied tofacitinib in patients with active moderate-to-severe RA in a 12 month Oral Standard and a six month Oral Step study. In the Oral Standard study, patients on stable methotrexate were treated with tofacitinib (5 or 10 mg twice daily) or 40 mg of the anti-TNF inhibitor adalimumab once every 2 weeks.[24] ACR20 response rates were 51.5, 52.6 and 47.2%, respectively, *versus* 28.3% for placebo-treated patients, demonstrating equivalence of tofacitinib to anti-TNF therapy. In the Oral Step study, patients had previously demonstrated an inadequate response to at least one TNF inhibitor and were treated with either 5 mg or 10 mg of tofacitinib twice a day.[20] At month 3, ACR20 response rates were 41.7 and 48.1%, respectively, demonstrating utility of tofacitinib in treatment-refractory patients. Adverse effects included an increased incidence of infections, a significant reduction in haemoglobin levels, increased blood lipid levels and neutropenia. In November 2012, the US FDA approved tofacitinib at 5 mg twice daily for treatment of moderately to severely active RA in patients that could not tolerate or had inadequate responses to methotrexate.[25]

5.2.1.3 Analysis of Tofacitinib in BioMAP Systems

BioMAP systems cover many different pathways where JAK signalling plays a role, including T cell proliferation, cytokine and growth factor signalling. Thus profiling of inhibitors can help identify potential biomarkers for the tested compounds, as well as providing mechanistic information on the relative roles of JAK kinases on various biological activities.

The effects of tofacitinib in BioMAP systems at concentrations of 14–370 nM are shown in Figure 5.2. These concentrations are relevant to human exposures. The reported average C_{max} for the 5 mg administration of tofacitinib is 50 ng ml^{-1} (160 nM), the 10 mg dose has an average C_{max} ~100 ng ml^{-1} (320 nM), while the higher 30 mg dose has a reported average C_{max} of 375 ng ml^{-1} (1.3 µM) in patients. Tofacitinib is highly active in the BioMAP BT system, a model of T cell-dependent B cell activation.[26] Tofacitinib reduces the levels of IgG, IL-6 and TNFα as well as IL-17A in this system, and also causes an increase in IL-2 levels. Inhibition of IgG, IL-6 and TNFα are consistent with efficacy in RA, as these are all clinical biomarkers for the disease.[27–29] Some inhibition of T cell and B cell proliferation is also observed (see proliferation endpoints in the SAg and BT systems, Figure 5.1). Among the other notable features of tofacitinib in this analysis is the potent inhibition of eotaxin-3 in the 4H system that consists of endothelial cells activated with IL-4 and histamine. This activity is consistent with inhibition of IL-4 signalling. IL-4 signalling in leukocytes has been shown to be coupled to JAK1 and JAK3. However, despite the expression of low but detectable levels of JAK3 in non-PBMC,[30] there does not appear to be any γc chain, suggesting that in these cells JAK3 does not functionally couple to the IL-4 receptor in the same way it does in PBMC.[31–33]

The BioMAP profile of tofacitinib at higher concentrations is shown in Figure 5.3. Interestingly, at these higher concentrations, probably supraphysiological, the profile of tofacitinib shows inhibition of a number of additional biomarkers associated with Th1 inflammation, including the chemokines Mig and IP-10, adhesion molecules, VCAM-1 and ICAM-1,

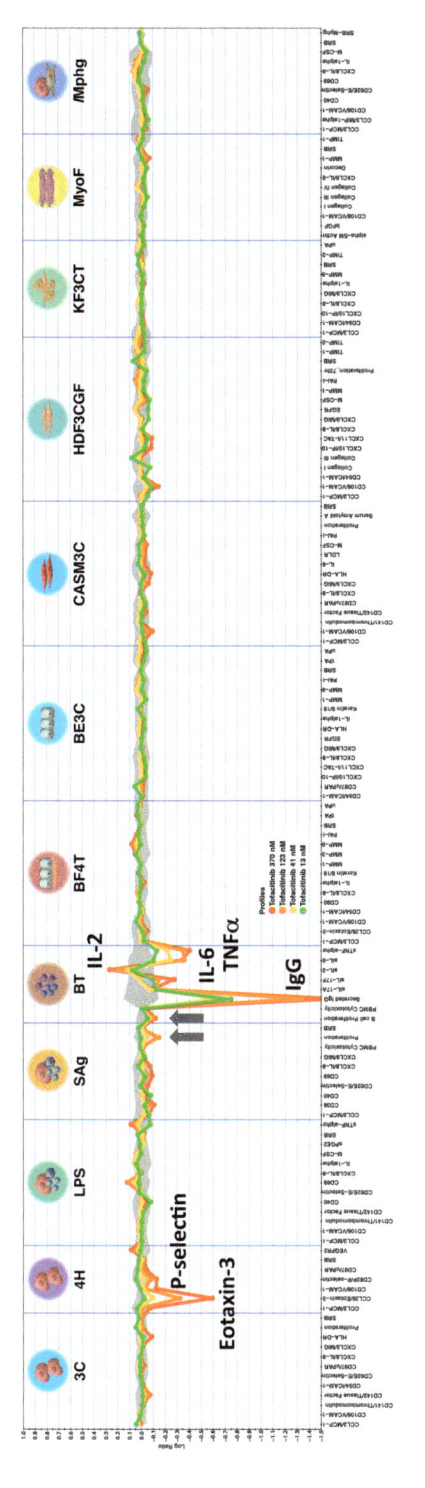

Figure 5.2 BioMAP profile of tofacitinib 41–370 nM. Tofacitinib was tested in 12 BioMAP model systems at 370 nM (red), 123 nM (orange), 41 nM (yellow) and 13 nM (green), as previously described.[4-6] The protein biomarkers measured are indicated along the x-axis. The y-axis shows the log expression ratios (log 10 [parameter value with drug/parameter value of control]) of the read-out protein levels relative to solvent controls.

Figure 5.3 BioMAP profile of tofacitinib 370 nM to 10 µM. Tofacitinib was tested in 12 BioMAP model systems at 10 µM (red), 3.3 µM (orange), 1.1 µM (yellow) and 0.37 µM (green), as described in Figure 5.2. The protein biomarkers measured are indicated along the x-axis. The y-axis shows the log expression ratios (log 10 [parameter value with drug/parameter value of control]) of the read-out protein levels relative to solvent controls.

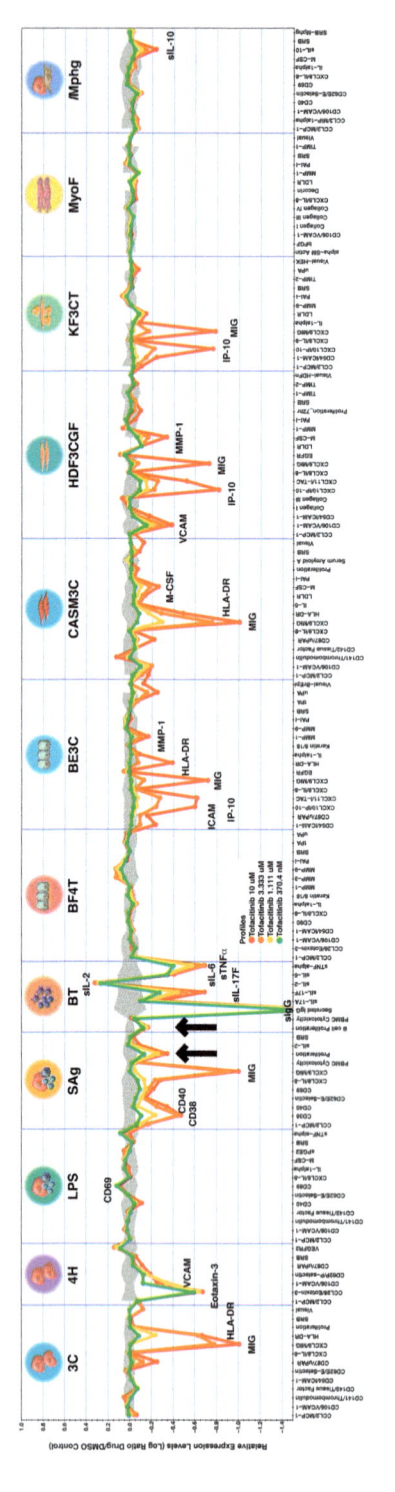

involved in leukocyte recruitment into inflammatory sites, HLA-DR and CD38. These activities are consistent with effects of tofacitinib on JAK1,-2 mediated IFN-g signalling (TH1), and are clinical biomarkers that correlate with disease in RA.[27,34–36] Immunomodulatory effects of tofacitinib include inhibition of HLA-DR, T cell activation (inhibition of most activities in the SAg system), IL-17A, F and IL-10, as well as upregulation of CD69 in the BioMAP LPS system. IL-17A has been shown to correlate with disease activity in RA patients.[37] The inhibition of IL-10 is of interest, as exogenously added IL-10 inhibits both CD69 in the LPS system as well as IL-2 in the BT system. It may be that the increases in these two biomarkers may be due to the loss of a feedback mechanism involving IL-10. Cross-talk between the TLR and JAK kinase signalling involving IFN-γ has been reported, consistent with this finding.[38]

Notably, many of the activities observed with tofacitinib at the higher concentrations are associated with clinical benefit for RA patients, as these are clinical biomarkers for RA that are associated with disease severity. This suggests that patients may continue to experience improvements in disease symptoms at higher doses of tofacitinib. This may have consequences, however. Tofacitinib inhibits JAK2 with a potency only 20 times less than inhibition of JAK3, although binding studies report that tofacitinib binds JAK3 and JAK2 with comparable efficiency (2.2 nM and 5 nM, respectively).[22,39] It has been suggested that inhibition of JAK2 should be avoided given its role in erythropoietin (EPO) and thrombopoietin (TPO) signalling. The potential for anaemia and other myeloid deficiencies in late progenitor cells (*e.g.* neutropenia) may be significant, particularly with prolonged treatment regimes. In clinical trials, anaemia and neutropenia are frequently reported during the first 6 months of treatment.[40] Indeed, data from phase III studies of tofacitinib report significant, dose-dependent decreases in mean neutrophil counts and include anaemia as one of the most frequently reported adverse events.

In conclusion, because JAK1 and possibly other JAK kinases have both structural and catalytic functions, which seem to differ with respect to their individual significance depending on the particular receptor-kinase complex, it may be difficult to predict or interpret cell-based results using only biochemical enzymatic inhibition data. This becomes especially important when one desires optimal and select JAK inhibitors for a particular therapeutic use. Testing compounds across a broad panel of primary human cell-based tissue and disease models, which can be correlated with clinical activity, can provide an alternative method for optimising compounds from this class.

5.2.2 SYK Kinase Inhibitors

5.2.2.1 Background – SYK Kinase

SYK kinase has been an attractive target for immune system disorders due to its role in immune cell signalling. SYK kinase is involved in the activation of a variety of immune cell types including macrophages, mast cells, B cells, neutrophils and dendritic cells,[41] and indeed has been suggested to function as a master regulator of Ig and FcR related cell signalling.[42] SYK kinase has been shown to play a role in B cell signalling through the B cell receptor, IgE-mediated mast cell degranulation, and FcγR-mediated macrophage phagocytosis.[43] Although SYK kinase knock-out mice present with perinatal lethality and show abnormal vascular development, mice lacking SYK kinase in the haematopoietic lineage are protected from autoantibody-induced arthritis.[44] These results and other studies have prompted interest in exploring SYK as a therapeutic target for RA.

Activation of SYK kinase is mediated by ITAM (immunoreceptor tyrosine-based activation motif)-containing receptors, such as Fc receptors and CD79 (Ig-a/Ig-b), the molecule that associates with surface BCR and controls B cell and mast cell activation responses. Phosphorylated ITAM domains serve as docking sites for the SH2 domains of the SYK family kinases, SYK and ZAP-70. Zap-70 has a somewhat analogous role in T cells, interacting with the TCR *via* the ITAM-containing CD3-zeta. SYK has also been shown to interact with other signalling molecules, including PLC-g, VAV, the p85 subunit of PI3K, LAT and LYN, to mediate downstream events. In macrophages, SYK associates with FcγR to regulate phagocytosis. Binding of FcγRs to their ligands induces activation and phosphorylation of ITAM motifs and recruitment of SYK. These biological activities support the development of inhibitors of SYK kinase for asthma/allergy as well as autoimmune conditions involving B cells.[42]

5.2.2.2 SYK and Fostamatinib

Fostamatinib was the first-in-class SYK kinase compound to enter clinical trials for RA (Figure 5.4). Interestingly, this molecule was developed in conjunction with efforts to identify small molecules in a high-throughput screening campaign using cultured human mast cells, and measuring inhibition of IgE-dependent mast cell degranulation. Fostamatinib has been evaluated in several clinical trials for RA, including phase II trials.[45–47] Treatment with 100 mg twice daily or 150 mg once daily resulted in ACR20 response rates of 67% and 57% *versus* 35% in placebo-treated patients.[47] In patients with active RA but inadequate response to either methotrexate[46] or anti-TNF inhibitors,[45] fostamatinib, given 100 mg twice daily, was not significantly more effective than placebo in anti-TNF inhibitor non-responders, but did show activity in patients taking background methotrexate. In December 2012, AstraZeneca and Rigel announced that in a phase IIb study in RA patients, fostamatinib as monotherapy was inferior to the anti-TNF inhibitor adalimumab.[48] While development of this molecule for RA has been discontinued, it is still undergoing clinical evaluation for the treatment of immune thrombocytopenic purpura (ITP), an autoimmune disease characterised by low platelet count (thrombocytopenia) from anti-platelet antibodies. In an open-label pilot study, three-quarters of ITP patients with refractory disease who had failed standard treatments

Fostamatinib R-788

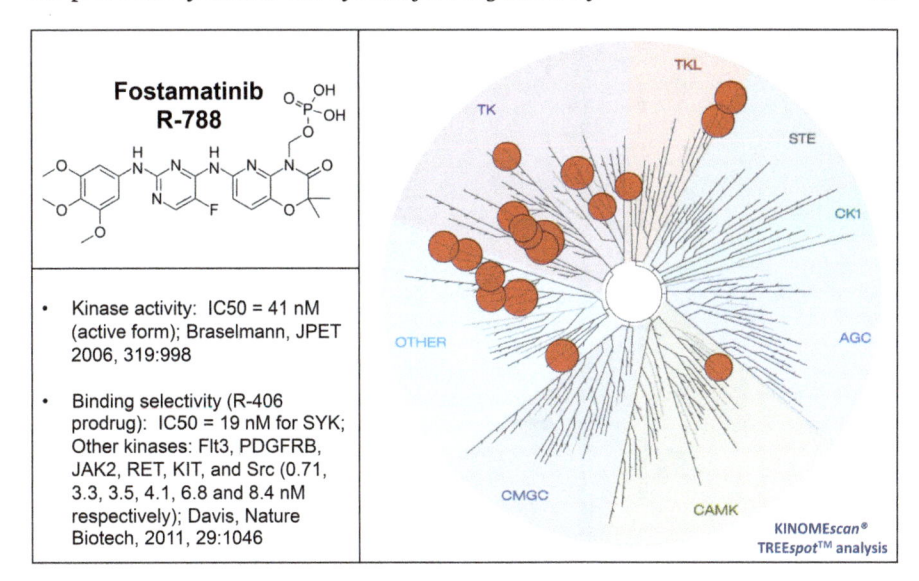

- Kinase activity: IC50 = 41 nM (active form); Braselmann, JPET 2006, 319:998

- Binding selectivity (R-406 prodrug): IC50 = 19 nM for SYK; Other kinases: Flt3, PDGFRB, JAK2, RET, KIT, and Src (0.71, 3.3, 3.5, 4.1, 6.8 and 8.4 nM respectively); Davis, Nature Biotech, 2011, 29:1046

Figure 5.4 Fofacitinib structure and kinase selectivity. Kinase binding data with R-406, the active metabolite of fostamatinib is from Davis et al.[22]

(rituximab, splenectomy or treatment with thrombopoietic agents), responded to treatment with fostamatinib.[49] Side effects of fostamatinib include gastrointestinal toxicity, neutropenia and modest increases in blood pressure as well as liver enzymes.[45,47] In humans, fostamatinib is highly bioavailable and rapidly absorbed.[50]

5.2.2.3 Analysis of Fostamatinib in BioMAP Systems

The BioMAP profile of fostamatinib is shown in Figure 5.5. Fostamatinib is broadly active, inhibiting B and T cell activation (SAg and BT systems), as well as monocyte activation (LPS system). Fostamatinib also affects endothelial cells (3C and 4H systems), epithelial cells (BE3C system), smooth muscle cells (CASM3C system), and to a lesser extent fibroblast-containing systems (HDF3CGF system). Inflammation-related activities include decreased chemokines eotaxin-3, IL-8, Mig, MIP1α and MCP-1; and inflammatory mediators PGE2, IL-1alpha, TNFα and IL-6. Inhibition of eotaxin-3 in the 4H system is consistent with inhibition of JAK kinases, a known secondary target of R406, the active metabolite of fostamatinib. Immunomodulatory activities include strong inhibition of activities in the BT system, a model of T cell-dependent B cell activation.[26] Activities consistent with tissue remodelling biology include inhibition of MMP3, uPA and uPAR.

Interestingly, the effects of fostamatinib on endothelial cells and reduction of PGE_2 in the LPS system, together with the observation that fostamatinib exposure is associated with increases in blood pressure, suggested

104   *Chapter 5*

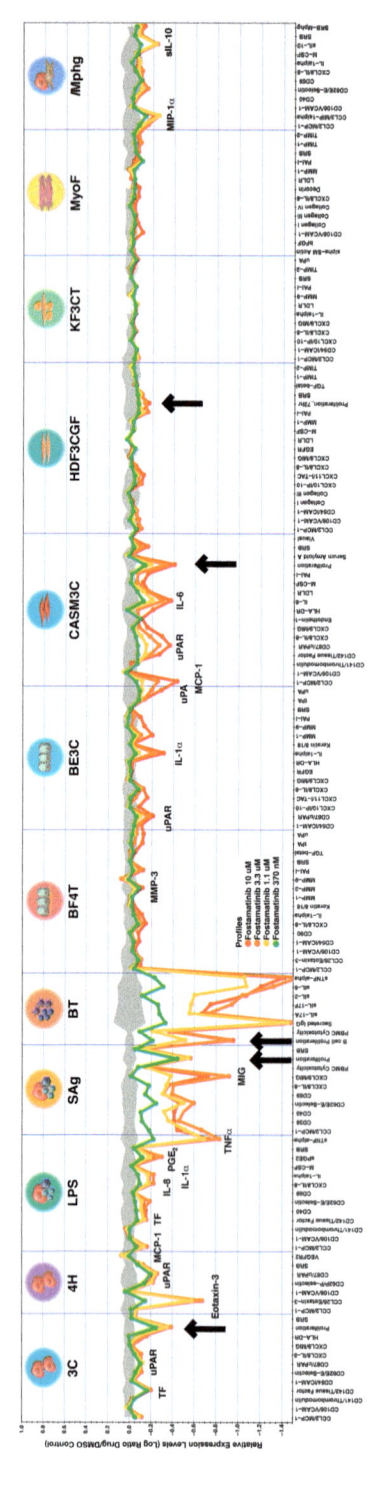

Figure 5.5   BioMAP profile of fostamatinib from 370 nM to 10 μM. Tofacitinib was tested in 12 BioMAP model systems at 10 μM (red), 3.3 μM (orange), 1.1 μM (yellow) and 0.37 μM (green), as described in Figure 5.2. The protein biomarkers measured are indicated along the *x*-axis. The *y*-axis shows the log expression ratios (log 10 [parameter value with drug/parameter value of control]) of the read-out protein levels relative to solvent controls.

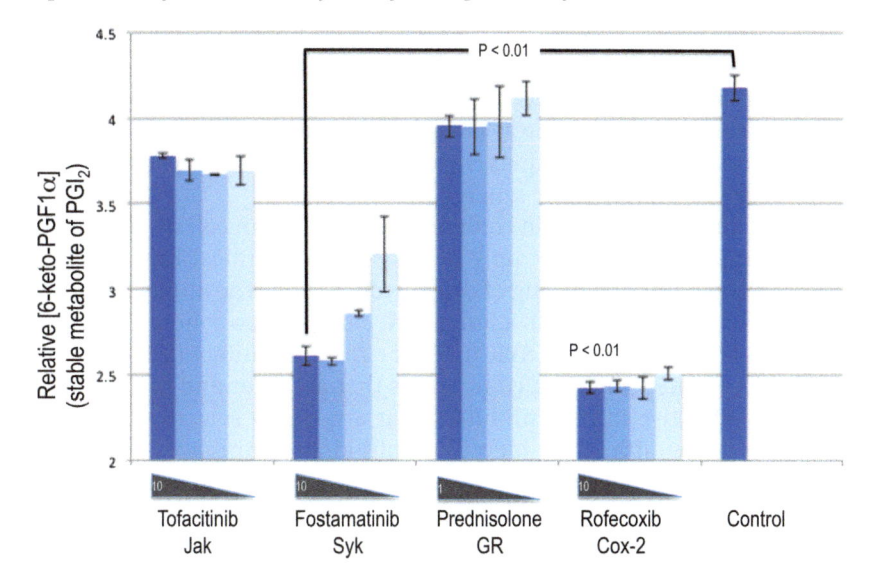

Figure 5.6 Effect of tofacitinib, fostamatinib, prednisolone and rofecoxib on 6-keto-PGF1α in the BioMAP LPS system. Tofacitinib, fostamatinib and rofecoxib at 10, 3.3, 1.1 and 0.37 μM or prednisolone at 1, 0.33, 0.11, 0.04 μM were tested in the BioMAP LPS system as described.[4] The levels of 6-keto-PGF1alpha, a stable metabolite of PGI_2 were measured. Compounds were tested in triplicate, and the average and standard deviation of samples are shown from a representative experiment.

that effects on another related prostaglandin, prostacyclin (PGI_2) should be assessed. As shown in Figure 5.6, fostamatinib significantly reduced the levels of prostacyclin measured in the LPS system. Prostacyclin is a vasodilator with anti-thrombotic and anti-proliferative effects. The effects of selective cyclooxygenase-2 inhibitors on PGI_2 has been proposed as one potential mechanism underlying the cardiovascular side effects of these drugs.[51]

The BioMAP profile of fostamatinib is consistent with a multi-kinase inhibitor. Given the dose-limiting side effects observed for fostamatinib in clinical trials, there is significant opportunity for identifying follow-up compounds with differential kinase selectivities that may be more effective and safe. It remains to be determined whether or not highly selective SYK kinase inhibitors will be sufficiently efficacious as several of the secondary targets of fostamatinib, including JAK and SRC-family kinases, could contribute to compound efficacy in immune system disorders.

5.3 Conclusions

While kinase inhibitors have been very successful in oncology indications for a number of years, only recently has a kinase inhibitor achieved regulatory approval in an inflammatory disease indication. The polypharmacy inherent in kinase inhibitors, due to the high degree of homology between family members, makes it challenging to identify safe and selective compounds. The limited knowledge about kinase-associated side effects, coupled with a poor understanding of pathway cross-talk, makes it difficult to predict the consequences of inhibiting kinases, particularly in light of pathway redundancy, feedback inhibition or synergy.

Profiling kinase inhibitors in complex primary human cell systems such as the BioMAP platform can not only reveal potential clinical biomarkers but provide a deeper understanding of drug mechanisms. The ability to correlate BioMAP profile activities with clinical effects, as shown here for the JAK inhibitor tofacitinib and the SYK inhibitor fostamatinib, will be useful for identifying second-generation compounds that are safer and more effective.

Clinically effective kinase inhibitors that are highly selective for a single target have been difficult to identify. Multi-kinase inhibitors have shown efficacy in RA, and now the challenge will be to identify compounds that are sufficiently effective, but with minimal side effects. Profiling these compounds in broad panels of primary human cell-based assays, such as those provided by BioMAP systems, can be helpful in assessing the biological activities that may be the result of inhibition of multiple targets, and determining their likely benefit to patients.

Acknowledgements

The authors would like to acknowledge Mark A. Polokoff for helpful discussions, and Daniel Trieber for the kinase binding data for tofacitinib and fostamatinib.

References

1. D. C. Swinney and J. Anthony, *Nat. Rev. Drug Discovery,* 2011, **10**, 507.
2. E. J. Kunkel, M. Dea, A. Ebens, E. Hytopoulos, J. Melrose, D. Nguyen, K. S. Ota, I. Plavec, Y. Wang, S. R. Watson, E. C. Butcher and E. L. Berg, *FASEB J.,* 2004, **18**, 1279.
3. E. J. Kunkel, I. Plavec, D. Nguyen, J. Melrose, E. S. Rosler, L. T. Kao, Y. Wang, E. Hytopoulos, A. C. Bishop, R. Bateman, K. M. Shokat, E. C. Butcher and E. L. Berg, *Assay Drug Dev. Technol.,* 2004, **2**, 431.
4. E. L. Berg, J. Yang, J. Melrose, D. Nguyen, S. Privat, E. Rosler, E. J. Kunkel and S. Ekins, *J. Pharmacol. Toxicol. Methods,* 2010, **61**, 3.
5. D. Xu, Y. Kim, J. Postelnek, M. D. Vu, D. Q. Hu, C. Liao, M. Bradshaw, J. Hsu, J. Zhang, A. Pashine, D. Srinivasan, J. Woods, A. Levin, A. O'Mahony, T. D. Owens, Y. Lou, R. J. Hill, S. Narula, J. DeMartino and J. S. Fine, *J. Pharmacol. Exp. Ther.,* 2012, **341**, 90.
6. G. Bergamini, K. Bell, S. Shimamura, T. Werner, A. Cansfield, K. Muller, J. Perrin, C. Rau, K. Ellard, C. Hopf, C. Doce, D. Leggate, R. Mangano, T. Mathieson, A. O'Mahony, I. Plavec, F. Rharbaoui, F. Reinhard,

M. M. Savitski, N. Ramsden, E. Hirsch, G. Drewes, O. Rausch, M. Bantscheff and G. Neubauer, *Nat. Chem. Biol.*, 2012, **8**, 576.
7. E. L. Berg, E. J. Kunkel, E. Hytopoulos and I. Plavec, *J. Pharmacol. Toxicol. Methods*, 2006, **53**, 67.
8. E. L. Berg, J. Yang and M. A. Polokoff, *J. Biomol. Screening*, 2013, **18**, 1260.
9. K. A. Houck, D. J. Dix, R. S. Judson, R. J. Kavlock, J. Yang and E. L. Berg, *J. Biomol. Screening*, 2009, **14**, 1054.
10. A. C. Melton, J. Melrose, L. Alajoki, S. Privat, H. Cho, N. Brown, A. M. Plavec, D. Nguyen, E. D. Johnston, J. Yang, M. A. Polokoff, I. Plavec, E. L. Berg and A. O'Mahony, *PloS One*, 2013, **8**, e58966.
11. J. J. O'Shea, S. M. Holland and L. M. Staudt, *New Engl. J. Med.*, 2013, **368**, 161.
12. N. K. Tonks, *FEBS J.*, 2013, **280**, 346.
13. J. Briscoe, D. Guschin, N. C. Rogers, D. Watling, M. Muller, F. Horn, P. Heinrich, G. R. Stark and I. M. Kerr, *Philos. Trans. R. Soc. Lond. B Biol. Sci.*, 1996, **351**, 167.
14. J. Briscoe, N. C. Rogers, B. A. Witthuhn, D. Watling, A. G. Harpur, A. F. Wilks, G. R. Stark, J. N. Ihle and I. M. Kerr, *EMBO J.*, 1996, **15**, 799.
15. L. D. Notarangelo, P. Mella, A. Jones, G. de Saint Basile, G. Savoldi, T. Cranston, M. Vihinen and R. F. Schumacher, *Hum. Mutat.*, 2001, **18**, 255.
16. P. Macchi, A. Villa, S. Giliani, M. G. Sacco, A. Frattini, F. Porta, A. G. Ugazio, J. A. Johnston, F. Candotti, J. J. O'Shea, *et al.*, *Nature*, 1995, **377**, 65.
17. E. Parganas, D. Wang, D. Stravopodis, D. J. Topham, J. C. Marine, S. Teglund, E. F. Vanin, S. Bodner, O. R. Colamonici, J. M. van Deursen, G. Grosveld and J. N. Ihle, *Cell*, 1998, **93**, 385.
18. S. J. Rodig, M. A. Meraz, J. M. White, P. A. Lampe, J. K. Riley, C. D. Arthur, K. L. King, K. C. Sheehan, L. Yin, D. Pennica, E. M. Johnson, Jr. and R. D. Schreiber, *Cell*, 1998, **93**, 373.
19. R. Kralovics, F. Passamonti, A. S. Buser, S. S. Teo, R. Tiedt, J. R. Passweg, A. Tichelli, M. Cazzola and R. C. Skoda, *New Engl. J. Med.*, 2005, **352**, 1779.
20. G. R. Burmester, R. Blanco, C. Charles-Schoeman, J. Wollenhaupt, C. Zerbini, B. Benda, D. Gruben, G. Wallenstein, S. Krishnaswami, S. H. Zwillich, T. Koncz, K. Soma, J. Bradley and C. Mebus, *Lancet*, 2013, **381**, 451.
21. J. Kremer, Z. G. Li, S. Hall, R. Fleischmann, M. Genovese, E. Martin-Mola, J. D. Isaacs, D. Gruben, G. Wallenstein, S. Krishnaswami, S. H. Zwillich, T. Koncz, R. Riese and J. Bradley, *Ann. Intern. Med.*, 2013, **159**, 253.
22. M. I. Davis, J. P. Hunt, S. Herrgard, P. Ciceri, L. M. Wodicka, G. Pallares, M. Hocker, D. K. Treiber and P. P. Zarrinkar, *Nat. Biotechnol.*, 2011, **29**, 1046.
23. J. M. Kremer, B. J. Bloom, F. C. Breedveld, J. H. Coombs, M. P. Fletcher, D. Gruben, S. Krishnaswami, R. Burgos-Vargas, B. Wilkinson, C. A. Zerbini and S. H. Zwillich, *Arthritis Rheum.*, 2009, **60**, 1895.

24. R. F. van Vollenhoven, R. Fleischmann, S. Cohen, E. B. Lee, J. A. Garcia Meijide, S. Wagner, S. Forejtova, S. H. Zwillich, D. Gruben, T. Koncz, G. V. Wallenstein, S. Krishnaswami, J. D. Bradley and B. Wilkinson, *New Engl. J. Med.,* 2012, **367**, 508.
25. A. Mullard, *Nat. Rev. Drug Discovery,* 2013, **12**, 87.
26. A. C. Melton, J. Melrose, L. Alajoki, S. Privat, H. Cho, N. Brown, A. M. Plavec, D. Nguyen, E. D. Johnston, J. Yang, M. A. polokoff, I. Plavec, E. L. Berg and A. O'Mahony, *PLoS One,* 2013, **8**, e58966.
27. N. Kutukculer, S. Caglayan and F. Aydogdu, *Clin. Rheumatol.,* 1998, **17**, 288–292.
28. G. Lin and J. Li, *Rheumatol. Int.,* 2010, **30**, 837.
29. B. de Paz, M. Alperi-Lopez, F. J. Ballina-Garcia, C. Prado, C. Gutierrez and A. Suarez, *J. Rheumatol.,* 2010, **37**, 2502.
30. J. W. Verbsky, E. A. Bach, Y. F. Fang, L. Yang, D. A. Randolph and L. E. Fields, *J. Biol. Chem.,* 1996, **271**, 13976.
31. K. Kotowicz, R. E. Callard, K. Friedrich, D. J. Matthews and N. Klein, *Int. Immunol.,* 1996, **8**, 1915.
32. S. Wery-Zennaro, M. Letourneur, M. David, J. Bertoglio and J. Pierre, *FEBS Lett.,* 1999, **464**, 91.
33. C. Doucet, C. Jasmin and B. Azzarone, *Oncogene,* 2000, **19**, 5898.
34. W. P. Kuan, L. S. Tam, C. K. Wong, F. W. Ko, T. Li, T. Zhu and E. K. Li, *J. Rheumatol.,* 2010, **37**, 257.
35. P. A. Klimiuk, S. Sierakowski, R. Latosiewicz, J. P. Cylwik, B. Cylwik, J. Skowronski and J. Chwiecko, *Ann. Rheum. Dis.,* 2002, **61**, 804.
36. R. J. Dolhain, P. P. Tak, B. A. Dijkmans, P. De Kuiper, F. C. Breedveld and A. M. Miltenburg, *Br. J. Rheumatol.,* 1998, **37**, 502.
37. S. A. Metawi, D. Abbas, M. M. Kamal and M. K. Ibrahim, *Clin. Rheumatol.,* 2011, **30**, 1201.
38. X. Hu, J. Chen, L. Wang and L. B. Ivashkiv, *J. Leukocyte Biol.,* 2007, **82**, 237.
39. M. W. Karaman, S. Herrgard, D. K. Treiber, P. Gallant, C. E. Atteridge, B. T. Campbell, K. W. Chan, P. Ciceri, M. I. Davis, P. T. Edeen, R. Faraoni, M. Floyd, J. P. Hunt, D. J. Lockhart, Z. V. Milanov, M. J. Morrison, G. Pallares, H. K. Patel, S. Pritchard, L. M. Wodicka and P. P. Zarrinkar, *Nat. Biotechnol.,* 2008, **26**, 127.
40. S. Busque, J. Leventhal, D. C. Brennan, S. Steinberg, G. Klintmalm, T. Shah, S. Mulgaonkar, J. S. Bromberg, F. Vincenti, S. Hariharan, D. Slakey, V. R. Peddi, R. A. Fisher, N. Lawendy, C. Wang and G. Chan, *Am. J. Transplant.,* 2009, **9**, 1936.
41. A. Mocsai, J. Ruland and V. L. Tybulewicz, *Nat. Rev. Immunol.,* 2010, **10**, 387.
42. R. Singh, E. S. Masuda and D. G. Payan, *J. Med. Chem.,* 2012, **55**, 3614.
43. C. Sedlik, D. Orbach, P. Veron, E. Schweighoffer, F. Colucci, R. Gamberale, A. Ioan-Facsinay, S. Verbeek, P. Ricciardi-Castagnoli, C. Bonnerot, V. L. Tybulewicz, J. Di Santo and S. Amigorena, *J. Immunol.,* 2003, **170**, 846.

44. Z. Jakus, E. Simon, B. Balazs and A. Mocsai, *Arthritis Rheum.*, 2010, **62**, 1899.
45. M. C. Genovese, A. Kavanaugh, M. E. Weinblatt, C. Peterfy, J. DiCarlo, M. L. White, M. O'Brien, E. B. Grossbard and D. B. Magilavy, *Arthritis Rheum.*, 2011, **63**, 337.
46. M. E. Weinblatt, A. Kavanaugh, M. C. Genovese, D. A. Jones, T. K. Musser, E. B. Grossbard and D. B. Magilavy, *J. Rheumatol.*, 2013, **40**, 369.
47. M. E. Weinblatt, A. Kavanaugh, M. C. Genovese, T. K. Musser, E. B. Grossbard and D. B. Magilavy, *New Engl. J. Med.*, 2010, **363**, 1303.
48. *AstraZeneca*, in [Press Release], 2013.
49. A. Podolanczuk, A. H. Lazarus, A. R. Crow, E. Grossbard and J. B. Bussel, *Blood*, 2009, **113**, 3154.
50. M. Baluom, E. B. Grossbard, T. Mant and D. T. Lau, *Br. J. Clin. Pharmacol.*, 2013, **76**, 78.
51. Y. Yu, E. Ricciotti, R. Scalia, S. Y. Tang, G. Grant, Z. Yu, G. Landesberg, I. Crichton, W. Wu, E. Pure, C. D. Funk and G. A. FitzGerald, *Sci. Transl. Med.*, 2012, **4**, 132ra154.

CHAPTER 6

Human in Vitro *ADMET and Prediction of Human Pharmacokinetics and Toxicity Liabilities at the Discovery Stage*

KATYA TSAIOUN

Safer Medicines Trust and Pharma Launcher, Boston, MA, USA
E-mail: katya@safermedicines.org

6.1 Introduction

Drug attrition that occurs in late clinical development or during post-marketing is a serious economic problem in the pharmaceutical industry.[1] The cost of drug approvals is approaching US$1 billion, and the cost of advancing a compound to phase 1 trials can reach up to US$100 million according to the *Tufts Center for the Study of Drug Development, Tufts University School of Medicine*.[2] The study also estimates a US$37000 direct out-of-pocket cost for each day a drug is in the development stage and opportunity costs of US$1.1 million in lost revenue.[2] Given these huge expenditures, substantial savings can accrue from early recognition of problems that would demonstrate a compound's potential to succeed in development.[3]

RSC Drug Discovery Series No. 41
Human-based Systems for Translational Research
Edited by Robert Coleman
© The Royal Society of Chemistry 2015
Published by the Royal Society of Chemistry, www.rsc.org

The costs associated with withdrawing a drug from the market are even greater. For example, terfenadine is both a potent hERG cardial channel ligand and is metabolised by the liver enzyme Cyp3A4. Terfenadine was frequently co-administered with Cyp3A4 inhibitors ketoconazole or erythromycin.[4] The consequent overload resulted in increases in plasma terfenadine to levels that caused cardiac toxicity,[5] resulting in the drug being withdrawn from the market[6] at an estimated cost of US$6 billion. Another example is the broad-spectrum antibiotic trovafloxacin, which was introduced in 1997 and soon became Pfizer's top seller. The drug was metabolically activated *in vivo* and formed a highly reactive metabolite causing severe drug-induced hepatotoxicity.[7] Trovafloxacin was black labelled in 1998,[8] costing Pfizer US$8.5 billion in lawsuits.[9] With the new ability to measure the impact of new molecules on cardiac ion channels such as hERG and other important ADMET (Absorption, Distribution, Metabolism, Excretion and Toxicity) parameters early in the discovery and development process, such liabilities are now recognised earlier, allowing for safer analogues to be advanced to more resource-consuming formal preclinical and clinical stages.

The purpose of preclinical ADMET, also referred to as early DMPK (drug metabolism and pharmacokinetics), is to reduce the risks similar to those described above, and avoid spending scarce resources on weak lead candidates and programs. This strategy allows drug development resources to be focused on fewer, but more likely to succeed drug candidates. In 1993, 40% of drugs failed in clinical trials because of pharmacokinetic (PK) and bioavailability problems in phase I clinical trials.[10] Since then, major technological advances have occurred in molecular biology and screening to allow major aspects of ADMET to be assessed earlier during the lead optimisation stage. By the late 1990s the pharmaceutical industry recognised the value of early ADMET assessment and began routinely employing it, with noticeable results. ADME and DMPK problems decreased from 40% to 11%.[4] Presently, a lack of efficacy and human toxicity are the primary reasons for failure.[11]

The terms 'drugability' and 'druglikeness' were first introduced by Dr Christopher Lipinski, who proposed 'Lipinski's Rule of 5' due to the frequent appearance of a number '5' in the rules.[12] The Rule of 5 has come to be a compass for the drug discovery industry.[13] It stipulates that small-molecule drug candidates should possess:

- a molecular weight less than 500 g mol^{-1}.
- a partition coefficient (log P – a measure of hydrophobicity) less than 5.
- no more than 5 hydrogen bond donors.
- no more than 10 hydrogen bond acceptors.

A compound with fewer than three of these properties is unlikely to become a successful orally bioavailable drug. But like every rule, it has notable exceptions. Such exceptions to Lipinski's Rule of 5 that have become marketed drugs are molecules taken up by active transport mechanisms, natural compounds, oligonucleotides and proteins.

The drug discovery industry is experiencing dramatic structural change and is no longer just the domain of traditional large pharmaceutical companies. Now venture capital funded start-ups, governments, venture philanthropy and other non-profit and academic organisations are important participants in the search for new drug targets, pathways and molecules. These organisations frequently form partnerships, sharing resources, capabilities, risks and rewards of drug discovery. Thus, it is becoming increasingly important to ensure that investors, donors and taxpayers' money is efficiently used so that new safe drugs for unmet medical needs may be delivered to the public. ADMET profiling has been proven to remove poor drug candidates from development and accelerate the discovery process.

Hence, *in vitro* and *in vivo* models remain mere approximations of the complex BBB and their relevance to human pharmacology must be carefully considered. The most appropriate method of conducting controlled experiments is to cross-compare the BBB passage of a series of compounds evaluated with both *in vitro* and *in vivo* models. This enables cross-correlations of pharmacokinetic data and the assessment of the predictive power of *in vitro* and *in vivo* tests.

6.2 The Science of ADMET

Regulatory authorities have relied upon *in vivo* testing to predict the behaviour of new molecules in the human body since the 1950s. Bioavailability, tissue distribution, pharmacokinetics, metabolism and toxicity are assessed typically in one rodent and one non-rodent species prior to administering a drug to a human to evaluate safety in a clinical trial (phase 1). Biodistribution is assessed using radioactively labelled compounds later in development because it is expensive both in terms of synthesising sufficient amounts of radioactively labelled compound and of performing the animal experiments.[14]

Pharmacodynamic (PD) effectiveness of test compounds is typically assessed initially through *in vitro* models such as receptor binding, followed by confirmation through *in vivo* efficacy models in mice or rats. The predictive ability depends on the therapeutic area and the animal model. Infectious disease and obesity models are considered to have the best predictive ability, whereas CNS and oncology animal models are generally the least predictive of human efficacy. Understanding the relationship between drug plasma and tissue concentrations (pharmacokinetics, or PK) and PD (termed PK/PD relationship) is crucial in determining the mechanism of action and metabolic stability of the molecule, which can explain and support efficacy results. *In vivo* pharmacokinetic (PK) studies in a variety of animal models are routinely used for lead optimisation to assess drug metabolism and absorption. It is important to note that there are significant differences in absorption and metabolism among species from animal studies, which may cause conflicting predictions of degradation pathways of new chemical entities (NCEs).

Toxicity and safety studies are performed in models that are relevant to the NCE's mode of action and therapeutic area. *In vivo* toxicity models are required for IND (Investigational New Drug) Application to the US Food and Drug Administration, European Medicines Agencies (EMA) and other regulatory authorities, but have substantial predictive weaknesses. In a retrospective study of 150 compounds from 12 large pharmaceutical companies, the combined animal toxicity study of rodents and non-rodents accurately predicted only slightly more than half of the human hepatotoxicity. This poor level of accuracy in animal toxicity studies caused large numbers of compounds to be removed from development without proceeding into clinical trials with the potential of demonstrating safety in human subjects.[15] The other ~50% whose toxicity could not be predicted was attributed to 'idiosyncratic human hepatotoxicity that cannot be detected by conventional animal toxicity studies'. Although it is widely recognised that mechanisms for toxicity are frequently quite different between species, animal testing remains the 'gold standard' for required regulatory and historical data reasons. The US FDA and other regulatory agencies are in the process of evaluating alternatives to animal testing, with the aim of developing models that are truly predictive of human mechanisms of toxicity, and limiting *in vivo* toxicology testing.

6.3 The ADMET Optimisation Loop

As already mentioned, historically ADMET studies were focused on *in vivo* assays. These are time- and resource-intensive, and generally low-throughput assays, resulting in their implementation later in the development process, when more resources are released to study the few molecules that have advanced to this stage. With the advent of *in vitro* high-throughput screening, molecular biology and miniaturisation technologies in the 1990s, early ADMET assays were developed to predict *in vivo* animal and human results, at a level of speed and cost-effectiveness appropriate for the discovery stage. This produced a major advance in the science of ADMET and has created a new paradigm that drug discovery programs follow in advancing compounds from hit to lead, from lead to advanced lead, and on to nominated clinical candidates. Now, early in the discovery phase, using human enzymes and human-origin cells, drug discovery programs are able to obtain highly actionable information about the drug-likeness of new molecules, the potential to reach target organ, and early indications of known human mechanisms of toxicities. ADMET assessment of varying complexity is currently routinely performed on compounds that have shown *in vitro* efficacy and in conjunction with or just prior to demonstrating early proof of principle *in vivo*.

The application of early ADMET is unique to each drug discovery program. The development path from discovery to IND is not straightforward and is dependent on the therapeutic area, route of administration, chemical series, and other parameters. Correspondingly, the importance of the various ADMET assays is based upon the specifics of the drug discovery program. ADMET assays can also be categorised into those that are routine and those reserved for more advanced profiling. This division is also a function of cost-effectiveness and the need for the specific information. For instance, data regarding induction of human liver enzymes and transporters are not relevant during the hit-to-lead phase and is normally obtained for fewer more advanced candidates.

In some cases the FDA requests data from *in vitro* ADMET assays. For example, *in vitro* drug–drug interaction (DDI) studies may now be conducted under guidance from FDA dated September 2006. The guidance document precisely outlines methods to conduct CYP-450 inhibition and induction and P-gp interaction studies.[16] This package is now typically included in an IND submission.

How should a discovery team employ early ADMET? The answer is not simple and formulaic, because it is a process. It is useful to start from the ultimate goal, which is usually a therapeutic drug for a specific patient population, and work backwards towards discovery. The drug discovery and development team first defines the target product profile (TPP), which includes indication, intended patient population, route of administration, acceptable toxicities, and ultimately will define the drug label. The TPP invariably will evolve during the life of the project, but having major parameters of TPP established initially maintains a collaboration and focus between disciplines such as biology and chemistry, discovery and development, preclinical and clinical groups. Once the TPP is identified, then major design elements of the phase 2 and 3 clinical trials can be outlined, leading to questions about the tolerability, toxicity and safety of the molecule. These parameters will then define the GLP toxicity studies in preclinical models, which will guide the team to the discovery and preclinical development data to be addressed in an early ADMET program.

How is this information implemented in the discovery phase? If a compound has high target receptor binding and biological activity in cells and in relevant *in vivo* animal models, what are the chances of it becoming a successful drug? A molecule needs to cross many barriers to reach its biological target. In order to obtain this goal, a molecule must be in solution and thus the first step is typically to assess the solubility of a compound. A solubility screen provides information about the NCE's solubility in fluids compatible with administration to humans. Chemical and metabolic stability is a further extension of the intrinsic properties of a molecule. Chemical stability in buffers, simulated gastric and intestinal fluids, and metabolic stability in plasma, hepatocytes or liver microsomes of different species can be measured to predict the rate of decay of a compound in the different environments encountered in the human body.

The second step is to define the absorption properties and the bioavailability of a molecule. Measurement of permeability across Caco-2 cell monolayers is a good predictor of human oral bioavailability. For CNS drugs, assessment of BBB penetration would be performed at this stage and is usually a key component of lead optimisation campaigns. Passive BBB permeability may be assessed using BBB-PAMPA assays which use phospholipid layers of different composition whereas potential for active uptake or efflux may be determined using *in vivo* models or cell lines naturally expressing endogenous human intestinal or BBB transporters (such as the Caco-2 cell line) or cell lines overexpressing specific transporters (such as MDCK-MDR1).

Measurement of binding to plasma proteins indicates the degree of availability of the free compound in the blood circulation. This is critical as only unbound drugs are able to reach the target and exert their pharmacological effects. Metabolism and DDI issues are discovered by screening for inhibition of cytochrome P450 liver enzymes (CYP450). All these assays allow chemists and biologists to obtain actionable information and provide a link between structure-activity (SAR) and structure-properties (SPR) relationships that drive decisions on selection of chemical series and molecules.

The next step is the involvement of DDIs and is required for advanced lead optimisation. The effect of drug transporters on permeability and the effect of drugs on transporter activity can be measured in Caco-2, MDCK-MDR1 or other models. P-gp interactions are particularly important for CNS drugs due to high expression of these efflux transporters in the human BBB. Early knowledge about these interactions is instrumental to the medicinal chemistry strategy and helps drive lead optimisation.

The effect of a compound on CYP450 metabolism can be identified by determining the 50% inhibitory concentration (IC_{50}) for each CYP450. These relationships between the NCE and metabolising enzymes need to be evaluated in the context of the human effective dose and maximum effective plasma concentrations. These human data are not normally available at early stages of discovery, but could be extrapolated from animal PK/PD results for compounds in more advanced stages of development. It is important to understand these transporter and CYP450 relationships for the following reasons.

1. The compound may affect the effective plasma concentrations of other concomitantly administered drugs if metabolised by the same CYPs (*i.e.* terfenadine).
2. If the parent drug is a CYP inducer, it may increase the clearance rate of concomitantly administered drugs which are metabolised by these CYPs. This may result in a decrease in these drugs' effective plasma concentrations, thus decreasing their pharmacological effect.
3. Metabolites formed *via* CYP metabolism may be responsible for undesirable side effects such as organ toxicity.
4. The metabolite of a compound may actually be responsible for a compound's efficacy, and not the parent compound. The metabolite may even have a better efficacy, safety and PK profile than its parent. As a result, metabolism can be exploited to produce a better drug which will impact the medicinal chemistry strategy.
5. The identification of drug-metabolising enzymes involved in the major metabolic pathways of a compound helps to predict the probable DDIs in humans. This information also may be used to design human clinical trials to detect unnecessary DDIs.

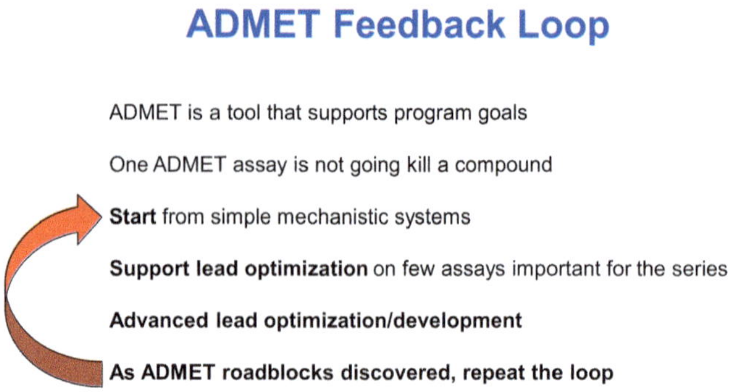

ADMET Feedback Loop

ADMET is a tool that supports program goals

One ADMET assay is not going kill a compound

Start from simple mechanistic systems

Support lead optimization on few assays important for the series

Advanced lead optimization/development

As ADMET roadblocks discovered, repeat the loop

Figure 6.1   ADMET feedback loop.

ADMET is a tool that supports overall program goals. Similar to the Rule of 5, which requires only three of the four conditions to be met, seldom will negative results from a single ADMET assay terminate a compound's development or the overall program. The results are more likely to alter the medicinal chemistry direction.

After assessing compounds in a few simple mechanistic systems such as plasma and liver microsomal stability screens in relevant species, the lead optimisation phase is started, which includes assays that identify potential liabilities. Finally, at the stage of advanced lead optimisation and development, more complex systems are used to more thoroughly understand a compound's metabolic fate and absorption mechanism to drive efficient development. As ADMET roadblocks are discovered, the cycle is repeated until the appropriate balance between efficacy and PK properties of the molecules is achieved (Figure 6.1).

6.4   Impact of Early Human Pharmacokinetics Prediction

Early ADMET provides the data necessary for selecting preclinical candidates by providing crucial information to medicinal chemists and accelerates the timelines for IND and subsequently NDA submission, which translates to lengthier commercialisation under patent protection and greater profits. For investors, this is a major parameter. For philanthropic organisations and governments, it means increasing the time of clinical benefit to the public. Data compiled by the Tufts Center for Drug Discovery have identified that for a typical, moderately successful proprietary drug (US$350 million annual sales), each day's delay equates to US$1.1 million in lost patent-protected revenues that provide the return on investment needed to fund drug discovery.[3] Further, shorter discovery and development timelines provide faster liquidity events for venture capital and angel investors. As drug discovery requires a longer commercialisation than any other form of product development, its slowness to produce returns is a major impediment for obtaining investment. Accelerating drug discovery and development should attract more investment in drug discovery research.

6.5 New Human ADMET Prediction Tools

ADMET technologies have become an active area of research as a necessity to accelerate the pace of drug discovery and development and the development of high-throughput and high-content technologies, as well as new cellular models such as stem cell research. Serious advances have been made in the last decade in areas of absorption, distribution and metabolism, while BBB and toxicity prediction has been more challenging to measure due to its complexity and multifactorial nature: there is a multitude of transporters that act on molecules in BBB and there is an even more challenging picture with regards to mechanisms of human organ toxicity. Another challenge is detection of human idiosyncratic toxicity. These mechanisms cause the most expensive, harmful and disheartening form of drug attrition – post-commercialisation toxicity. Many idiosyncratic drug reactions are due to formation of short-lived reactive metabolites that bind covalently to cell proteins.[17] The extent to which a compound will generate these metabolites can now in some cases only be detected before a compound is administered to humans. Other mechanisms of human toxicity can now be observed early in discovery with the help of the new technologies and are briefly described in the following section.

6.5.1 The Blood-Brain Barrier Challenge

Although lack of efficacy and unexpected toxicity are the major causes of drug failure in clinical trials, a prime determinant is the ability of a drug to penetrate biological barriers such as cell membranes, intestinal walls, or the BBB. For drugs that target the CNS such as stroke, *in vitro* efficacy combined with the inability to penetrate the BBB typically result in poor *in vivo* efficacy in patients. Another obstacle caused by BBB permeability is that many drugs not intended as CNS therapeutics cause neurotoxicity. So early knowledge of this important drug property is essential not only for CNS programs, but also for other therapeutic indications.

The delivery of systemically administered drugs to the brain of mammals is limited by the BBB as it effectively isolates the brain from the blood because of the presence of tight junctions connecting the endothelial cells of the brain vessels. In addition, specific metabolising enzymes and efflux pumps such as P-glycoprotein (P-gp) and the multi-drug-resistance protein (MRP), located within the endothelial cells of the BBB, actively pump exogenous molecules out of the brain.[18,19] This is one of the reasons for CNS drugs having a notoriously high failure rate.[20] In recent years, 9% of compounds that entered phase 1 survived to launch and only 3–5% of CNS drugs were commercialised.[20] More than 50% of this attrition resulted from failure to demonstrate efficacy in phase 2 studies. Over the last decade, phase 2 failures have increased by 15%. Compounds with demonstrated efficacy against a target *in vitro* and in animal models frequently proved to be ineffective in humans. Many of these failures occur due to the inability to reach the CNS targets, such as in stroke due to lack of BBB permeability. For drugs targeted to reduce damage from a stroke, the delivery method, BBB permeability, and drug metabolism and clearance can provide life or death to a patient if the drug is not delivered to the target tissue in its active form in a matter of hours from the event.

Due to the extraordinary cost of drug development, it is highly desirable to have effective, cost-efficient and high-throughput tools to measure BBB permeability before proceeding to expensive and time-consuming animal BBB permeability studies or human clinical trials. With *in vitro* tools available, promising drug candidates with ineffective BBB penetration may be improved by removing structural components that mediate interaction(s) with efflux proteins, and/or lowering binding to brain tissue at earlier stages of development to increase intrinsic permeability.[21]

The development of drugs targeting the CNS requires precise knowledge of the drug's brain penetration. Ideally, this information would be obtained as early as possible to focus resources on compounds most likely to reach the target organ. The physical transport and metabolic composition of the BBB is highly complex. Numerous *in vitro* models have been designed to study kinetic parameters in the CNS, including non-cerebral peripheral endothelial cell lines, immortalised rat brain endothelial cells, primary cultured bovine, porcine or rat brain capillary endothelial cells and co-cultures of primary brain capillary cells with astrocytes.[22-24] *In vitro* BBB models must be carefully assessed for their capacity to reflect accurately the passage of drugs into the CNS *in vivo*.

Alternatively, several *in vivo* techniques have been used to estimate BBB passage of drugs directly in laboratory animals. *In vivo* transport across the BBB was first studied in the 1960s using the early indicator diffusion method (IDM) of Crone.[25] Other *in vivo* techniques were later proposed including brain uptake index (BUI) measurement,[14] the *in situ* brain perfusion method,[26,27] autoradiography and intracerebral microdialysis.[28] Unfortunately these methods have limitations, including the need for sophisticated equipment, technical expertise, mathematical modelling, species differences, invasiveness, and low throughput, and render them unsuitable for use during early stages of drug discovery and development.

Artificial membrane permeability assays (PAMPA and BBB-PAMPA) offer a cost-effective and high-throughput method of screening for passively absorbed compounds but do not predict active transport in or out of the brain.

The development of a new co-culture-based model of human BBB able to predict passive and active transport of molecules into the CNS has recently been reported.[29] This new model consists of primary cultures of human brain capillary endothelial cells co-cultured with primary human glial cells.[22,29] The advantages of this system include:

i. made of human primary culture cells.
ii. avoids species, age and inter-individual differences because the two cell types are removed from the same person.
iii. expresses functional efflux transporters such as P-gp, MRP-1, 4,5 and BCRP.

This model has potential for assessment of permeability of drug and specific transport mechanisms, which is not possible in PAMPA or other cell models due to incomplete expression of active transporters.

One important step in development of any *in vitro* model is to cross-correlate *in vitro* and *in vivo* data in order to validate experimental models and to assess the predictive power of the techniques.[30] The human BBB model was validated against a 'gold standard' *in vivo* model and has shown an excellent *in vitro–in vivo* correlation.[29,31] In this carefully designed *in vivo–in vitro* correlation study the authors reported the evaluation of the BBB permeabilities for a series of compounds studied correlatively *in vitro* using a human BBB model and *in vivo* with quantitative PET imaging.[29] Six clinical PET tracers with different molecular size ranges (Figure 6.2) and degree of BBB penetration were used including [^{18}F]-FDOPA and [^{18}F]-FDG, ligands of amino acid and glucose transporters, respectively. The findings demonstrate that the *in vitro* co-culture model of human BBB has important features of the BBB *in vivo* including low paracellular permeability, well-developed tight junctions, functional expression of important known efflux transporters and is suitable for discriminating between CNS and non-CNS compounds. To further demonstrate the relevance of the *in vitro* human system, drug permeation into the human brain was evaluated using PET imaging in parallel to the assessment of drug permeability across the *in vitro* model of the human BBB. *In vivo* plasma-brain exchange parameters used for comparison were determined previously in humans by PET using a kinetic analysis of the radiotracer binding. 2-[^{18}F]fluoro-A-85380 and [^{11}C]-raclopride show absent or low cerebral uptake with the distribution volume under 0.6. [^{11}C]-flumazenil, [^{11}C]-befloxatone, [^{18}F]-FDOPA and [^{18}F]-FDG show a cerebral uptake with the distribution volume above 0.6. The *in vitro* human BBB model discriminates compounds similar to *in vivo* human brain PET imaging analysis. This data illustrates the close relationship between *in vitro* and *in vivo* pharmacokinetic data (r^2 0.90, $p < 0.001$) (Figure 6.2). Past *in vivo–in vitro* studies often did not have good correlations for substances transported into or out of the brain *via* active transport. Presumably this is due to experiments being performed either with models that did not have adequate expression of active human transporters (such as PAMPA or MDCK cells) or using too high concentrations of compounds *in vitro*, which are known to saturate the

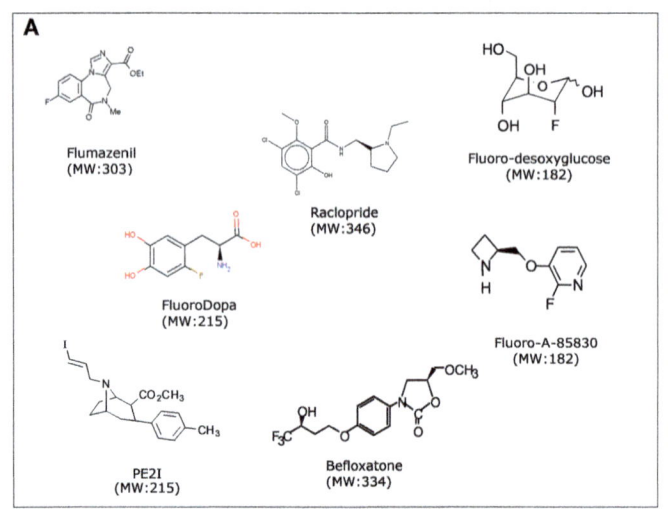

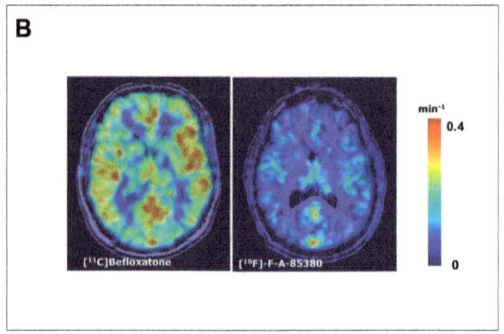

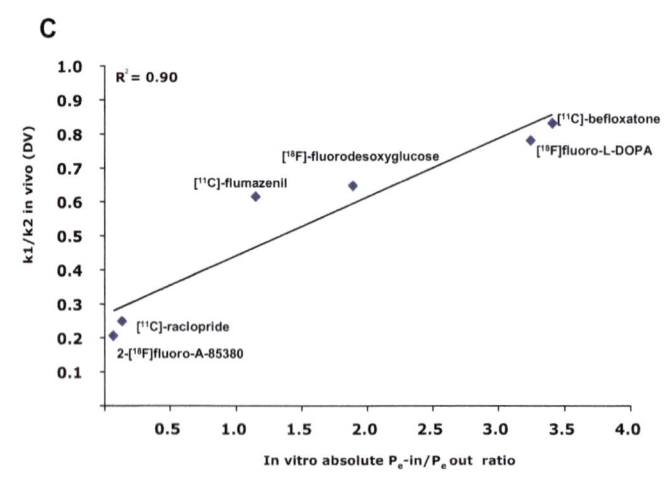

Figure 6.2 *In vitro-in vivo* drug transport correlation (figures from Josserand et al.[29]). (A) Chemical structures of radioligands investigated and used clinically. (B) Typical imaging data. Co-registered PET-MRI images representing the k1 obtained in human after intravenous injection of

transporters. Using the radioactive labelled probes and the small amounts of compounds avoids these issues.

In conclusion, this *in vitro* human BBB model offers great potential for both being developed into a reproducible screen for passive BBB permeability and determining active transport mechanisms. Due to its high-throughput potential, the model may provide testing of large numbers of compounds of pharmaceutical importance for CNS diseases. Validation work is in progress in which activity of transporters that are important in CNS BBB are being assessed in a functional assay and compared between Caco-2 and hBBB models.[31]

6.5.2 Mechanisms of Human Toxicity

Idiosyncratic hepatotoxicity or drug-induced liver injury (DILI) occurs in only one out of about 10 000 patients and is usually statistically impossible to discover during clinical trials. In spite of its name, which is defined as 'a rare event with undefined mechanism', some mechanisms have now been identified, including mitochondrial toxicity and the formation of reactive metabolites. Another mechanism of human toxicity that is not limited to the liver, but may also affect lung, spleen, and heart tissues, is phospholipidosis.

6.5.2.1 Genotoxicity

Genotoxicity of drugs is an important concern to the regulatory authorities. The FDA recommends a number of *in vitro* and *in vivo* tests to measure the mutagenic potential of chemical compounds, including the Ames test in *Salmonella typhimurium*. GreenScreen GC, a new, high-throughput assay that links the regulation of the human GADD45a gene to the production of green fluorescent protein (GFP) has become available. The assay relies on the DNA damage-induced upregulation of the RAD54 gene in yeast measured using a promoter–GFP fusion reporter.[32] The test is more specific and sensitive for

---

[^{11}C]-befloxatone (left) and [^{18}F]-F-A-85380 (right). The PET images representing the k1 are as follows: PET image obtained at 1 min post injection (mean value between 30 s and 90 s) is considered as independent to the receptor binding. This image (in Bq/mL) is corrected from the vascular fraction (Fv in Bq/mL, considered as 4% of the total blood concentration at 1 min) and divided by the arterial plasma input function (AUC$_{0-1\ min}$ of the plasma concentration, in Bq∗min mL^{-1}). The resulting parametric image, expressed in min^{-1}, represent an index of the k$_1$ parameter of the radiotracer. (C) *In vivo* distribution volume (DV) as function of the *in vitro* P_e-out/P_e-in ratio (Q). Regression line was calculated, and correlation was estimated by the two-tailed non-parametric Spearman test. [^{11}C]PE2I radioligand was not plotted in the figure because the *in vivo* K1/k2 parameter in human is not available.

genotoxicity than those currently used such as the Ames and mouse lymphoma tests.

6.5.2.2 Mitochondrial Toxicity

Mitochondrial toxicity is increasingly implicated in drug-induced idiosyncratic toxicity. Many of the drugs that have been withdrawn from the market due to organ toxicity have been found to be mitochondrial toxicants.[33] Mitochondrial toxicants injure mitochondria by inhibiting respiratory complexes of the electron chain, inhibiting or uncoupling oxidative phosphorylation, inducing mitochondrial oxidative stress, or inhibiting DNA replication, transcription or translation.[34]

Toxicity testing of drug candidates is usually performed in immortalised cell lines that have been adapted for rapid growth in a reduced oxygen atmosphere. Their metabolism is often anaerobic by glycolysis, despite having functional mitochondria and an adequate oxygen supply. Alternatively, normal cells generate ATP for energy consumption aerobically by mitochondrial oxidative phosphorylation. The anaerobic metabolism of transformed cell lines is less sensitive to mitochondrial toxicants, causing systematic under-reporting in toxicity testing.[34,35] To address this issue, HepG2 and NIH/3T3 cells can be grown in media in which glucose is replaced by galactose.[33] The change in sugar results in the metabolism of the cell to possess a respiratory substrate that is both more similar to normal cells and sensitive to mitochondrial toxicants without reducing sensitivity to non-mitochondrial toxicants (Figure 6.3).

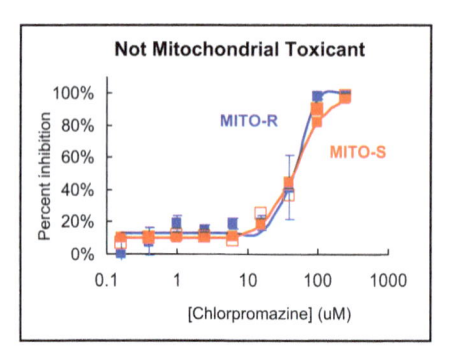

Figure 6.3 Effect of antimycin A, a compound known to be toxic to mitochondria (A) and imipramine (B) on parent HepG2 cells (Mito-R – blue) and a HepG2 cell line that has been developed to become sensitive to mitochondrial toxicants (Mito-S – red). Figure from Annand R., *et al.*, PPTOXII: Role of Environmental Stressors in the Developmental Origins of Disease, December 7–10, 2009, Miami Beach, FL.

Acetaminophen-Glutathione Adduct

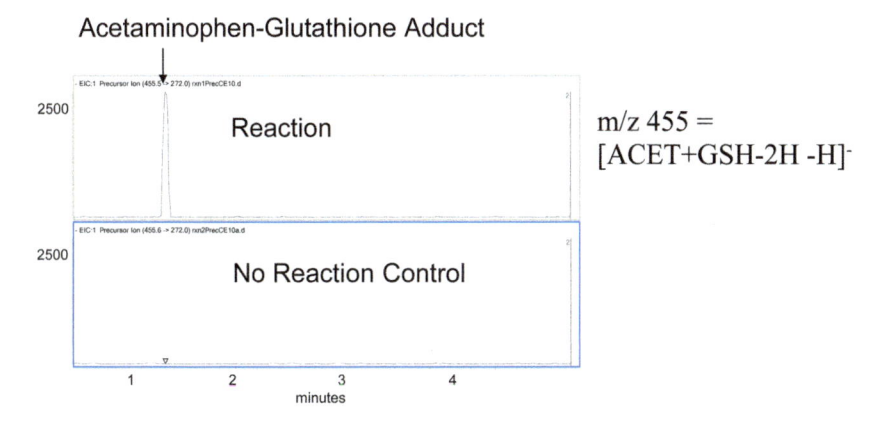

[Chromatogram: Reaction showing peak at ~1.5 min, intensity 2500]

m/z 455 = [ACET+GSH-2H -H]⁻

[Chromatogram: No Reaction Control, intensity 2500]

minutes (1, 2, 3, 4)

Figure 6.4 Formation of reactive metabolites of acetaminophen. Acetaminophen was incubated with microsomes and glutathione in the presence and absence of NADPH. An adduct of glutathione with acetaminophen was formed in the presence of NADPH. When NADPH was absent (No Reaction Control), no adduct was formed. Figure from Annand R., *et al.*, PPTOXII: Role of Environmental Stressors in the Developmental Origins of Disease, December 7–10, 2009, Miami Beach, FL.

6.5.2.3 Reactive Metabolites Formation

Another property of compounds that can cause idiosyncratic toxicity is their ability to form reactive intermediates.[36] Formation of short-lived reactive metabolites is known to be the mechanism of toxicity of some compounds such as acetaminophen.[37] The formation of reactive metabolites can be identified by incubating test compounds with liver microsomes and adding glutathione to trap the reactive intermediates, which are then identified by LC/MS/MS (Figure 6.4). Conversion of more than 10% of the test agent to reactive intermediates indicates that the compound may be implicated in idiosyncratic toxicity.

6.5.2.4 Phospholipidosis

Phospholipidosis is a lysosomal storage disorder and can be caused by drugs that are cationic amphiphiles.[38] The disorder is considered to be mild and can often self-resolve. However, drugs that cause phospholipidosis can also produce organ damage, and thus this disorder is a concern to the regulatory agencies.[38] A cell-based assay for phospholipidosis has been developed[39] that involves accumulation of a fluorescent phospholipid, resulting in an increase of fluorescence in the lysosomes of cells that have been treated with drugs that cause phospholipidosis (Figure 6.5). If phospholipidosis is absent, the phospholipid is degraded and fluorescence does not increase. Increases in fluorescence are normalised to cell numbers because many of these drugs are also cytotoxic (Figure 6.5).

Phospholipidosis of Compounds in HepG2 Cells

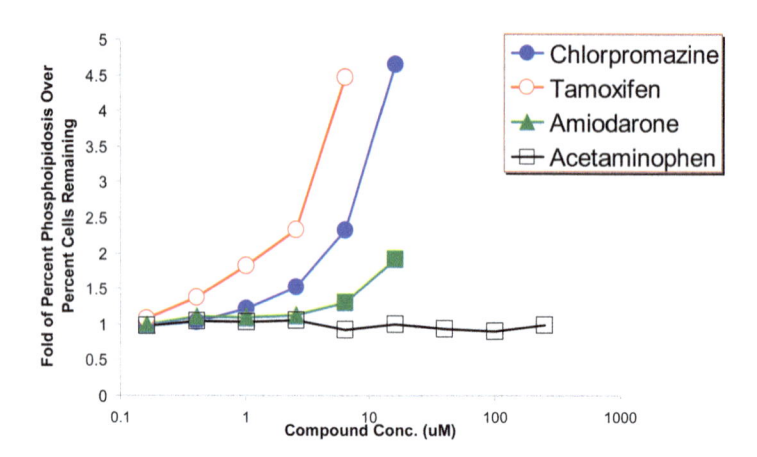

Figure 6.5 Drug-induced phospholipidosis (PLD) is determined by measuring the accumulation of a fluorescent phospholipid in cells treated with increasing drug concentrations. Fluorescence is measured and normalised to cell number. Fluorescence is increased in cells treated with compounds that are known to cause PLD (chlorpromazine, tamoxifen, amiodarone), but it is not increased in cells treated with a compound that is known not to cause PLD (acetaminophen). Figure from Annand R., et al., PPTOXII: Role of Environmental Stressors in the Developmental Origins of Disease, December 7–10, 2009, Miami Beach, FL.

6.5.2.5 High-Content Toxicology: The Present and Future of Predictive Toxicology

As already mentioned, drug safety is a major concern for the pharmaceutical industry, with greater than 30% of drug candidates failing in clinical trials as a result of toxicity.[3,10] Furthermore, there are numerous examples of drugs that have been withdrawn from the market or given black box warnings as a result of side effects not identified in clinical trials. Developing and commercialising a drug is a large financial commitment and failure at this stage can be catastrophic for a company. To address this problem, there has been a significant drive to incorporate multiparametric toxicity assessment at much earlier stages in the drug discovery and development process.

It is well recognised that animal models are often not reflective of all mechanisms of human toxicity. This is corroborated by a large percentage of drugs that fail in the clinic due to toxicity, not identified in lengthy preclinical animal trials. Human hepatotoxicity, as well as hypersensitivity and cutaneous reactions, are particularly difficult to identify during mandated preclinical studies. Only 50% of drugs found to be hepatotoxic in clinical studies showed concordance with animal toxicity results.[15,40]

Initiatives such as ECVAM, ICCVAM, EPA ToxCast and NC3Rs are currently addressing this problem by identifying alternatives to animal safety testing. The cosmetics industry is at the forefront of this work and an EU ban of the sale of cosmetics tested on animals was implemented in 2013.

The introduction of more relevant and sophisticated *in vitro* human systems is essential to overcome these issues and will enable higher throughput screening assays to be implemented earlier and more cost effectively. The widespread use of *in vitro* methods has to some extent been hampered by their relatively poor predictive ability, as traditionally only single markers of toxicity have been investigated. However in many cases, drug toxicity is a highly complex process, which can manifest itself *via* multiple mechanisms. It has been recognised that predictions of toxicity can only be improved by investigating a broad panel of markers and their relationship to each other.

6.5.3 The Power of Multiparametric Screening in Toxicology

Technologies such as high-content screening (HCS) have transformed cell biology and enabled subtle changes in multiple cellular processes to be tracked within the same cell population. The technique uses either fluorescently labelled antibodies or dyes to stain specific areas of the cell, which have critical roles in cell health or the maintenance of cellular function. The impact of concentration-dependent and time-dependent drug exposure on these cellular processes can be investigated and related to specific toxicological or efficacious responses. The ability to analyse multiple end points simultaneously, yet selectively, is a major advantage and as well as being more sensitive, allows greater predictivity and an improved mechanistic understanding over traditional single end point measurements.[41]

The power of HCS in toxicity assessment was illustrated in two key papers authored by scientists at Pfizer,[41,42] where a panel of six to eight key toxicity markers were identified and used to predict human hepatotoxicity. The articles highlight the improved predictive power of HCS over existing conventional *in vitro* toxicity assays and over traditional preclinical animal tests. HCS technology is now routinely used for *in vitro* toxicity assessment in large pharmaceutical companies. The most recent example of such approach was published by scientists from Lundbeck.[43] They describe a validation of a novel HCS assay based on six parameters (nuclei counts, nuclear area, plasma membrane integrity, lysosomal activity, mitochondrial membrane potential (MMP), and mitochondrial area) using ~100 drugs of which the clinical hepatotoxicity profile is known. They found a 100-fold toxicity index between the lowest toxic concentration and the therapeutic C_{max} as optimal for classification of compounds as hepatotoxic or non-hepatotoxic. Most parameters by themselves appeared to have ~50% sensitivity and ~90% specificity. Drugs that had activities in two or more parameters at a concentration below 100-fold their C_{max} are typically hepatotoxic, whereas non-hepatotoxic drugs typically hit less than two parameters within that 100-fold toxicity index. This assay is employed in discovery projects to prioritise novel compound series during hit-to-lead, to steer away from a DILI risk during lead optimisation, for risk assessment towards candidate selection and to provide guidance on safe human exposure levels.

6.6 Bridging the Gap between *in Vitro* and *in Vivo*

It is critical to link the *in vitro* HCS information to animal or human pathology and establish a relationship to the *in vivo* data. The patterns observed in the key toxicity markers are often characteristic of specific mechanisms of known pathology.

Multiparametric systems utilise powerful techniques such as HCS imaging and combine them with databases of clinical information for drugs with known human and preclinical toxicological profiles. Using such systems, the toxicological profiles generated by such technologies can be compared with known drugs for which animal or clinical data are available. Specific changes in cellular function observed for particular mechanisms of toxicity can then be recognised and used to predict for unknown NCEs.

It is well recognised that toxicological events can be organ-specific, time-dependent and concentration-dependent. Multiple cell models that represent general cytotoxicity (such as HepG2, H9C2 cells) or organ-specific toxicity (using primary hepatocytes as a model for hepatotoxicity or stem cell-derived models) have been successfully utilised with imaging. Depending on the cell model, the cellular health parameters include the following: cell cycle arrest, nuclear size, oxidative stress, stress kinase activation, DNA damage response, DNA fragmentation, mitochondrial potential and mass, mitosis marker, cytoskeletal disruption, apoptosis, steatosis, phospholipidosis, ROS generation, hypertrophy and general cell loss. To assess early- and late-stage toxic responses, it is very useful to measure the exposure to drug candidates at multiple time points. Dose-dependent effects are typically investigated by exposing the cells to multiple physiologically relevant concentrations of the compound.

Data may be represented as AC_{50} (concentration at which average response is 50% of control activity), AC_{90}, AC_{20} or minimum toxic concentration for each cell health parameter, and the collection of AC values over the entire cell feature set comprises the response profile. A variety of visual and quantitative data mining tools including classifiers, correlation analysis and cluster analysis may be used to analyse the compound profiles (Figure 6.6). Using such approaches, there are a number of different ways in which the data can be interpreted:

1. Similarity profile plots can identify potential mechanisms of actions by correlating unknown test compound response with known control compounds where the mechanisms of action are already known.

Human in Vitro *ADMET and Prediction of Human Pharmacokinetics* 127

1. HCS Imaging System

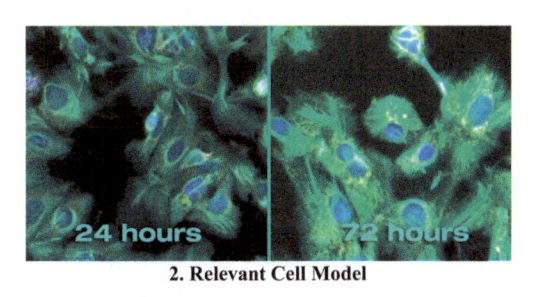

2. Relevant Cell Model

3. Cell Health Biomarker Panels

4. Bioinformatics Data Analysis
5. Ranking and Safety Alerts
6. Informed Decisions

Figure 6.6 Typical high-content imaging system components.

2. The relative toxicity of compounds in a series can be predicted by the classifiers and used to rank compounds for prioritisation of the most promising candidates.
3. The most potent or earliest cellular response can be extracted from the data, which may highlight the optimal end point(s) for designing higher throughput systems to investigate SAR within a series.

In summary, HCS technology is an example of a novel approach which identifies time-dependent, sub-lethal effects on cell health and function. HCS technology has a number of advantages over existing single end point assays and has the ability to predict mechanistic outcomes by correlating with known compound profiles and pathologies. Toxicological response is influenced by many factors including dose administration (including tissue exposure levels), time of exposure and/or accumulation in specific cells. Many of these factors are influenced by the pharmacokinetics of the drug administered and its effect on absorption, distribution, metabolism and excretion. Considering ADME data in conjunction with multiparametric measurements of *in vitro* toxicology is likely to be an important consideration in the future direction of predictive toxicology. Incorporation of human PK parameters prediction using PBPK modelling helps to ensure that *in vitro* cytotoxicity is relevant to projected tissue exposure.

By expanding such technologies to incorporate different mechanisms and cell models and incorporating knowledge of predicted human exposure, such systems will continue to improve the reliability of the classification and will become a part of routine drug discovery and development process in years to come.

6.7 Conclusions and Future Directions

Much of the progress in the field of ADMET profiling has occurred in the last 15 years. This progress has decreased the proportion of drug candidates failing in clinical trials for ADME reasons, providing optimism in an otherwise declining productivity in drug discovery. The principal barrier now is the toxicity portion of ADMET. The prediction of human-specific toxicology must be improved.

Cell-based assays using established cell lines and co-cultures have been used to determine toxicity to various organs, but many of these cell lines have lost some of the physiological activities present in normal cells. HepG2 cells, for instance, have greatly reduced levels of metabolic enzymes. Primary human hepatocytes can be used but are expensive, suffer from high donor-to-donor variability, and maintain their characteristics for only a short time. Three-dimensional models have been developed for cell-based therapies including micropatterned co-cultures of human liver cells that maintain the phenotypic functions of the human liver for several weeks.[44] This development should provide more accurate information about toxicity when used in ADMET screening and could be extended to other organ-specific cells leading to integrated tissue models in the 'human on a chip'.[45] The potential of stem cells to differentiate into cell lines of many different lineages may be exploited to develop human and animal stem cell-derived systems for major organ systems.[46]

HCS has been used for early cytotoxicity measurement since 2006 and provides great optimism.[41] This method has been optimised for hepatocytes and is more predictive of hepatotoxicity than other currently available methods and in the future could be applied to cells of other organs.

Molecular profiling is another alternative and is defined as any combination or individual application of mRNA expression, proteomic, toxicogenomic or metabolomic measurements that characterise the state of a tissue.[47] This approach has been applied in an attempt to develop profiles or signatures of certain toxicities. Molecular profiles, in conjunction with agents that specifically perturb cellular systems, have been used to identify patterns of changes in gene expression and other parameters at sub-toxic drug concentrations that might be predictive of hepatotoxicity, including idiosyncratic hepatotoxicity.[48] In the future, larger data sets, high-throughput gene disruptions, and more diverse profiling data will lead to more detailed knowledge of disease pathways and will facilitate in target selection and the construction of detailed models of cellular systems for use in ADMET screening to identify toxic compounds early in the discovery process. The combination of *in silico*, *in vitro* and *in vivo* methods and models into multiple content databases, data mining, human-on-a-chip systems, and predictive modelling algorithms, visualisation tools, and high-throughput data analysis solutions can be integrated to predict a system's ADMET properties. Such models are starting to be built and should be widely available within the next decade.[49] The use of these tools will lead to a greater understanding of the interactions of drugs with their targets and predict their toxicities.

To conclude, the future should provide a decrease in late-stage development failures and withdrawals of marketed drugs, faster timelines from discovery to market, and reduced development costs through the reduction of late-stage failures.

References

1. K. I. Kaitin, *Clin. Pharm. Ther.*, 2008, **83**, 210.
2. J. A. DiMasi, R. W. Hansen and H. G. Grabowski, *J. Health Econ.*, 2003, **22**, 151.
3. I. Kola and J. Landis, *Nat. Rev. Drug Discovery*, 2004, **3**, 711.
4. P. K. Honig, R. L. Woosley, K. Zamani, D. P. Conner and L. R. Cantilena Jr, *Clin. Pharm. Ther.*, 1992, **52**, 231.
5. P. K. Honig, D. C. Wortham, K. Zamani, D. P. Conner, J. C. Mullin and L. R. Cantilena, *J. Am. Med. Assoc.*, 1993, **269**, 1513.
6. US Food and Drug Administration. (1998). *Seldane and generic terfenadine withdrawn from market*. FDA Talk Paper T98–10. Available at http://www.fda.gov/bbs/topics/answers/ans00853.html.
7. P. Ball, L. Mandell, Y. Niki and G. Tillotson, *Drug Safety*, 1999, **21**, 407–421.
8. US Food and Drug Administration, *Public Health Advisory. Trovan (trovafloxacin/alatroflocacin mesylate)*, June 9, 1999, Available at http://www.fda.gov/cder/news/trovan.

9. J. Stephens, *'Panel Faults Pfizer in '96 Clinical Trial in Nigeria'*. The Washington Post: p. A01, May 7, 2006, Available at http://www.washingtonpost.com/wp-dyn/content/article/2006/05/06/AR2006050601338.html.
10. H. Kubinyi, *Nat. Rev. Drug Discovery,* 2003, **2**, 665.
11. D. Schuster, C. Laggner and T. Langer, *Curr. Pharm. Des.,* 2005, **11**, 3545.
12. C. A. Lipinski, F. Lombardo, B. W. Dominy and P. J. Feeney, *Adv. Drug Delivery Rev.,* 2001, **46**, 3.
13. C. A. Lipinski, *J. Pharmacol. Toxicol. Method,* 2000, **44**, 235.
14. A. H. Schinkel, J. J. Smit, O. van Tellingen, J. H. Beijnen, E. Wagenaar, L. van Deemter, C. A. Mol, M. A. van der Valk, E. C. Robanus-Maandag, H. P. te Riele, A. J. M. Berns and P. Borst, *Cell,* 1994, **77**, 491.
15. R. S. Schinkel and A. Minn, *Cell. Mol. Biol.,* 1999, **45**, 15.
16. O. Hurko and J. L. Ryan, *NeuroRx,* 2005, **2**, 671.
17. X. Liu and C. Chen, *Curr. Opin. Drug Discovery Dev.,* 2005, **8**, 505.
18. I. Megard, A. Garrigues, S. Orlowski, S. Jorajuria, P. Clayette, E. Ezan and A. Mabondzo, *Brain Res.,* 2002, **927**, 153.
19. D. J. Begley, D. Lechardeur, Z. D. Chen, C. Rollinson, M. Bardoul, F. Roux, D. Scherman and N. J. Abbott, *J. Neurochem.,* 1996, **67**, 988.
20. M. A. Deli, C. S. Abraham, Y. Kataoka and M. Niwa, *Cell. Mol. Neurobiol.,* 2005, **25**, 59.
21. C. Crone, *Acta Physiol. Scand.,* 1963, **58**, 292.
22. W. H. Oldendorf, *Brain Res.,* 1970, **24**, 372.
23. Y. Takasato, S. I. Rapoport and Q. R. Smith, *Am. J. Physiol.,* 1984, **247**, H484.
24. A. Kakee, T. Terasaki and Y. Sugiyama, *J. Pharmacol. Exp. Ther.,* 1996, **277**, 1550.
25. W. F. Elmquist and R. J. Sawchuk, *Pharm. Res.,* 1997, **14**, 267.
26. H. Olson, G. Betton, D. Robinson, K. Thomas, A. Monro, G. Kolaja, P. Lilly, J. Sanders, G. Sipes, W. Bracken, M. Dorato, K. Van Deun, B. P. B. Smith and A. Heller, *Regul. Toxicol. Pharmacol.,* 2000, **32**, 56.
27. US Food and Drug Administration Website. *Guidance for Industry Drug Interaction Studies – Study Design, Data Analysis, and Implications for Dosing and Labeling September, 2006* Available at: http://www.fda.gov/CDER/guidance/6695dft.
28. J. Uetrecht, *Chem. Res. Toxicol.,* 2008, **21**, 84.
29. V. Josserand, H. Pélerin, B. de Bruin, B. Jego, B. Kuhnast, F. Hinnen, F. Ducongé, R. Boisgard, F. Beuvon, F. Chassoux, C. Daumas-Duport, E. Ezan, F. Dollé, A. Mabondzo and B. Tavitian, *J. Pharmacol. Exp. Ther.,* Jan 2006, **316**, 79.
30. W. M. Pardridge, D. Triguero, J. Yang and P. A. Cancilla, *J. Pharmacol. Exp. Ther.,* 1990, **253**, 884.
31. M. Jacewicz, A.-C. Guyot, R. Annand, J. Gilbert, A. Mabondzo, K. Tsaioun. *Contribution of Transporters to Permeability across Cell Monolayers. Comparison of Three Models.* Apredica, CEA. American Association of Pharmaceutical Scientists, poster presentation, Atlanta, GA, November 2008.

32. J. A. Dykens and Y. Will, *Drug Discovery Today*, 2007, **12**, 777.
33. S. Rodríguez-Enríquez, O. Juárez, J. S. Rodríguez-Zavala and R. Moreno-Sánchez, *Eur. J. Biochem.*, 2001, **268**, 2512.
34. L. D. Marroquin, J. Hynes, J. A. Dykens, J. D. Jamieson and Y. Will, *Toxicol. Sci.*, 2007, **97**, 539.
35. D. P. Williams and B. K. Park, *Drug Discovery Today*, 2003, **8**, 1044.
36. Z. X. Liu and N. Kaplowitz, *Exp. Opin. Drug Metab. Toxicol.*, 2006, **2**, 493.
37. O. Ademuyiwa, R. Agarwal, R. Chandra and J. R. Behari, *Chem. Biol. Interact.*, 2009, **179**, 314.
38. M. Natalie, S. Margino, H. Erik, P. Annelieke, V. Geert and V. Philippe, *Toxicol In Vitro*, 2009, **23**, 217.
39. H. Olson, G. Betton, J. Stritar and D. Robinson, *Toxicol. Lett.*, 1998, **102–103**, 535.
40. P. J. O'Brien, W. Irwin, D. Diaz, E. Howard-Cofield, C. M. Krejsa, M. R. Slaughter, B. Gao, N. Kaludercic, A. Angeline, P. Bernardi, P. Brain and C. Hougham, *Arch. Toxicol.*, 2006, **80**, 580.
41. J. J. Xu, P. V. Henstock, M. C. Dunn, A. R. Smith, J. R. Chabot and D. de Graaf, Cellular Imaging Predictions of Clinical Drug-Induced Liver Injury, *Toxicol. Sci.*, 2008, **105**, 97.
42. M. Persson, A. F. Løye, T. Mow and J. J. Hornberg, *J. Pharmacol. Toxicol. Methods,* 2013 Nov–Dec, **6** 8(3), 302–313.
43. P. W. Hastwell, L.-L. Chai, K. J. Roberts, T. W. Webster, J. S. Harvey, R. W. Rees and R. M. Walmsley, *Mutat. Res.*, 2006, **607**, 160.
44. S. R. Khetani and S. N. Bhatia, *Nat. Biotechnol.*, 2008, **26**, 120.
45. K. Viravaidya and M. L. Shuler, *Biotechnol. Prog.*, 2004, **20**, 590.
46. L. Yang, M. H. Soonpaa, E. D. Adler, T. K. Roepke, S. J. Kattman, M. Kennedy, E. Henckaerts, K. Bonham, G. W. Abbott, R. M. Linden, L. J. Field and G. M. Keller, *Nature*, 2008, **453**, 524.
47. R. B. Stoughton and S. H. Friend, *Nat. Rev. Drug Discovery*, 2005, **4**, 345.
48. N. Kaplowitz, *Nat. Rev. Drug Discovery*, 2005, **4**, 489.
49. S. Ekins, Y. Nikolsky and T. Nikolskaya, *Trends Pharmacol. Sci.*, 2005, **26**, 202.

CHAPTER 7

'Body-on-a-Chip' Technology and Supporting Microfluidics

A. S. T. SMITH[a], C. J. LONG[a], C. MCALEER[a], X. GUO[a], M. ESCH[b], J. M. PROT[b], M. L. SHULER*[b], AND J. J. HICKMAN*[a]

[a]NanoScience Technology Center, University of Central Florida, Orlando, FL 32826, USA; [b]Biomedical Engineering, Cornell University, Ithaca, NY, USA
*E-mail: jhickman@mail.ucf.edu, mls50@cornell.edu

7.1 Introduction

Government approval of a novel therapeutic agent is the desired outcome of a complex research and development process often lasting 10–15 years and costing roughly £750 million (US$ 1.2 billion).[1,2] A significant percentage of this cost can be attributed to the high rate of attrition of compounds subjected to pharmaceutical testing and evaluation. For every 10 000 compounds identified with possible therapeutic potential, typically only one will be determined to be both safe for use in humans and to have medicinal value, and be approved to become a licensed medicine.[3]

A major contribution to this high rate of attrition is the relative inadequacy of current test beds for preclinical evaluation protocols. Due to the inherent complexity of living systems, drug trials on animal models often produce data that is difficult to interpret. More importantly, animal models have, on numerous occasions, been shown to be poor predictors of drug effects in humans.[4] Data collected between 1991 and 2000 highlights that only 11% of compounds found to be beneficial during preclinical evaluation in animals

---

RSC Drug Discovery Series No. 41
Human-based Systems for Translational Research
Edited by Robert Coleman
© The Royal Society of Chemistry 2015
Published by the Royal Society of Chemistry, www.rsc.org

demonstrated medicinal value in clinical trials and went on to become registered treatments.[5]

In vitro cell culture models represent simplified systems in which to study drug effects. The use of single cell types or defined co-cultures makes interpretation of the data easier and understanding the effects of drug treatment more straightforward. Moreover, because a relatively small number of cells are necessary for most cultures, human primary and stem cells can be obtained and differentiated for use in experimentation, removing the issue of animal–human incompatibility. However, these models often suffer from incomplete maturation of cells, giving abnormal phenotypes in culture, and the lack of interconnected tissues limits understanding of cell–cell and tissue–tissue interplay in response to drug treatment. Assessment of a cell's functional output is also problematic in such models because typical *in vitro* assays focus on evaluating cell survivability and/or analysis of phenotypic biomarkers, and do not therefore have the capacity to measure and record correct physiological function.

In an attempt to generate a middle ground between standard *in vitro* and *in vivo* experimentation, recent research efforts have focused on the production of complex, and yet well characterised, integrated multi-organ *in vitro* models for high-content drug screening. This chapter will discuss the development of these models and describe how they could be applied to future preclinical drug screening protocols in an attempt to streamline the process and provide better evaluation of how these compounds will behave in humans. Current technology relating to novel functional single-organ *in vitro* assays which accurately recapitulate the functional behaviour of native tissues will also be discussed. Such models represent exciting possibilities for future integration into multi-organ models, moving the field closer to the development of a complete 'body-on-a-chip' system capable of demonstrating accurate human responses to novel physical, chemical or pathological challenges in a controlled *in vitro* environment.

7.2 Multi-Organ *in Vitro* Models

Given the dynamic, complex and interactive nature of living systems, it is evident that single cell-type culture models are inadequate as predictors of *in vivo* responses to pharmacological treatment. To account for indirect modes of activity, *via* non-target organ metabolic processes, multi-organ systems are necessary to provide investigators with accurate predictions of whole body responses. Relatively simple studies, using physical barriers between cell populations, have verified that inter-organ connectivity can be achieved *in vitro*, *via* release of biomolecules into a shared medium.[6,7] The field is now moving toward the application of microfluidic fabrication techniques to generate similar multi-organ platforms possessing correct physiological scales, sheer stresses, liquid-cell ratios and fluid residence times in the various organ compartments.[3] Although *in vitro* organ models are not a perfect mimic of *in vivo* systems, the increased predictive power of integrated model systems, in relation to modelling real tissue behaviour, could become a valuable tool for studying time-dependent multi-organ interactions in a controlled yet dynamic and complex biological environment. To date, interaction between two organ systems has been demonstrated by several groups,[8–10] and interaction among tissues within three and four organ models has also been reported.[11–14] A model of dynamic interaction between more than four organs has not been demonstrated, however cutting edge designs are currently being developed which will theoretically be able to support as many as 10 organ systems within a single platform (Figure 7.1).

The concept for a 'body-on-a-chip' microphysiological system comes from the desire to emulate a physiologically-based pharmacokinetic (PBPK) pharmacodynamic (PD) mathematical model using a physical model. PBPK

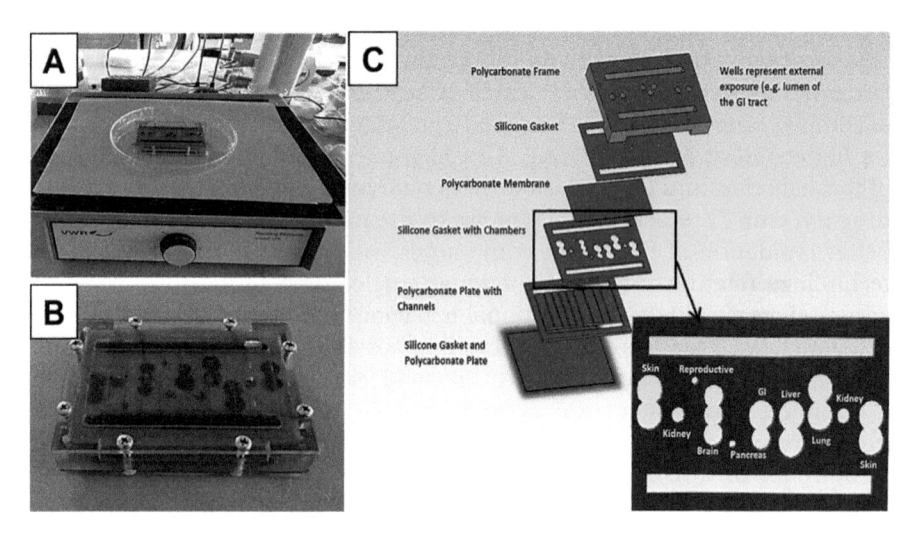

Figure 7.1 Overview of a 10 chamber *in vitro* model based on a simplified three chamber system using gravity driven flow.[17] A: Prototype unit on a rocker platform. B: Close-up inverted view of prototype unit after dye distribution study. C: Exploded schematic showing compartmentalisation design for 10 chamber system within this single unit. Barrier tissues (*e.g.* GI tract, skin, *etc.*) are made by plating cells and allowing them to grow into a layer on a pretreated polycarbonate membrane (0.4 μm pore size), which was sandwiched between a chamber gasket and another gasket or a polycarbonate channel plate creating chambers for the cells. The design provides an apical and a basolateral side for each chamber, allowing passage of chemicals across the barrier. Non-barrier tissues are 3D constructs made by re-suspending cells in hydrogels or plating them on a polymeric scaffold. The 3D tissue constructs are placed directly onto a polycarbonate membrane that is again sandwiched between a chamber gasket and another gasket or polycarbonate channel plate creating the chambers for the cells.

models are designed to predict the time-dependent distribution of a chemical and its metabolite in the body. A PD model predicts the response of a tissue or organ to a time-dependent concentration of a chemical in that organ, including its spatial distribution. A PBPK-PD model is aimed at predicting the response of the body to exposure to drugs or chemicals.[15] The primary limitation of such mathematical models is the current lack of detailed knowledge pertaining to all of the biological mechanisms associated with the adsorption, distribution, metabolism, elimination and biological responses of tissues to a drug or chemical, and its metabolites. Because living cells are used in physical 'body-on-a-chip' devices, it is not necessary to know, *a priori*, the mechanism of toxicity and the specific kinetic parameters involved within a system, because the cells present within the model are potentially capable of performing the full complement of metabolic processes they perform normally *in vivo*. The tissue constructs therefore contain the reactions and interactions not known to the modeller, but important in the body. Experiments with complete 'body-on-a-chip' devices should produce data that could be compared with those obtained from mathematical PBPK models, to provide a basis from which a mechanism of toxicity may be hypothesised and tested. By combining PBPK-PD models with an experimental device, the quality of predictions and depth of understanding could be increased.[16,17]

'Body-on-a-chip' devices are scaled based on the PBPK model according to organ mass ratios, with the fluid residence times per organ or tissue matching those *in vivo* for most organs in the body. As for a PBPK model, all tissue parts must be included either explicitly or implicitly in the physical model. The tissues that are not explicitly included within the device, even where there is no reaction or absorption, must be included because the residence times in these organ 'chambers' alter the dynamics of exposure to chemicals and metabolites. The tissues that are not explicitly included within such devices are often grouped together, and their volumes are included in a compartment labelled 'other tissue compartment'.[14,18–20] This compartment can be divided into 'rapidly perfused' and 'slowly perfused' tissue compartments to adapt the flow of a given medium within a multi-organ model to better mimic blood flow, dispersal and accumulation as it occurs *in vivo*. Cell types derived from barrier tissues, such as the gastrointestinal tract and skin, can be plated on support sheets possessing microscale pores to produce engineered biological membranes with apical and basolateral sides, and allow controlled passage of chemicals across the barrier tissue into the recirculating blood surrogate. In this manner, selective barriers can be developed to restrict the movement of certain metabolites and factors, while permitting passage of others in a biologically relevant manner.

While engineering on a biologically relevant scale is desirable as a means to improve the physiological relevance of a multi-organ model, the small volume of blood surrogate and low cell numbers make analysis of soluble metabolites, such as markers of cell health and functionality, problematic. Furthermore, the physical dimensions and complexity of such devices restrict operator access, making removal of medium for analysis difficult. Although removal of medium for analysis is possible with careful design, the small cell populations inherent to devices of this nature increases the required sensitivity of any analytical assays employed. For this reason, optical and electrical assessment of cell viability and function *in situ* is highly desirable in the development of 'body-on-a-chip' devices, because these techniques permit acquisition of real-time data in a non-invasive manner. Cell survival, as a measure of drug toxicity, can be assessed using simple live/dead assays or trypan blue exclusion methodologies.[13,14] The use of fluorescent reporter molecules, autofluorescent substrates and analysis of the spectral signatures of specific compounds of interest can also be used to provide insight into the biochemical pathways involved in producing any observed toxic effects.[18,19,21,22] Advances in genetic manipulation technology, particularly using stem cells,[23] also holds exciting possibilities for 'body-on-a-chip' devices because fluorescent reporters bound to promoters for specific biochemical markers can be used to study whether drug treatment causes alterations in phenotypic expression patterns.[24]

Integration of microelectrode arrays into multi-organ devices provides the means to drastically increase the analytical power of such systems. Measurement of the electrophysiological properties of cell types such as neurons, cardiomyocytes and skeletal muscle permits direct functional assessment of these cells in real time.[25-27] Electrical impedance measurements are attractive as a means to measure cell viability in a continuous and non-invasive manner,[28] and the use of microelectrodes could also be implemented for measurement of a wide variety of relevant metabolites such as lactate.[29] In addition, application of technologies designed to measure physical displacement, such as flexible posts[30] or microscale cantilever arrays,[31] could facilitate measurement of force generation in cardiac, skeletal and smooth muscle tissues, providing analytical assessment of physical maturation within multi-organ models.

Although some of these technologies have yet to be incorporated into 'body-on-a-chip' devices, the described analytical techniques are all amenable to integration with such platforms. Successful development of a multi-organ system, utilising human cell sources, and employing such a diverse array of analytical tools as described above should provide investigators with the means to generate highly accurate and informative predictions of whole body responses during preclinical drug development trials.

7.3 Functional Single-Organ *in Vitro* Models

The successful development of integrated multi-organ *in vitro* systems is predicated on the validation of functional, single-organ (or tissue) assays which can be adapted and incorporated into extant 'body-on-a-chip' platforms. Many organ analogues have already been developed with such potential and are discussed in more detail below. Due to space limitations, the systems discussed here do not represent a complete list of tissues for which *in vitro* models are currently being developed. Instead, the highlighted systems represent those organs with the greatest level of *in vitro* development, or those of greatest interest to the drug development industry.

7.3.1 Heart/Cardiac Tissue

Ex vivo experimental devices to study mammalian heart function began with the pioneering work of Oscar Langendorff in the late 19th century, in which perfused whole hearts were used to investigate physiology, pathology and contraction.[32] This method remains a valuable tool for studying the effects of physical and chemical challenges on heart contractility,[33] however, the tissue and supporting systems are far too large for incorporation into chip-based assays. Force measurements have been performed on single cardiomyocytes through measuring deflection of elastic substrates,[34] pillars[35] or attached beads[36] in response to cardiomyocyte contraction, or by directly measuring force *via* a piezoelectric force transducer attached to the cell.[37] While the Langendorff method is too large for a chip-based system, single-cell models are too small, lacking the complexity of interactions of multiple cells, and are inadequate for testing heart functional contraction.[38] Microtissues composed of sufficient numbers of cells to mimic the tissue, while remaining small enough for incorporation into microscale devices, are more appropriate for mimicking heart function. Microtissue systems for measuring cardiomyocyte force generation have been created using elastomeric PDMS in the form of cantilevers,[39–41] films attached at one end,[17,18] films mounted on posts,[42,43] and using hydrogel materials for cantilevers[44] and posts.[45]

The measurement of the electrical activity of cardiomyocyte cultures is commonly performed using microelectrode arrays, in which an array of microscale electrodes records voltage changes due to depolarisation and repolarisation of the cell membranes during activity at various positions on the culture surface. A review of this technology has been performed, in which specific applications of microelectrode arrays in biomedical research are discussed in detail.[46] Of particular interest for the study of cardiomyocytes is an extension of the microelectrode array technology in which cardiomyocytes are patterned on the surface to produce defined structures and conduction paths on the array of electrodes (Figure 7.2). This extension has been used to measure the effects of pharmaceuticals on the conduction velocity and action potential refractory period in cardiomyocyte culture.[25]

7.3.2 Lung

The air–liquid interface between the air in the lung alveoli and the capillaries surrounding the alveolar structures in the lung is composed primarily of a bilayer of epithelial and endothelial cells, along with the surfactant and mucus produced by specialised cells. This interface acts to allow the exchange of oxygen and carbon dioxide, and as a physical barrier between inhaled air and the rest of the body.[47] With the goal of producing devices for

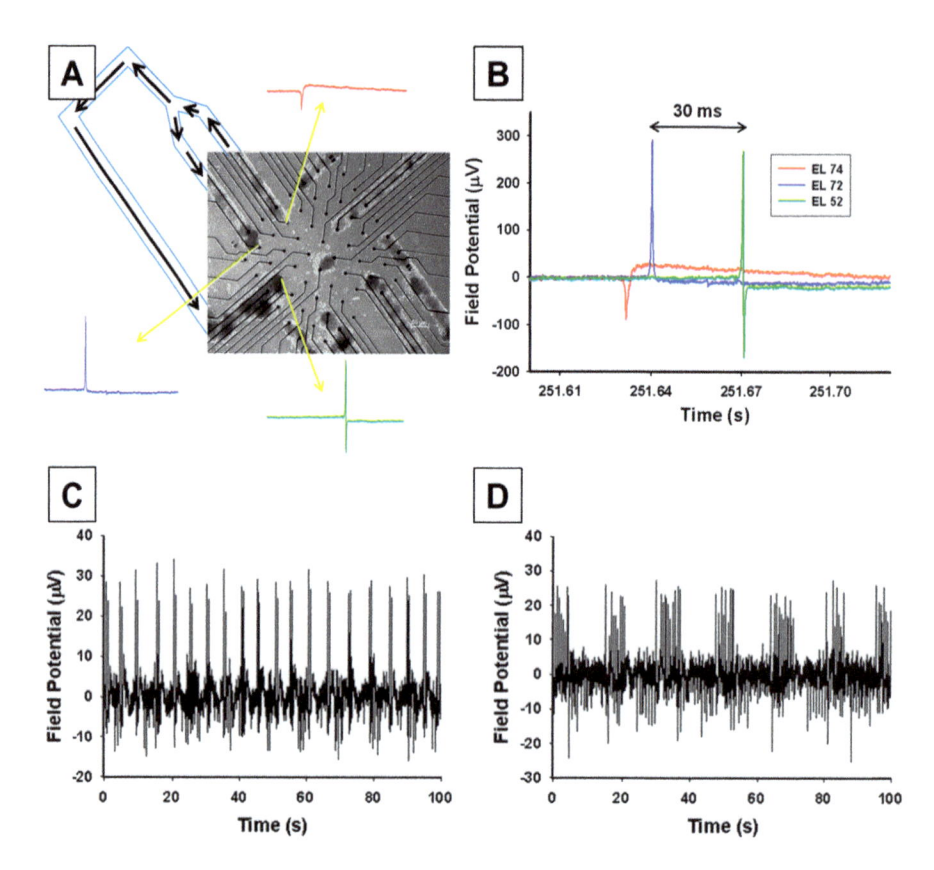

Figure 7.2 Microelectrode arrays (MEAs) can be used to measure cardiac electrical conductance *in vitro*. A: A phase contrast image of cardiomyocytes cultured on a chemically patterned MEA substrate (distance between electrodes = 200 μm). The remainder of the chemical pattern is illustrated artificially in this image to highlight the complete conduction pathway. Representative extracellular signals recorded on specific electrodes are also included. B: An overlay of the recorded traces from A, showing the temporal relationship of the signals. In such experiments, conduction velocity is determined based on the time delay in the signal and the distance between the recording electrodes. In this experiment, the calculated conduction velocity was 0.190 ms^{-1}. C: Such recording equipment can also be used to assess response of cultured cardiomyocytes to drug treatment. In the illustrated experiment, the effect of the HERG channel antagonist sparfloxacin was assessed. Field potential recordings before addition of sparfloxacin (C) indicated stable beating frequencies and synchronous activity. Field potential recording after the addition of 2 μM sparfloxacin (D) showed burst-like activity with higher intra-burst frequencies and a lack of synchronicity between the electrodes. Data reprinted with permission from Natarajan *et al.*[25]

biomedical testing, several systems have been developed that model various portions of lung physiology. In one system, in which an epithelial layer separated air and liquid compartments, the air–liquid interface was shown to be an important component in producing physiological conditions in terms of epithelial layer integrity and surfactant production.[48] In another study, a model of the airway architecture was produced to study airway occlusion and small airway clearance. In this device, the surfactant was shown to be a critical component when looking at damage to the lung epithelium.[49] A similar device added the action of mechanical stretching to combine both mechanical and fluid stresses on the epithelial cells to further provide physiological similarity.[50]

The bilayer of epithelial cells and endothelial cells acting as an air–liquid interface was recreated in a PDMS device that incorporated mechanical stretch to simulate the mechanical forces caused by breathing.[51] During evaluation of responses to bacteria, neutrophils, and pathogen phagocytosis, the mechanical forces from stretching were found to be an important component for recreating correct physiological behaviour. This system has since been advanced to model drug-induced pulmonary oedema, and has demonstrated that mechanical forces during breathing are crucial in the development of increasing vascular leakage during drug-induced toxicity.[52]

PDMS, which is commonly used in microfluidic devices, has extremely high gas permeability, preventing the control and measurement of gas concentrations in the air and liquid microenvironments in lung devices produced using this material. Because the primary function of the lung is to regulate dissolved gas concentrations in the blood by controlling the exchange of oxygen and carbon dioxide between the air and liquid, this limitation of PDMS devices is especially problematic. In the design of a lung device to allow the control and measurement of dissolved gas concentrations, a silicon-based microfluidic device was designed for the liquid flow side using iterative computational fluid dynamics and experimental flow visualisation. In this process, the microfluidic paths and internal structures were optimised to produce uniform flow patterns.[53]

7.3.3 Gastrointestinal (GI) Tract

In vitro models of the gastrointestinal (GI) tract epithelium have been used to estimate the bioavailability of orally administered drugs for many years.[54–56] The standard approach to estimating the bioavailability of drugs is to culture monolayers of the Caco-2 human cell line on porous membranes and measure the transport of the drug from the apical to the basolateral side of the cell culture. Caco-2 cells provide the main routes of intracellular and paracellular substance transport. To make the model more authentic, goblet cells and M-cells have been included in the cell culture system.[10,12,57] Exposure of Caco-2 cells to physiological, peristaltic motion has also been shown to improve cell differentiation and *in vitro* tissue function, leading to the development of absorptive, mucus-producing, enteroendocrine and Paneth cell types in culture.[58] Such cultures also show improved drug metabolising activity compared with standard cultures. In addition to transport studies, measurements of changes in transepithelial resistance (TER) across the Caco-2 cell layer provide an estimate of a drug's potential to damage the integrity of the GI tract epithelium.

To simulate the first-pass metabolism that occurs as a result of the portal circulation that delivers digested substances directly to the liver, *in vitro* models of the GI tract have been combined with models of the liver in co-culture systems that are either static or dynamic in nature[59] (Scheme 7.1). These systems have the ability to mimic modulating effects of the GI tract epithelium on liver tissue and *vice versa*. When a static GI tract/liver co-culture system was challenged with benzo[a]pyrene at the apical side of the GI tract tissue, the tissue modulated the response of the liver tissue to toxic metabolites that were generated in both tissues by transporting them to the apical side of the epithelial cell layer.[60] The same system showed that cytochrome P450 (CYP) 1A1/2 induction was enhanced in both Caco-2 cells and HEPG2 cells when these cell types were co-cultured within the same system.[60] These experiments were conducted under static conditions and the authors note that cell viability was enhanced when a perfusion system was added.[60] A co-culture system that was constructed using microfluidic fabrication technology demonstrated the power of the GI tract epithelium to modulate the response of the liver tissue to acetaminophen.[61] In this system, tubing connected the basolateral chamber of the GI tract module with a microfluidic chip that contained HEPG2/C3A cells. When acetaminophen was administered through a separate microfluidic circuit that was supplied to the apical side of the GI tract tissue, the toxicity to the liver tissue was ameliorated by the presence of Caco-2 cells. A slightly different perfused system was used by Brand *et al.*[62] to test the ability of a peroxovanadium compound to increase the glucose consumption of HEPG2 cells within a co-culture system in which the compound travelled across the GI tract epithelium first. To increase the authenticity of the tissue's reaction to challenges by drugs, tissue slices have been incorporated into microfluidic co-culture systems.[63] These systems demonstrate that individualised medicine with a patient's tissues is in principle possible. Within the co-culture system, the primary bile acid chenodeoxycholic acid (CDCA) induced the expression of fibroblast growth factor 15 (FGF15) in the intestinal tissue slice, which resulted in a stronger down-regulation of the enzyme, cytochrome P450 7A1 (CYP7A1), in the liver slice. This down-regulation was less prominent when the liver slice was exposed to CDCA in a single tissue.[63] Other groups have developed microfluidic models of the GI tract, but have not connected them with other tissue models.[64,65] Given the widespread use of human Caco-2 cells in modelling GI tract functionality *in vitro*, incorporation of a human-based model of this tissue into a multi-organ platform is an exciting possibility for the near future.

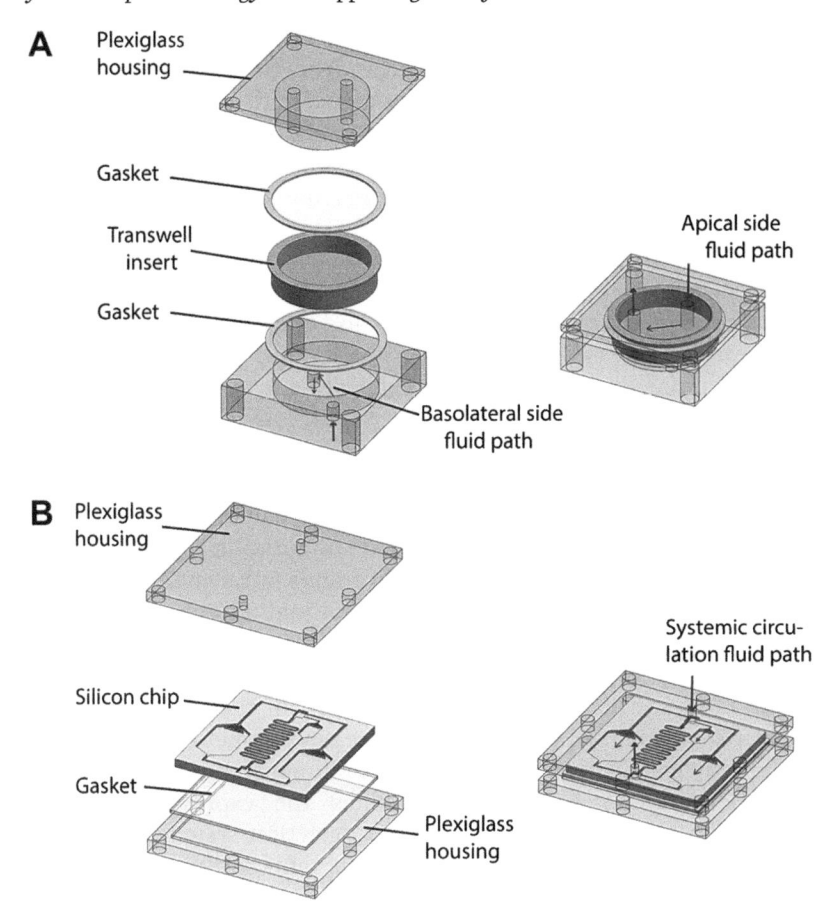

'Body-on-a-Chip' Technology and Supporting Microfluidics 141

Scheme 7.1 A schematic representation of a microfluidic chamber for the culture of gastrointestinal epithelial cells (Caco-2). A transwell insert is sandwiched between two Plexiglass housings which determine the final sizes of the apical and basolateral chambers. The illustrated chamber has been used in conjunction with the chip shown in B to simulate the oral uptake and the first-pass metabolism of acetaminophen.[58] B: A schematic representation of a body-on-a-chip device containing four organ compartments (liver, bone marrow, adipose and kidney). This device was designed to mimic physiological fluid residence times within each organ compartment. It has been microfabricated in silicon using standard photolithography and dry etching techniques. The Plexiglass chip housing provides microfluidic access to the chip.

7.3.4 Liver

The liver plays a major role in maintenance of system homeostasis through a variety of critical metabolic functions such as glucose storage, hormone synthesis and drug metabolism. Hepatocytes represent 60% of the total cell mass of the liver, the other 40% being primarily sinusoids (blood vessels), fat-storing stellate cells, Kuppfer cells and endothelial cells,[66] and are involved in the detoxification of drugs through a complex enzymatic system.[67] This phenomenon can be summarised in three main phases: phase 1 involving functionalisation by CYP450 enzymes, phase 2 where conjugation through sulphotransferase, glycosyltransferase and glutathione S-transferase enzymes occurs, and phase 3 which corresponds to the excretion of the compound through different transporters.[68] Given the determinant function in drug metabolism that the liver performs, and that hepatotoxicity is the primary cause of drug withdrawal, reproduction of a liver model is of great interest for the pharmaceutical industry,[69] and critical in the development of accurate multi-organ systems.

Primary human hepatocytes are considered the gold standard in drug metabolism studies. However, these cells rapidly lose their specific functions when cultivated in a standard 2D tissue culture environment.[70] Several strategies have been developed to improve the maintenance of hepatocyte functionality, such as co-culture with non-parenchymal cells and the utilisation of 3D matrices in a sandwich configuration.[68,71] More recently, microfabrication has become an attractive alternative method to produce platforms capable of cultivating liver cells and maintaining their functional maturation *in vitro*. Dynamic cultures with controlled shear rates, patterned cell surfaces, and reduced cell numbers and media volumes have been the main advantages achieved through application of these technologies. For example, recently published work details the development of a micro-device, utilising Matrigel, which confers a three-dimensional environment on cultured HEPG2/C3A cells, and facilitates the expression of CYP450 enzymes.[72] An *in situ* fluorescence optical detection system was combined with this micro-device, allowing for the successful real-time monitoring of CYP activity. Current work is now concentrated on the integration of primary hepatocytes with non-parenchymal cells, cultivated in a 3D scaffold,[73] to better mimic the physiological situation.

Dynamic culture and microenvironments have been shown to have no negative effect on cell survival when compared to static well plates in culture. Moreover, it has been reported that, when cultivated in a microfluidic bioreactor, hepatic cells increase expression of major drug metabolism enzymes at the genetic and protein levels. Such results demonstrate the relevance of this tool in improving cell metabolism and detection of toxic responses that have remained hidden in well plate studies, and therefore highlight its importance in the development of next-generation toxicity and pharmacological *in vitro* liver assays.[74,75]

It has been clearly demonstrated that these new *in vitro* tools are beneficial for hepatic culture, given their ability to address the loss of cellular function often observed in classical well plate studies. Development of medium- or high-throughput assays for the extrapolation of predictive *in vitro* results, and comparison to *in vivo* observations, remains a substantial challenge in the development of liver analogues toward their application in preclinical drug development studies.

7.3.5 Kidney

Modelling of renal function is vitally important to the development of a physiological multi-organ system for drug discovery, because the kidney's role in blood filtration and waste removal has significant impacts on drug, metabolite and waste concentrations *in vivo*. To that end, kidney transport barrier function has been modelled *in vitro* by culturing renal tubular epithelial cells on a porous membrane and measuring water and Na$^+$ uptake in response to hormonal stimulation.[76] In these experiments, cell culture under physiologically relevant flow rates was found to produce a fully differentiated and polarised kidney epithelium. Three-dimensional models of normal and diseased kidney tissue have also been produced, capable of the controlled uptake of 6-carboxy-fluorescein.[77] Uptake activity subsequently ceased following treatment with a specific uptake inhibitor, indicating the functional ability of these constructs to remove circulating compounds from the blood surrogate.

While these studies were performed using rodent cell types, recent work has utilised primary human proximal tubular epithelial cells in the development of a dynamic flow model of renal function.[78] This model has demonstrated that exposure of these cells to physiologically relevant shear stresses results in enhanced cellular polarisation and cilia formation as well as improved albumin transport, glucose reabsorption and brush border alkaline phosphatase activity. Furthermore, such cultures were found to produce more biologically representative responses to cisplatin-induced toxicity compared with cells maintained under standard culture conditions.

In addition to primary cultures, movement has been made toward the development of functional human renal cells from stem cell sources. Recent work has shown that treatment of human stem cells with a growth factor cocktail can induce cells to express the glomeruli markers WT-1 and rennin.[79] Furthermore, treatment of human stem cells with activin A, retinoic acid and BMP-4 or BMP-7 has been shown to induce their differentiation into renal precursor cells,[80] however, their functional ability was not determined. Recently, investigators have produced human stem cell-derived cells capable of cAMP uptake in response to hormone treatment, GGT activity and ammonia production; all functional responses characteristic of the proximal tubular cells of the kidney. Incorporation of such human cells into 'organ-on-a-chip' systems has yet to be established however, and is an important consideration for further development of physiologically relevant multi-organ systems for drug development.

7.3.6 Adipose Tissue

Adipose tissue compartments are a necessary consideration in the design of 'body-on-a-chip' models because fat can have a substantial effect on the distribution and accumulation of drugs and their metabolites within *in vivo* systems. In most platforms to date, fat tissues are included implicitly, rather than through incorporation of a specified tissue compartment containing adipose tissue. However, modelling the fat compartment explicitly may have advantages. In experiments with the environmental toxin naphthalene, it has been shown that the presence of differentiated 3T3-L1 adipocytes reduces the toxic response of co-cultured lung and liver cells to the toxin. Adipocytes presumably ameliorate damage to co-cultured lung cells through absorption of toxins and damaging compounds, such as hydrogen peroxide, and prevention of glutathione (GSH) depletion before re-synthesis is activated.[19] Development of a 3D adipose tissue model derived from human adipogenic stem cells has recently been reported, and such cultures represent exciting potential models for investigating obesity and diabetes *in vitro*.[81] The validation of such tissues *in vitro* makes the integration of human adipose tissue models into multi-organ platforms a viable possibility for the near future.

7.3.7 Central Nervous System (CNS)/Peripheral Nervous System (PNS)

Development of micro-fabricated platforms designed to emulate neurobiological tissues has facilitated novel experimental designs in which the precise spatio-temporal control of the cellular environment is possible.[82] *In vitro* cultures of many neural modalities have been developed and tested,[83] including nociceptive, neuromotor, autonomic, and those responsible for circadian rhythms and osmoregulatory pathways. Attempts to construct a nervous system on a chip have led to the development of several breakthrough technologies arising from multidisciplinary fields.

Originally, primary neurons or glia from animal tissues were the major cell source for developing *in vitro* models. However, there is an increasing demand for human-based *in vitro* systems for use in disease and drug screening studies.[84] Due to the limited availability of primary human neural tissue, and the rapid growth of stem cell technology, *in vitro* nervous systems utilising human stem cells as the tissue source are becoming increasingly common. Embryonic stem cells,[85,86] adult stem cells[87,88] and iPSCs (induced pluripotent stem cells)[89] have all been used to derive various nervous system cell types for use in the development of novel *in vitro* systems.[90] The generation of motoneurons[87,91] (Figure 7.3) and mid-brain dopamine neurons[92–94] is well established, while the differentiation of cortical neurons,[95,96] sensory neurons,[97,98] oligodendrocytes[88,99] and astrocytes[100,101] have also been reported. Even the differentiation of blood–brain barrier (BBB) components such as pericytes, brain endothelial cells and smooth muscle cells has started to emerge.[102–105] The derivation of patient-specific iPSCs opens up a particularly exciting avenue of research focused on the generation of patient-specific disease models, essential for aetiological study, drug screening and therapy design.[106–108]

Addition of cells to *in vitro* systems in a spatially controlled manner has become feasible due to the introduction of several technologies. By combining SAMs (self-assembled monolayers) with advanced photolithography methods, the surface area designated for cell adhesion and proliferation can be designed and engineered with high resolution.[109,110] One type of

SAM surface, trimethoxysilyl propyldiethylenetriamine (DETA-silane), has been shown to support neuronal, endothelial and cardiac cell growth,[111-114] and has been employed in generating high-resolution *in vitro* patterned circuits of embryonic hippocampal neurons.[115] Moreover, DETA-silane substrates have been shown to promote guided axonal/dendritic process extension (polarised) at the level of a single neuron.[110]

Microfluidic devices have also been developed to achieve the directed placement of neurons and glia, combining microfabrication and surface micropatterning technologies.[116,117] Compartmentalised chambers not only separate the growth of axons and dendrites by controlled placement of microstructures such as micro-grooves, but also offer the flexibility to cultivate multiple cell types, with each component capable of independent manipulation. The incorporation of multi-electrode array (MEA) technology to such chips allows researchers to analyse and monitor these controlled neural networks with high resolution and in a high-throughput manner.[118]

The electrical activity of neurons and their synaptic connections form the functional basis for all neural networks. Glass pipette patch-clamp analysis is a state-of-the-art technique that allows investigators to monitor the intrinsic electrophysiological and synaptic properties of neurons.[119,120] However, this technique is labour intensive, has low data throughput, and requires a high degree of operator training. MEAs, in contrast, have the ability to monitor network activity by measuring local field potentials from multiple extracellular sites,[121-123] but specific ion channel activities are difficult to delineate using these multiplexed signals.[124] Planar patch-clamp array chips (a combination of patch clamp and MEA technology) enable the simultaneous high-resolution electrophysiological interrogation of individual neurons at multiples sites in synaptically connected neuronal networks.[125] The high-resolution and high-throughput features provide the potential for utilising this technology in screening drugs with a broad range of targets and assessing responses at both the single cell and neural network levels.[123,126]

In order to accurately model the nervous system, multiple models have been established, aimed at addressing varied neuronal functions. The different nervous subsystems and the current state of development of their *in vitro* analogues are discussed in the following paragraphs.

Brain networks: Both cortical networks and hippocampal networks have been intensively investigated using MEA technologies.[122,127-131] Neural pathways between cortical and hippocampal neurons have been investigated in microfluidic platforms utilising organotypic brain slices from rodents.[132] Microfluidic systems to study neurotransmitter release and subsequent neurotransmission have also been reported.[133] Such 2D systems have obvious advantages for 'body-on-a-chip' technologies due to their simple configurations and ease of access to the cultured cells; however, multiple 3D neural cell culture systems are also being developed.[134]

Neuromuscular junctions: Development of *in vitro* models of the neuromuscular junction (NMJ) have been the focus of extensive research efforts, due to the breakdown of this structure being involved in the development of

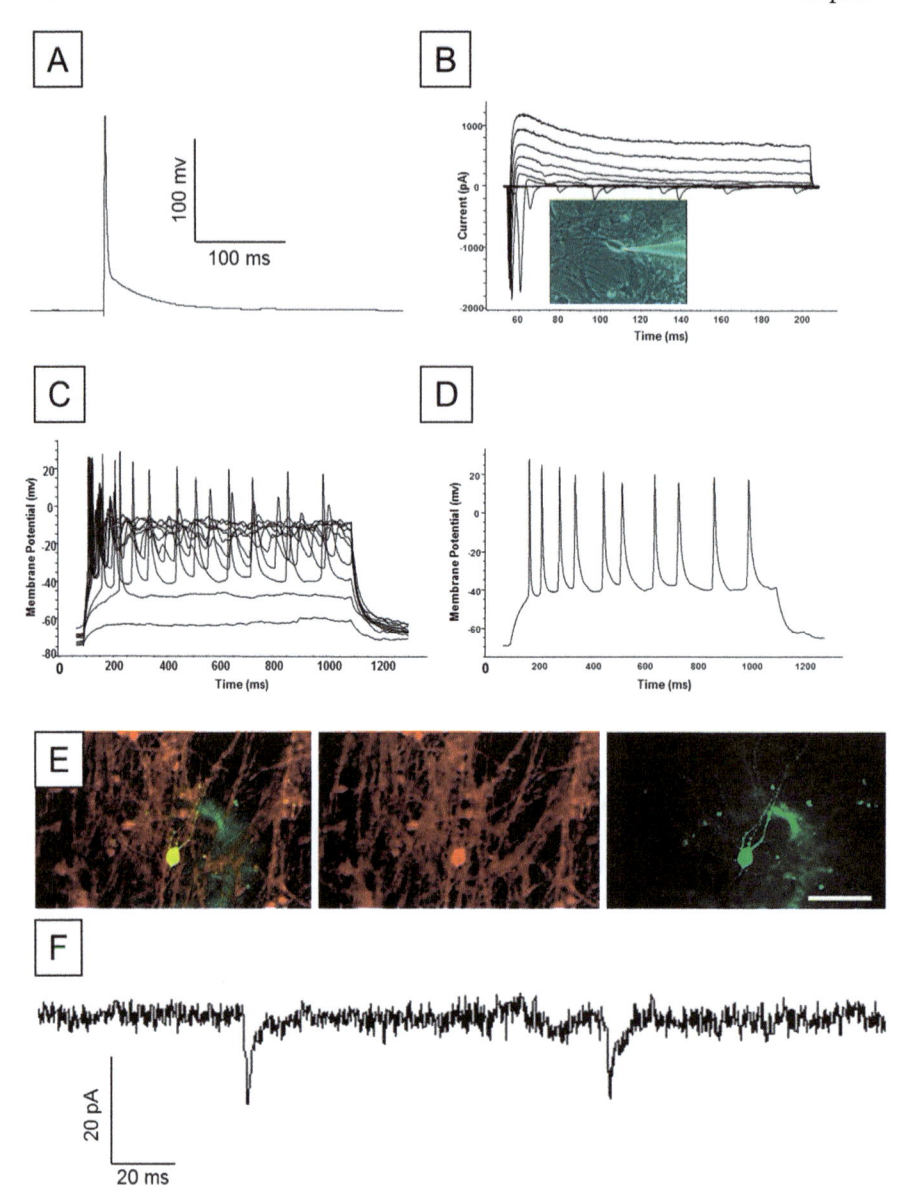

Figure 7.3 Characterisation of human motoneurons developed for use in multi-organ *in vitro* models for drug development protocols. A. An example trace of an action potential (AP) elicited when the cell received a saturated stimulus (2 ms 500 pA inward current). B. An example trace of active Na$^+$ and K$^+$ currents from a voltage clamp recording of a differentiated neuron. The insert picture indicates the recorded neuron. C. An example of a repetitive firing action potential from a current clamp recording of a differentiated motoneuron. D. An example trace of repetitive firing action potentials from C. E. Choline

multiple disease states (such as amyotrophic lateral sclerosis) and its importance in understanding the effects of spinal cord injury. Measurement of functional neuromuscular transmission *in vitro* was first reported in 1970 using chick muscle cells and motoneurons.[135] Since then, the ability to promote NMJ formation *in vitro* has been demonstrated using both rodent[136–140] and human[9,141] cell sources. Motoneurons maintained in co-culture with skeletal muscle myotubes display normal responses to neurotransmitter application,[140] as well as characteristic ionic currents and action potential firing.[9,140,141] Immunocytochemical staining has provided evidence for the co-localisation of pre- and post-synaptic structures *in vitro*,[9,139–141] and dual patch-clamp recordings have demonstrated the depolarisation of cultured myotube membranes through stimulation of associated motoneurons, providing evidence of functional neuromuscular transmission between these cell types.[139]

Incorporation of innervated skeletal muscle cultures into a multi-organ 'body-on-a-chip' platform prevents the application of complex and invasive patch-clamp protocols for assessing NMJ functionality. Simplified systems are required which will allow functional data to be recorded continuously, and in real time, without necessitating the removal of cells from the multi-organ platform for interrogation. The use of micro-fabricated silicon cantilevers (Figure 7.4) enables the assessment of myotube contraction profiles through measurement of cantilever deflection *via* a laser and photo-detector system.[31] Incorporation of motoneurons into this culture model allows for the investigation of functional neuromuscular transmission through addition of motoneuron synaptic stimulants, such as glutamate, and the recording of subsequent myotube contractions.[142] Such a culture system represents a far simpler, less invasive technique for interrogating *in vitro* synapses and has important implications for the *in vitro* study of peripheral neuropathies as well as testing of novel therapeutic agents. The relative simplicity of this model makes it amenable to integration with more complex culture systems. It represents an exciting possible model for incorporation into 'body-on-a-chip' platforms in the near future.

Myelination: Models capable of inducing axonal myelination *in vitro* are another area of intense interest because the phenomenon is central to both neural regeneration and certain disease states, including multiple sclerosis (MS) and peripheral neuropathy. Myelination models of the PNS have been shown to promote the *in vitro* myelination of motor neurons[143] and sensory

---

acetyltransferase immunostaining (red) to confirm the identity of a recorded cell as a motoneuron. The cell was labelled with Alexa 488 dye (green) from the intracellular solution during patch clamp analysis. Scale bar = 50 µm. F. Spontaneous excitatory post-synaptic currents from an induced motoneuron, indicating functional synapse formation between the patched cell and another neuron(s) in the culture. Data reprinted with permission from Guo *et al*.[87]

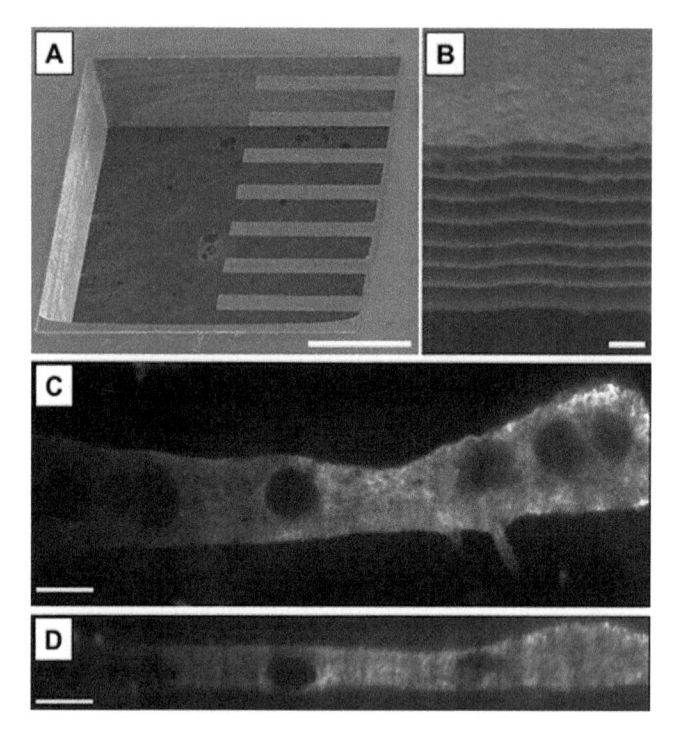

Figure 7.4 Microscale silicon cantilevers for measurement of skeletal muscle contractile force. Deflection of a cantilever in response to the contraction of a myotube cultured on its surface can be measured using a laser and photodetector and converted into a direct measurement of force within the cell. A. SEM micrograph of a silicon cantilever array at low magnification and taken at 50° from normal. Scale bar = 500 μm. B. High magnification image of one cantilever from A used to measure cantilever thickness. The image was taken at 50° from normal, scale bar = 1 μm. C. Confocal image of a myotube cultured on a silicon cantilever (cantilever not visible) and stained for myosin heavy chain. Scale bar = 20 μm. D. A reconstructed three-dimensional image of the cell in C, illustrating the depth of the cultured myotube. Scale bar = 20 μm. Data reprinted with permission from Wilson et al.[31]

neurons[144] when co-cultured with Schwann cells. In the CNS, *in vitro* systems capable of myelinating both hippocampal neurons and spinal motor neurons using oligodendrocytes have likewise been reported.[145–147]

Sensory systems: Sensory neuron activity is vital for accurate analysis of nervous system function,[148–150] because it is a crucial part of the reflex arc, and the damage or otherwise malfunction of these cells is associated with many diseases including peripheral neuropathy. The isolation and culture of rat dorsal root ganglion (DRG) neurons is a well-established technique,[151] and such cells are able to form functional connections with intrafusal muscle fibres in culture when maintained on bioengineered surfaces and bioMEMs devices.[152] Establishment of an *in vitro* reflex arc would be an invaluable tool for both basic and applied biomedical research, and has so far generated a series of *in vitro* biological systems using either primary rat cells or human stem cell-derived components.[9,152–154]

Blood–brain barrier: The blood–brain barrier (BBB) is a biological construct that selectively restricts the passage of blood-derived substances to the CNS, and plays a critical role in maintaining CNS homeostasis and neuronal function. The BBB consists primarily of three cell types: microvascular endothelial cells on the luminal side of the vascular wall, and pericytes and astrocytes (glial cells) lining the extra-luminal surface. An *in vitro* model of the BBB is crucial in understanding its physiology and addressing complex issues relating to drug delivery, as well as pathogenesis of neurological diseases.[155–157] The most common and widely utilised BBB model was developed based on a simplified view of the barrier, and utilized a monolayer of specialised brain microvascular endothelial cells in a *trans*-well apparatus.[158] The addition of extra-luminal astrocytes (in juxtaposition to the endothelial monolayer) have been shown to facilitate the formation of more stringent inter-endothelial tight junctions that more closely resembles that of the BBB *in situ*.[159] However, commercially available membranes used in *trans*-well studies are more than 10 µm thick, with uneven pore distribution, resulting in a high hydraulic resistance between the two compartments and preventing close interaction between the cell populations. Ultra-thin nanofabricated membranes can address these shortcomings because they constitute a highly porous surface with controlled pore dimensions and can be less than 3 µm thick.[160–162] Different membranes have demonstrated endothelial and astrocyte biocompatibility, and can potentially be used in other devices as a substitute for commercial systems.

To allow for endothelial exposure to physiological shear stresses, a purpose-built cone and plate viscometer has been used to enable the endothelial exposure to flow *in vitro*.[163] Recently, some artificial capillary-like structures, such as hollow fibres (generally made of thermoplastic polymers such as polypropylene, polysulfone, *etc.*), have been used to develop 3D BBB models. These dynamic *in vitro* 3D models (DIV-BBB models) allow brain microvascular endothelial cells to be co-cultured with abluminal astrocytes while being subjected to quasi-physiological pulsatile laminar shear stress.[164] Such a complex system represents a BBB model that closely resembles the *in vivo* tissue both functionally and anatomically. A human DIV-BBB model has also been developed based on co-cultures of human microvascular endothelial cells from 'normal' and drug-resistant epileptic brain tissue, with human brain astrocytes from epilepsy patients or controls.[165] This humanised DIV-BBB model was then upgraded to allow the *trans*-endothelial trafficking of immune cells, mimicking neuro-inflammation.[166] To better mimic BBB functionality *in vivo*, further issues need to be addressed, such as the use of a more biologically relevant hollow fibre material, the incorporation of an active transport system, and the utilisation of patient-derived cells for disease-specific models.

7.3.8 Skeletal Muscle

In vitro skeletal muscle culture from mammalian sources was first established in the early 1970s.[167] Since that time, it has evolved into a multifaceted methodology for high-throughput drug discovery and regenerative medicine. The field has expanded vastly to include the utilisation of multiple cell sources, development of specific media formulations to achieve different levels of maturation, and designs for direct measurement of *in vitro* muscle functionality. Primary and stem cell sources have expanded to include, but are not limited to, human,[9,168] rabbit,[169,170] avian[171,172] and, most predominantly, rodent tissues.[138,173–176] Growth and differentiation media containing animal sera have given way to serum-free media formulations that are fully defined, and therefore suitable for human-based drug discovery. Silicon- and PDMS-based devices used to measure *in vitro* physiological parameters of muscle are the most widely used, and incorporate either polymer cantilevers,[40] silicon cantilevers[177–179] (Figure 7.4) or flexible post-based designs.[180,181] In all cases, these models rely on the ability to measure substrate displacement or deformation in response to muscle fibre contraction and converting this data mathematically into a measurement of force output. Each system possesses inherent advantages and limitations with respect to their capacity to successfully integrate into multi-organ 'body-on-a-chip' platforms. Post-based and polymer cantilever designs are easily fabricated but rely on optical imaging for analysis, which will probably be difficult to accomplish in the three-dimensional configuration of a multi-organ system. Silicon-based cantilever strategies are relatively expensive to fabricate. However, they allow for the incorporation of recording electrodes and piezoelectric functionality which make them ideal candidates for 'body-on-a-chip' systems.

7.3.9 Blood Surrogate

The task of developing a suitable media system to adequately support multiple organ models within a single platform becomes increasingly difficult as more cell types are added. Successfully identifying a cocktail of growth factors and small molecules capable of maintaining the viability and, most importantly, the functionality of multiple cell types *in vitro* is a costly and time-consuming process. Critically, the varying rates of maturation of different cell types and their changing nutritional and developmental requirements at multiple maturation stages often necessitates the implementation of a dynamic medium replacement regime where specific factors are substituted in and out as the cells mature.[9] In the past, many investigators have relied on the use of animal sera to provide the necessary trophic support to single and multiple cell types. However, the use of such solutions is incompatible with proposed high-throughput *in vitro* drug assessment assays because they represent an undefined and variable component within the culture environment capable of significantly altering culture conditions between experimental repeats and thereby confounding the collected results.[182]

Certain co-culture models have been developed utilizing serum-free, defined media compositions. For example, successful establishment of muscle and motoneurons, and the promotion of functional synaptic connections, has been reported *in vitro* using a defined media system.[9,141] Furthermore, this media system has been adapted to promote the functional maturation of a variety of excitatory cell types including cardiomyocytes,[25] sensory neurons[152,183] and hippocampal neurons.[184] As mentioned previously, dynamic co-cultures of four cell types have been established and successful maintenance of four human cell types (liver, lung, kidney and adipose) in a serum-free system has been reported.[11] Establishment of a defined medium or media combination capable of successfully maintaining more than four cell types *in vitro* has yet to be reported but it remains a critical rate-limiting step in the advancement of multi-organ platform development.

7.4 Microfluidics

Microfluidic technologies have supported the development of physiologically-based 'body-on-a-chip' devices in several ways. First, they provide the opportunity to fine-tune liquid-to-cell ratios. Because surface-to-volume ratios are high within the devices, the liquid-to-cell ratios approach physiological values. Second, the low Reynolds numbers typically achieved in a microfluidic system enable precise control of the shear stresses. Third, the physical and chemical microenvironments can be crafted to establish defined conditions and structures. Furthermore, microfabrication technologies allow the design of devices with excellent control over tissue chamber dimensions. Scaling the tissue chambers and distributing the blood surrogate in a physiologically relevant manner improves the predictive power of efficacy and toxicity measurements conducted with such devices.

The fluid dynamics and transport phenomena inside microfluidic systems can also be analysed theoretically. In simple cases an analytical solution may be obtained but it is more typical to use a computational method. The flow inside a microfluidic channel can be analysed by solving the Navier–Stokes equation, assuming incompressible fluid:

$$\rho \frac{\partial \vec{u}}{\partial t} = -\rho \vec{u} \nabla \vec{u} - \nabla p + \eta \nabla^2 \vec{u} \tag{7.1}$$

(rate of change of momentum) = (convective force)
+ (pressure force) + (viscous force)

where u is the velocity field, ρ is the density, η is the viscosity, and p is the pressure. In the simple case of a long cylindrical tube with radius R, the steady-state flow rate can be described by following equation:

$$Q = \frac{\pi R^4}{8\eta}\left(-\frac{\mathrm{d}p}{\mathrm{d}x}\right) = \frac{\pi R^4}{8\eta}\frac{\Delta p}{L} \tag{7.2}$$

where Δp is the pressure drop across the channel with the length of L, and η is viscosity. Eqn (7.2) relates the volumetric flow rate with the pressure drop and the geometry of the channel (the radius and the length). An analogy to an electrical circuit can be made by relating the flow rate (Q) to the current (I), and the pressure drop (Δp) to the voltage drop (ΔV); then an equation similar to Ohm's law can be written:

$$\Delta V = I \cdot R (\text{electrical wire}), \Delta p = Q \cdot R (\text{fluidic channel}) \quad (7.3)$$

In fluid dynamics, the resistance R is dictated mainly by the geometry of the channel. In the case of a rectangular channel with a high aspect ratio ($w \gg h$), the resistance can be described by the following equation:

$$R = \frac{12 \mu L}{w h^3} \quad (7.4)$$

Using the analogy of an electrical circuit to describe a microfluidic system allows a network of microfluidic channels to be analysed in a similar manner.[185,186] For example, the fluidic resistance of serially connected microfluidic channels is the same as the sum of fluidic resistance of all channels. In a similar way, the reciprocal of the fluidic resistance of parallel-connected microfluidic channels is the same as the sum of the reciprocal of the resistance of all channels.

Microfluidic systems allow researchers to precisely control the parameters that define the microenvironment of *in vivo* tissues, enabling parametric study of the relationship between environment and the cellular behaviour. Young *et al.*[187] used a microfluidic device to study the adhesion properties of endothelial cells in the presence of various extracellular matrix (ECM) proteins and fluidic shear stresses. The fluidic shear stress in a parallel plate is determined by the geometry of the channel and the fluid velocity in the following manner:

$$\tau = \frac{6 \mu Q}{w h^2} \quad (7.5)$$

where μ is the viscosity, Q is the flow rate, w is the width and h is the height of the channel. A unique advantage of using microfluidics was demonstrated in a study that used a microscale channel with a tapered profile, generating a wide range of shear stresses.[188] Using microfluidic channels, it has also been determined that endothelial cells require a minimum initial shear stress *in vitro* to enable their growth and development to confluent layers that express VE-cadherin, a standard marker for adherence junctions.[189]

While polydimethylsiloxane (PDMS) is the most popular material for microfluidic systems and carries many advantages, the material properties of the PDMS surface are not ideal.[190] There are concerns that the adsorption of compounds from the medium may interfere with an accurate experimental outcome. ECM proteins or hydrogels resemble the natural *in vivo* tissue environment better than PDMS,[191] and silicon-based devices have been used in combination with these materials to construct a variety of 'body-on-a-chip' devices, including a device that consists of a microfluidic chamber filled with a hydrogel to provide a three-dimensional scaffold for hepatocyte culture.[3,13] Hydrogels are more porous than PDMS, allowing molecular diffusion inside the scaffold matrix similar to diffusion in tissues.[192] Many efforts have been directed toward fabricating hydrogels into a microfluidic scaffold.[193,194] Other hydrogels, such as polyethylene glycol diacrylate (PEG-DA)[195] and fibrin,[196] have also been used to create microfluidic devices. While these approaches have focused on creating rather simple representations of the blood vasculature, an alternative approach has focused on recreating a more complex vasculature network. In this case, a sacrificial moulding technique was used to create an interconnected network inside a scaffold made of collagen[196] or PDMS.[197]

7.5 Concluding Remarks

The development of complex, integrated multi-organ systems, interconnected by a physiologically relevant microfluidic network in a defined common medium, has great potential for streamlining the drug development process. Analysis of interlinked functional human assays in response to novel therapeutics provides the means to replace animals in this process. In doing so, it provides investigators with more biologically relevant, high information content models capable of being maintained in a defined and fully-controlled environment for both acute and, more importantly, chronic applications. Further work is required to develop the systems covered in this chapter, as well as models for organs and tissues not addressed, toward a point where these technologies can be effectively integrated to produce a viable single product. However, single-organ models are already being used to test and predict drug responses in numerous human tissues, and further development of these models toward more complex interconnected systems has been demonstrated in prototype devices and will probably be realised in the near future.

Acknowledgements

We would like to acknowledge funding by the National Center for Advancing Translational Sciences (NCATS) of the National Institutes of Health (NIH) Microphysiological Systems UH2TR000516.

References

1. J. A. Dimasi, *Clin. Pharmacol. Ther.*, 2001, **69**, 286.
2. J. A. DiMasi and H. Grabowski, *Manag. Decision Econ.*, 2007, **28**, 469.
3. M. B. Esch, T. L. King and M. L. Shuler, *Ann. Rev. Biomed. Eng.*, 2011, **13**, 55.
4. R. Greek and A. Menache, *Int. J. Med. Sci.*, 2013, **10**, 206.
5. I. Kola and J. Landis, *Nat. Rev. Drug Discovery*, 2004, **3**, 711.
6. A. P. Li, C. Bode and Y. Sakai, *Chem. Biol. Interact.*, 2004, **150**, 129.

7. I. Wagner, E. M. Materne, S. Brincker, U. Sussbier, C. Fradrich, M. Busek, F. Sonntag, D. A. Sakharov, E. V. Trushkin, A. G. Tonevitsky, R. Lauster and U. Marx, *Lab Chip,* 2013, **13**, 3538.
8. L. Shintu, R. Baudoin, V. Navratil, J. M. Prot, C. Pontoizeau, M. Defernez, B. J. Blaise, C. Domange, A. R. Pery, P. Toulhoat, C. Legallais, C. Brochot, E. Leclerc and M. E. Dumas, *Anal. Chem.,* 2012, **84**, 1840.
9. X. Guo, M. Gonzalez, M. Stancescu, H. H. Vandenburgh and J. J. Hickman, *Biomaterials,* 2011, **32**, 9602.
10. G. J. Mahler, M. L. Shuler and R. P. Glahn, *J. Nutr. Biochem.,* 2009, **20**, 494.
11. C. Zhang, Z. Zhao, N. A. Abdul Rahim, D. van Noort and H. Yu, *Lab Chip,* 2009, **9**, 3185.
12. F. Antunes, F. Andrade, F. Araujo, D. Ferreira and B. Sarmento, *Eur. J. Pharm. Biopharm.,* 2013, **83**, 427.
13. J. H. Sung and M. L. Shuler, *Lab Chip,* 2009, **9**, 1385.
14. A. Sin, K. C. Chin, M. F. Jamil, Y. Kostov, G. Rao and M. L. Shuler, *Biotechnol. Prog.,* 2004, **20**, 338.
15. M. Danhof, E. C. de Lange, O. E. Della Pasqua, B. A. Ploeger and R. A. Voskuyl, *Trends Pharmacol. Sci.,* 2008, **29**, 186.
16. J. H. Sung, A. Dhiman and M. L. Shuler, *J. Pharm. Sci.,* 2009, **98**, 1885.
17. J. H. Sung, C. Kam and M. L. Shuler, *Lab Chip,* 2010, **10**, 446.
18. K. Viravaidya, A. Sin and M. L. Shuler, *Biotechnol. Prog.,* 2004, **20**, 316.
19. K. Viravaidya and M. L. Shuler, *Biotechnol. Prog.,* 2004, **20**, 590.
20. N. Bhargava, M. Das, M. Edwards, M. Stancescu, J. F. Kang and J. J. Hickman, *In Vitro Cell Dev. Biol. Anim.,* 2010, **46**, 685.
21. B. Ma, G. Zhang, J. Qin and B. Lin, *Lab Chip,* 2009, **9**, 232.
22. D. A. Tatosian and M. L. Shuler, *Biotechnol. Bioeng.,* 2009, **103**, 187.
23. L. Gerrard, D. Zhao, A. J. Clark and W. Cui, *Stem Cells,* 2005, **23**, 124.
24. H. Xu, W. L. Kraus and M. L. Shuler, *Biotechnol. Bioeng.,* 2008, **101**, 12767.
25. A. Natarajan, M. Stancescu, V. Dhir, C. Armstrong, F. Sommerhage, J. J. Hickman and P. Molnar, *Biomaterials,* 2011, **32**, 4267.
26. K. Varghese, P. Molnar, M. Das, N. Bhargava, S. Lambert, M. S. Kindy and J. J. Hickman, *PLoS One,* 2010, **5**, e8643.
27. C. G. Langhammer, M. K. Kutzing, V. Luo, J. D. Zahn and B. L. Firestein, *Biotechnol. Prog.,* 2011, **27**, 891.
28. J. Z. Xing, L. Zhu, S. Gabos and L. Xie, *Toxicol. In Vitro,* 2006, **20**, 995.
29. I. A. Ges and F. Baudenbacher, *Biosens. Bioelectron.,* 2010, **26**, 828.
30. H. Vandenburgh, *Tissue Eng. Part B Rev.,* 2010, **16**, 55.
31. K. Wilson, M. Das, K. J. Wahl, R. J. Colton and J. Hickman, *PLoS One,* 2010, **5**, e11042.
32. O. Langendorff, *Pflüg. Archiv. Eur. J. Physiol.,* 1895, **61**, 291.
33. R. M. Bell, M. M. Mocanu and D. M. Yellon, *J. Mol. Cell. Cardiol.,* 2011, **50**, 940.
34. J. G. Jacot, A. D. McCulloch and J. H. Omens, *Biophys. J.,* 2008, **95**, 3479.
35. Y. Tanaka, K. Morishima, T. Shimizu, A. Kikuchi, M. Yamato, T. Okano and T. Kitamori, *Lab Chip,* 2006, **6**, 230.

36. S. Yin, X. Zhang, C. Zhan, J. Wu, J. Xu and J. Cheung, *Biophys. J.*, 2005, **88**, 1489.
37. G. Lin, R. E. Palmer, K. S. Pister and K. P. Roos, *IEEE Trans. Biomed. Eng.*, 2001, **48**, 996.
38. T. Kaneko, K. Kojima and K. Yasuda, *Analyst*, 2007, **132**, 892.
39. J. Kim, J. Park, K. Na, S. Yang, J. Baek, E. Yoon, S. Choi, S. Lee, K. Chun and S. Park, *J. Biomech.*, 2008, **41**, 2396.
40. A. Agarwal, J. A. Goss, A. Cho, M. L. McCain and K. K. Parker, *Lab Chip*, 2013, **13**, 3599.
41. A. Grosberg, P. W. Alford, M. L. McCain and K. K. Parker, *Lab Chip*, 2011, **11**, 4165.
42. P. W. Alford, A. W. Feinberg, S. P. Sheehy and K. K. Parker, *Biomat*, 2010, **31**, 3613.
43. T. Boudou, W. R. Legant, A. Mu, M. A. Borochin, N. Thavandiran, M. Radisic, P. W. Zandstra, J. A. Epstein, K. B. Margulies and C. S. Chen, *Tissue Eng. Part A*, 2012, **18**, 910.
44. V. Chan, J. H. Jeong, P. Bajaj, M. Collens, T. Saif, H. Kong and R. Bashir, *Lab Chip*, 2012, **12**, 88.
45. J. Dahlmann, A. Krause, L. Moller, G. Kensah, M. Mowes, A. Diekmann, U. Martin, A. Kirschning, I. Gruh and G. Drager, *Biomat*, 2013, **34**, 940.
46. I. L. Jones, P. Livi, M. K. Lewandowska, M. Fiscella, B. Roscic and A. Hierlemann, *Anal. Bioanal. Chem.*, 2011, **399**, 2313.
47. A. Tam, S. Wadsworth, D. Dorscheid, S. F. Man and D. D. Sin, *Ther. Adv. Respir. Dis.*, 2011, **5**, 255.
48. E. Bellas, K. G. Marra and D. L. Kaplan, *Tissue Eng. Part C Methods*, 2013, **19**, 745.
49. D. Huh, H. Fujioka, Y. C. Tung, N. Futai, R. Paine, J. B. Grotberg and S. Takayama, *Proc. Natl. Acad. Sci. U.S.A.*, 2007, **104**, 18886.
50. N. J. Douville, Y. C. Tung, R. Li, J. D. Wang, M. E. El-Sayed and S. Takayama, *Anal. Chem.*, 2010, **82**, 2505.
51. D. Huh, B. D. Matthews, A. Mammoto, M. Montoya-Zavala, H. Y. Hsin and D. E. Ingber, *Science*, 2010, **328**, 1662.
52. D. Huh, D. C. Leslie, B. D. Matthews, J. P. Fraser, S. Jurek, G. A. Hamilton, K. S. Thorneloe, M. A. McAlexander and D. E. Ingber, *Sci. Trans. Med.*, 2012, **4**, 159ra147.
53. C. Long, C. Finch, M. Esch, W. Anderson, M. Shuler and J. Hickman, *Ann. Biomed. Eng.*, 2012, **40**, 1255.
54. P. Artursson, K. Palm and K. Luthman, *Adv. Drug. Delivery. Rev.*, 2001, **46**, 27.
55. K. M. Hillgren, A. Kato and R. T. Borchardt, *Med. Res. Rev.*, 1995, **15**, 83.
56. A. Quaroni and J. Hochman, *Adv. Drug Delivery Rev.*, 1996, **22**, 3.
57. G. J. Mahler, M. B. Esch, E. Tako, T. L. Southard, S. D. Archer, R. P. Glahn and M. L. Shuler, *Nat. Nanotechnol.*, 2012, 7, 264.
58. H. J. Kim and D. E. Ingber, *Integr. Biol.*, 2013, **5**, 1130–1140.
59. Y. Y. Lau, Y. H. Chen, T. T. Liu, C. Li, X. Cui, R. E. White and K. C. Cheng, *Drug Metab. Dispos. Biol. Fate Chem.*, 2004, **32**, 937.

60. S. H. Choi, O. Fukuda, A. Sakoda and Y. Sakai, *Mater. Sci. Eng. C,* 2004, **24**, 333.
61. G. J. Mahler, M. B. Esch, R. P. Glahn and M. L. Shuler, *Biotechnol. Bioeng.,* 2009, **104**, 193.
62. R. M. Brand, T. L. Hannah, C. Mueller, Y. Cetin and F. G. Hamel, *Ann. Biomed. Eng.,* 2000, **28**, 1210.
63. P. M. van Midwoud, M. T. Merema, E. Verpoorte and G. M. Groothuis, *Lab Chip,* 2010, **10**, 2778.
64. Y. Imura, Y. Asano, K. Sato and E. Yoshimura, *Anal. Sci. Int. J. Japan Soc. Anal. Chem.,* 2009, **25**, 1403.
65. S. Sant, S. L. Tao, O. Z. Fisher, Q. Xu, N. A. Peppas and A. Khademhosseini, *Adv. Drug Delivery Rev.,* 2012, **64**, 496.
66. A. Blouin, R. P. Bolender and E. R. Weibel, *J. Cell Biol.,* 1977, **72**, 441.
67. A. Dash, W. Inman, K. Hoffmaster, S. Sevidal, J. Kelly, R. S. Obach, L. G. Griffith and S. R. Tannenbaum, *Exp. Opin. Drug Metab. Toxicol.,* 2009, **5**, 1159.
68. N. J. Hewitt, M. J. Lechon, J. B. Houston, D. Hallifax, H. S. Brown, P. Maurel, J. G. Kenna, L. Gustavsson, C. Lohmann, C. Skonberg, A. Guillouzo, G. Tuschl, A. P. Li, E. LeCluyse, G. M. Groothuis and J. G. Hengstler, *Drug Metab. Rev.,* 2007, **39**, 159.
69. A. Guillouzo, *Environ. Health Perspect.,* 1998, **106**(Suppl 2), 511.
70. E. LeCluyse, A. Madan, G. Hamilton, K. Carroll, R. DeHaan and A. Parkinson, *J. Biochem. Mol. Toxicol.,* 2000, **14**, 177.
71. E. L. LeCluyse, *Eur. J. Pharm. Sci.,* 2001, **13**, 343.
72. J. H. Sung, J. R. Choi, D. Kim and M. L. Shuler, *Biotechnol. Bioeng.,* 2009, **104**, 516.
73. R. Kostadinova, F. Boess, D. Applegate, L. Suter, T. Weiser, T. Singer, B. Naughton and A. Roth, *Toxicol. Appl. Pharmacol.,* 2013, **268**, 1.
74. J. M. Prot, O. Videau, C. Brochot, C. Legallais, H. Benech and E. Leclerc, *Int. J. Pharm.,* 2011, **408**, 67.
75. J. M. Prot, A. Bunescu, B. Elena-Herrmann, C. Aninat, L. C. Snouber, L. Griscom, F. Razan, F. Y. Bois, C. Legallais, C. Brochot, A. Corlu, M. E. Dumas and E. Leclerc, *Toxicol. Appl. Pharm.,* 2012, **259**, 270.
76. K. J. Jang and K. Y. Suh, *Lab Chip,* 2010, **10**, 36.
77. B. Subramanian, D. Rudym, C. Cannizzaro, R. Perrone, J. Zhou and D. L. Kaplan, *Tissue Eng Part A,* 2010, **16**, 2821.
78. K. J. Jang, A. P. Mehr, G. A. Hamilton, L. A. McPartlin, S. Chung, K. Y. Suh and D. E. Ingber, *Integr. Biol.: Quantitative Biosci. Nano Macro,* 2013, **5**, 1119.
79. M. Schuldiner, O. Yanuka, J. Itskovitz-Eldor, D. A. Melton and N. Benvenisty, *Proc. Natl. Acad. Sci. U. S. A.,* 2000, **97**, 11307.
80. C. A. Batchelder, C. C. Lee, D. G. Matsell, M. C. Yoder and A. F. Tarantal, *Differentiation,* 2009, **78**, 45.
81. E. Bellas, K. G. Marra and D. L. Kaplan, *Tissue Eng. Part C Methods,* 2013, **19**, 745.
82. A. M. Taylor and N. L. Jeon, *Curr. Opin. Neurobiol.,* 2010, **20**, 640.

83. K. Ballanyi, *Isolated central nervous system circuits*, Humana Press, New York, 2012.
84. J. Janne, H. Johan and B. Petter, *J. Cell. Physiol.*, 2009, **219**, 513.
85. M. A. Johnson, J. P. Weick, R. A. Pearce and S.-C. Zhang, *J. Neurosci.*, 2007, **27**, 3069.
86. J. Sharp, J. Frame, M. Siegenthaler, G. Nistor and H. S. Keirstead, *Stem Cells*, 2010, **28**, 152.
87. X. Guo, K. Johe, P. Molnar, H. Davis and J. Hickman, *J Tissue Eng. Regen. Med.*, 2010, **4**, 181.
88. H. Davis, X. Guo, S. Lambert, M. Stancescu and J. J. Hickman, *ACS Chem. Neurosci.*, 2012, **3**, 31.
89. B.-Y. Hu, J. P. Weick, J. Yu, L.-X. Ma, X.-Q. Zhang, J. A. Thomson and S.-C. Zhang., *Proc. Natl. Acad. Sci. U. S. A.*, **107**, 4335.
90. V. Selvaraj, P. Jiang, O. Chechneva, U. G. Lo and W. Deng, *Front. Biosci.*, 2012, **17**, 65.
91. X. J. Li, Z. W. Du, E. D. Zarnowska, M. Pankratz, L. O. Hansen, R. A. Pearce and S. C. Zhang, *Nat. Biotechnol.*, 2005, **23**, 215.
92. A. L. Perrier, V. Tabar, T. Barberi, M. E. Rubio, J. Bruses, N. Topf, N. L. Harrison and L. Studer, *Proc. Natl. Acad. Sci. U. S. A.*, 2004, **101**, 12543.
93. T. C. Schulz, S. A. Noggle, G. M. Palmarini, D. A. Weiler, I. G. Lyons, K. A. Pensa, A. C. Meedeniya, B. P. Davidson, N. A. Lambert and B. G. Condie, *Stem Cells*, 2004, **22**, 1218.
94. J. Kim, *Nature*, 2002, **418**, 50.
95. R. Nat., *J. Cell. Mol. Med.*, **15**, 1429.
96. Y. Shi, P. Kirwan and F. J. Livesey, *Nat. Protoc.*, 2012, **7**, 1836.
97. O. Pomp, I. Brokhman, I. Ben-Dor, B. Reubinoff and R. S. Goldstein, *Stem Cells*, 2005, **23**, 923.
98. I. Brokhman, L. Gamarnik-Ziegler, O. Pomp, M. Aharonowiz, B. E. Reubinoff and R. S. Goldstein, *Differentiation*, 2008, **76**, 145.
99. G. I. Nistor, M. O. Totoiu, N. Haque, M. K. Carpenter and H. S. Keirstead, *Glia*, 2005, **49**, 385.
100. S. J. Davies, C. H. Shih, M. Noble, M. Mayer-Proschel, J. E. Davies and C. Proschel, *PLoS One*, 2011, **6**, e17328.
101. L. Emdad, S. L. D'Souza, H. P. Kothari, Z. A. Qadeer and I. M. Germano, *Stem Cells Dev.*, 2012, **21**, 404.
102. H. Chaudhury, E. Raborn, L. C. Goldie and K. K. Hirschi, *Cells Tissues Organs*, 2012, **195**, 41.
103. P. Dore-Duffy, *Curr. Pharm. Design*, 2008, **14**, 1581.
104. P. Dore-Duffy, A. Katychev, X. Wang and E. Van Buren, *J. Cerebr. Blood F Met.*, 2006, **26**, 613.
105. C. Cheung and S. Sinha, *J. Mol. Cell. Cardiol.*, 2011, **51**, 651.
106. N. Egawa, S. Kitaoka, K. Tsukita, M. Naitoh, K. Takahashi, T. Yamamoto, F. Adachi, T. Kondo, K. Okita, I. Asaka, T. Aoi, A. Watanabe, Y. Yamada, A. Morizane, J. Takahashi, T. Ayaki, H. Ito, K. Yoshikawa, S. Yamawaki, D. Suzuki, H. Watanabe, T. Hioki, K. Kaneko, K. Makioka, H. Okamoto,

A. Takuma, S. Tamaoka, K. Hasegawa, T. Nonaka, M. Hasegawa, A. Kawata, M. Yoshida, T. Nakahata, R. Takahashi, M. C. N. Marchetto, F. H. Gage, S. Yamanaka and H. Inoue, *Sci. Transl. Med.*, 2012, **4**, 145ra104.
107. B. Bilican, A. Serio, S. J. Barmada, A. L. Nishimura, G. J. Sullivan, M. Carrasco, H. P. Phatnani, C. A. Puddifoot, D. Story, J. Fletcher, I.-H. Park, B. A. Friedman, G. Q. Daley, D. J. A. Wyllie, G. E. Hardingham, I. Wilmut, S. Finkbeiner, T. Maniatis, C. E. Shaw and S. Chandran, *Proc. Natl. Acad. Sci. U. S. A.*, 2012, **109**, 5803.
108. G. L. Boulting., 3491793, Harvard University, 2011.
109. D. Brambley, B. Martin and P. D. Prewett, *Adv. Mater. Opt. Electr.*, 1994, **4**, 55.
110. D. A. Stenger, J. J. Hickman, K. E. Bateman, M. S. Ravenscroft, W. Ma, J. J. Pancrazio, K. Shaffer, A. E. Schaffner, D. H. Cribbs and C. W. Cotman, *J. Neurosci. Methods*, 1998, **82**, 167.
111. M. Das, P. Molnar, H. Devaraj, M. Poeta and J. J. Hickman, *Biotechnol. Prog.*, 2003, **19**, 1756.
112. M. Das, N. Bhargava, C. Gregory, L. Riedel, P. Molnar and J. J. Hickman, *In Vitro Cell Dev. B*, 2005, **41**, 343.
113. A. E. Schaffner, J. L. Barker, D. A. Stenger and J. J. Hickman, *J. Neurosci. Methods*, 1995, **62**, 111.
114. M. Das, P. Molnar, C. Gregory, L. Riedel, A. Jamshidi and J. J. Hickman, *Biomaterials*, 2004, **25**, 5643.
115. M. S. Ravenscroft, K. E. Bateman, K. M. Shaffer, H. M. Schessler, D. R. Jung, T. W. Schneider, C. B. Montgomery, T. L. Custer, A. E. Schaffner, Q. Y. Liu, Y. X. Li, J. L. Barker and J. J. Hickman, *J. Am. Chem. Soc.*, 1998, **120**, 12169.
116. W. C. Chang and D. W. Sretavan, *Langmuir*, 2008, **24**, 13048.
117. A. M. Taylor, S. W. Rhee, C. H. Tu, D. H. Cribbs, C. W. Cotman and N. L. Jeon, *Langmuir*, 2002, **19**, 1551.
118. J. Erickson, A. Tooker, Y. C. Tai and J. Pine, *J. Neurosci. Methods*, 2008, **175**, 1.
119. E. Neher and B. Sakmann, *Nature*, 1976, **260**, 799.
120. O. P. Hamill, A. Marty, E. Neher, B. Sakmann and F. J. Sigworth, *Pflugers Arch.*, 1981, **391**, 85.
121. A. Stett, U. Egert, E. Guenther, F. Hofmann, T. Meyer, W. Nisch and H. Haemmerle, *Anal. Bioanal. Chem.*, 2003, **377**, 486.
122. G. J. Brewer, M. D. Boehler, A. N. Ide and B. C. Wheeler, *J. Neurosci. Methods*, 2009, **184**, 104.
123. A. F. M. Johnstone, G. W. Gross, D. G. Weiss, O. H. U. Schroeder, A. Gramowski and T. J. Shafer, *Neurotoxicology*, 2010, **31**, 331.
124. I. Jones, P. Livi, M. Lewandowska, M. Fiscella, B. Roscic and A. Hierlemann, *Anal. Bioanal. Chem.*, 2011, **399**, 2313.
125. C. Py, M. Martina, G. A. Diaz-Quijada, C. C. Luk, D. Martinez, M. W. Denhoff, A. Charrier, T. Comas, R. Monette, A. Krantis, N. I. Syed and G. A. R. Mealing, *Front. Pharmacol.*, 2011, **2**, 51.
126. J. M. Nagarah, *Front. Pharmacol.*, 2011, **2**, 74.

127. M. Gandolfo, A. Maccione, M. Tedesco, S. Martinoia and L. Berdondini, *J Neural Eng.*, 2010, **7**, 056001.
128. E. Marconi, T. Nieus, A. Maccione, P. Valente, A. Simi, M. Messa, S. Dante, P. Baldelli, L. Berdondini and F. Benfenati, *PLoS One*, 2012, **7**, e34648.
129. M. Geissler and A. Faissner, *J. Neurosci. Methods*, 2012, **204**, 262.
130. H. A. Johnson, A. Goel and D. V. Buonomano, *Nat. Neurosci.*, 2010, **13**, 917.
131. H. V. Oviedo, I. Bureau, K. Svoboda and A. M. Zador, *Nat. Neurosci.*, 2010, **13**, 1413.
132. Y. Berdichevsky, K. J. Staley and M. L. Yarmush, *Lab Chip*, 2010, **10**, 999.
133. C. A. Croushore and J. V. Sweedler, *Lab Chip*, 2013, **13**, 1666.
134. D. K. Cullen, J. A. Wolf, V. N. Vernekar, J. Vukasinovic and M. C. LaPlaca, *Crit. Rev. Biomed. Eng.*, 2011, **39**, 201.
135. G. D. Fischbach, *Science*, 1970, **169**, 1331.
136. M. P. Daniels, B. T. Lowe, S. Shah, J. Ma, S. J. Samuelsson, B. Lugo, T. Parakh and C. S. Uhm, *Microsc. Res. Tech.*, 2000, **49**, 26.
137. B. Faraut, A. Ravel-Chapuis, S. Bonavaud, M. Jandrot-Perrus, M. Verdière-Sahuqué, L. Schaeffer, J. Koenig and D. Hantaï, *Eur. J. Neurosci.*, 2004, **19**, 2099.
138. M. Das, J. W. Rumsey, C. A. Gregory, N. Bhargava, J. F. Kang, P. Molnar, L. Riedel, X. Guo and J. J. Hickman, *Neuroscience*, 2007, **146**, 481.
139. J. A. Umbach, K. L. Adams, C. B. Gundersen and B. G. Novitch, *PLoS One*, 2012, 7.
140. G. B. Miles, D. C. Yohn, H. Wichterle, T. M. Jessell, V. F. Rafuse and R. M. Brownstone, *J. Neurosci.*, 2004, **24**, 7848.
141. X. Guo, M. Das, J. Rumsey, M. Gonzalez, M. Stancescu and J. Hickman, *Tissue Eng. Part C Methods*, 2010, **16**, 1347.
142. A. S. T. Smith, C. J. Long, K. Pirozzi and J. J. Hickman, *Technology*, 2013, **1**, 37.
143. J. W. Rumsey, M. Das, M. Stancescu, M. Bott, C. Fernandez-Valle and J. J. Hickman, *Biomaterials*, 2009, **30**, 3567.
144. M. Stettner, K. Wolffram, A. K. Mausberg, C. Wolf, S. Heikaus, A. Derksen, T. Dehmel and B. C. Kieseier, *J. Neurosci. Methods.*, 2013, **214**, 69.
145. Z. Chen, Z. Ma, Y. Wang, Y. Li, H. Lü, S. Fu, Q. Hang and P.-H. Lu, *Brain Res.*, 2010, **1309**, 9.
146. H. Davis, M. Gonzalez, N. Bhargava, M. Stancescu, J. J. Hickman and S. Lambert, *J. Biomat. Tissue Eng.*, 2012, **2**, 206.
147. A. Gardner, P. Jukkola and C. Gu, *Nat. Protoc.*, 2012, **7**, 1774.
148. D. J. Joseph, P. Choudhury and A. B. Macdermott., *J. Neurosci. Methods*, **189**, 197.
149. S. Copray, R. Liem, I. J. Mantingh-Otter and N. Brouwer, *Muscle Nerve*, 1996, **19**, 1401.
150. J. Liu, J. W. Rumsey, M. Das, P. Molnar, C. Gregory, L. Riedel, J. J. Hickman and J. D. Sato, *In Vitro Cell Dev-An*, 2008, **44**, 162.
151. A. K. Hall., in *Current Protocols in Neuroscience*, John Wiley & Sons, Inc., Editon edn., 2001.

152. J. W. Rumsey, M. Das, A. Bhalkikar, M. Stancescu and J. J. Hickman., *Biomaterials*, **31**, 8218.
153. M. Das, J. W. Rumsey, N. Bhargava, M. Stancescu and J. J. Hickman., *Biomaterials*, **31**, 4880.
154. X. Guo, J. E. Ayala, M. Gonzalez, M. Stancescu, S. Lambert and J. J. Hickman, *Biomaterials*, 2012, **33**, 5723.
155. D. Hallier-Vanuxeem, P. Prieto, M. Culot, H. Diallo, C. Landry, H. Tähti and R. Cecchelli, *Toxicol In Vitro*, 2009, **23**, 447.
156. P. Naik and L. Cucullo, *J. Pharm. Sci.*, 2012, **101**, 1337.
157. L. Cucullo, B. Aumayr, E. Rapp and D. Janigro, *Curr. Opin. Drug Discovery Dev.*, 2005, **8**, 89.
158. V. Berezowski, C. Landry, S. Lundquist, L. Dehouck, R. Cecchelli, M. P. Dehouck and L. Fenart, *Pharm. Res.*, 2004, **21**, 756.
159. G. W. Goldstein, *Ann. N. Y. Acad. Sci.*, 1988, **529**, 31.
160. G. Shayan, N. Felix, Y. Cho, M. Chatzichristidi, M. L. Shuler, C. K. Ober and K. H. Lee, *Tissue Eng. Part C Methods*, 2012, **18**, 667.
161. G. Shayan, Y. S. Choi, E. V. Shusta, M. L. Shuler and K. H. Lee, *Eur. J. Pharm. Sci.*, 2011, **42**, 148.
162. S. Harris and M. Shuler, *Biotechnol. Bioprocess Eng.*, 2003, **8**, 246.
163. J. C. F. Dewey, S. R. Bussolari, J. M. A. Gimbrone and P. F. Davies, *J. Biomech. Eng.*, 1981, **103**, 177.
164. S. Santaguida, D. Janigro, M. Hossain, E. Oby, E. Rapp and L. Cucullo, *Brain Res.*, 2006, **1109**, 1.
165. L. Cucullo, M. Hossain, E. Rapp, T. Manders, N. Marchi and D. Janigro, *Epilepsia*, 2007, **48**, 505.
166. L. Cucullo, N. Marchi, M. Hossain and D. Janigro, *J. Cerebr. Blood F Met.*, 2011, **31**, 767.
167. R. Bischoff, *Anat. Rec.*, 1974, **180**, 645.
168. C. A. Powell, B. L. Smiley, J. Mills and H. H. Vandenburgh, *Am. J. Physiol.*, 2002, **283**, C1557.
169. K.-T. Hartner and D. Pette, *Eur. J. Biochem.*, 1990, **188**, 261.
170. N. Rubinstein, K. Mabuchi, F. Pepe, S. Salmons, J. Gergely and F. Sreter, *J. Cell Biol.*, 1978, **79**, 252.
171. D. Bader, T. Masaki and D. A. Fischman, *J. Cell Biol.*, 1982, **95**, 763.
172. H. Vandenburgh and S. Kaufman, *Science*, 1979, **203**, 265.
173. M. Das, J. W. Rumsey, N. Bhargava, M. Stancescu and J. J. Hickman, *Biomaterials*, 2009, **30**, 5392.
174. K. Naumann and D. Pette, *Differentiation*, 1994, **55**, 203.
175. T. A. Rando and H. M. Blau, *J. Cell Biol.*, 1994, **125**, 1275.
176. U. Wehrle, S. Düsterhöft and D. Pette, *Differentiation*, 1994, **58**, 37.
177. K. Wilson, P. Molnar and J. J. Hickman, *Lab Chip*, 2007, 7, 920.
178. K. Shimizu, H. Sasaki, H. Hida, H. Fujita, K. Obinata, M. Shikida and E. Nagamori., *Micro Electro Mechanical Systems, 2009. MEMS 2009. IEEE 22nd International Conference on*, 2009.
179. E. Defranchi, E. Bonaccurso, M. Tedesco, M. Canato, E. Pavan, R. Raiteri and C. Reggiani, *Microsc. Res. Technol.*, 2005, **67**, 27.

180. H. Vandenburgh, J. Shansky, F. Benesch-Lee, V. Barbata, J. Reid, L. Thorrez, R. Valentini and G. Crawford, *Muscle Nerve,* 2008, **37**, 438.
181. G. Supinski, D. Nethery, T. M. Nosek, L. A. Callahan, D. Stofan and A. DiMarco, *Am. J. Physiol.,* 2000, **278**, R891.
182. J. van der Valk, D. Brunner, K. De Smet, A. Fex Svenningsen, P. Honegger, L. E. Knudsen, T. Noraberg, J. Lindl, A. Price, M. L. Scarino and G. Gstraunthaler, *Toxicol In Vitro,* 2010, **24**, 1053.
183. X. Guo, S. Spradling, M. Stancescu, S. Lambert and J. J. Hickman, *Biomaterials,* 2013, **34**, 4418.
184. A. E. Schaffner, J. L. Barker, D. A. Stenger and J. J. Hickman, *J. Neurosci. Methods,* 1995, **62**, 111.
185. A. Ajdari, *CR Physique,* 2003, **5**, 539.
186. K. W. Oh, K. Lee, B. Ahn and E. P. Furlani, *Lab Chip,* 2012, **12**, 515.
187. E. W. Young, A. R. Wheeler and C. A. Simmons, *Lab Chip,* 2007, **7**, 1759.
188. E. Gutierrez and A. Groisman, *Anal. Chem.,* 2007, **79**, 2249.
189. M. B. Esch, D. J. Post, M. L. Shuler and T. Stokol, *Tissue Eng. Part A,* 2011, **17**, 2965.
190. E. Berthier, E. W. Young and D. Beebe, *Lab Chip,* 2012, **12**, 1224.
191. M. C. Cushing and K. S. Anseth, *Science,* 2007, **316**, 1133.
192. T. Frisk, S. Rydholm, H. Andersson, G. Stemme and H. Brismar, *Electrophoresis,* 2005, **26**, 4751.
193. N. W. Choi, M. Cabodi, B. Held, J. P. Gleghorn, L. J. Bonassar and A. D. Stroock, *Nat. Mater.,* 2007, **6**, 908.
194. Y. Zheng, J. Chen, M. Craven, N. W. Choi, S. Totorica, A. Diaz-Santana, P. Kermani, B. Hempstead, C. Fischbach-Teschl, J. A. Lopez and A. D. Stroock, *Proc. Natl. Acad. Sci. U. S. A.,* 2012, **109**, 9342.
195. M. P. Cuchiara, A. C. Allen, T. M. Chen, J. S. Miller and J. L. West, *Biomaterials,* 2010, **31**, 5491.
196. A. P. Golden and J. Tien, *Lab Chip,* 2007, **7**, 720.
197. L. M. Bellan, S. P. Singh, P. W. Henderson, T. J. Porri, H. G. Craighead and J. A. Spector, *Soft Matter,* 2009, **5**, 1354.

CHAPTER 8

Utility of Human Stem Cells for Drug Discovery

SATYAN CHINTAWAR[a], MARTIN GRAF[b], AND ZAMEEL CADER*[a]

[a]Nuffield Department of Clinical Neurosciences, Weatherall Institute of Molecular Medicine, John Radcliffe Hospital, University of Oxford, Oxford OX3 9DU, UK; [b]Roche Pharmaceutical Research and Early Development, Discovery Technologies, Roche Innovation Center Basel, 124 Grenzacherstrasse, CH 4070 Basel, Switzerland
*E-mail: zameel.cader@ndcn.ox.ac.uk

8.1 Introduction

The last few decades have seen significant progress in basic, translational and clinical research, raising new hopes for the prevention, treatment and cure of serious illnesses. Advances in understanding the molecular mechanism of disease have opened up many avenues for the discovery of novel and innovative medicines. Investment in drug research and development (R&D) has increased considerably over the past decades but resultant new medicines have not increased proportionately. It has been estimated that the average time of development of a new drug, from project initiation to launch, has increased from 9.7 years during the 1990s to 13.9 years from 2000 onwards. Total R&D expenditures and the cost of developing a new drug have increased substantially while the rate of introduction of new molecular entities (NMEs) has at best remained approximately constant because of their failure in late-phase clinical trials.[1]

If drug discovery and development is to become more efficient and commercially viable, current practice needs to change considerably. To reduce high attrition rates and increased cost, it would be crucial to identify efficacy and safety issues as early as possible during the drug discovery process. Including models with better prediction of efficacy and safety of candidate molecules at the earliest stages would increase the likelihood of compounds advancing through end-stage clinical trials. Human induced pluripotent stem cell modelling potentially provides such a platform, where assays capture more of the physiological complexity of human tissue and the genetic background of the relevant patient population while remaining amenable to high-throughput screening (see Figure 8.1). This chapter

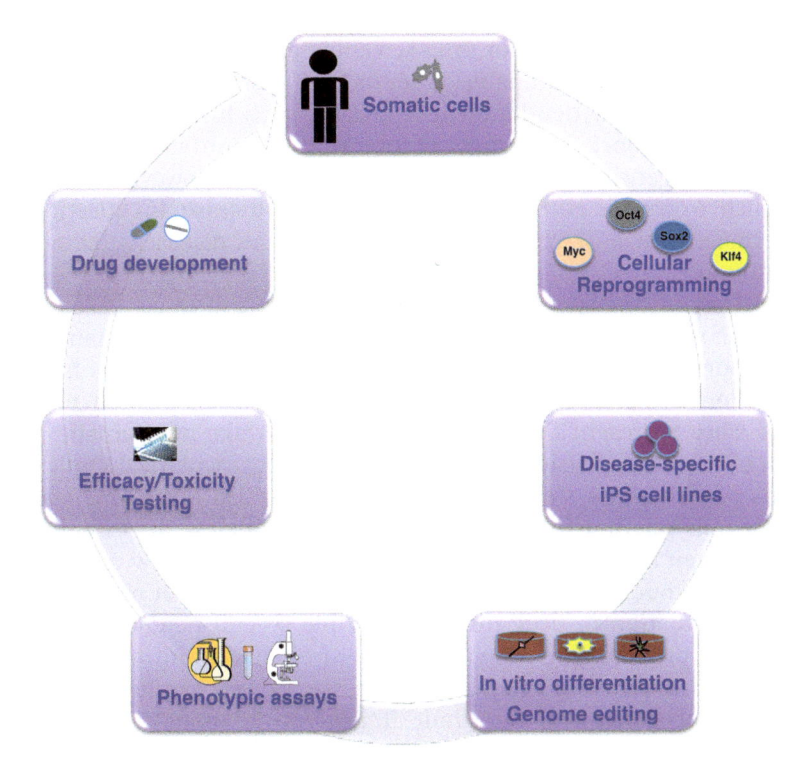

Figure 8.1 The application of human induced pluripotent stem cell (iPSC) technology for drug discovery. Easily accessible cells from patients, such as skin fibroblasts or blood, are reprogrammed to derive disease- and patient-specific iPS cell lines. These cell lines can serve as a source to derive disease-affected cell types that can be used as *in vitro* models for phenotypic assays and target identification studies. These assays are then screened in high or moderate throughput for efficacy and safety assessment of candidate molecules to derive novel effective and safer medicines.

discusses the current progress on establishing human stem cell-based disease models for drug discovery and reflects on hurdles and challenges ahead to realise the full potential of this very promising technology.

8.2 Existing Approaches to Drug Discovery

There are about 37 trillion cells in the human body, with over 200 different subtypes. While a disease process probably affects multiple cell types in multiple tissues to varying degrees, it is usually possible to identify a particular tissue or cell bearing the brunt of disease. Access to human primary cells from affected tissue can thus offer real insights into disease mechanism and serve a role in drug discovery but supply is understandably scarce. Additional issues include limited proliferative abilities and donor-to-donor variability, challenging their use as reliable and sustainable cellular models for drug screening. The workhorses for drug screening are instead immortalised cell lines that are easily expandable, a reproducible cell source, and routinely used for broad assay development. Typically, such lines are engineered to over-express or knock out a target of interest in order to maximise the assay window or to generate a disease-relevant phenotype. However it is unlikely that such a system sufficiently resembles the disease in question, and furthermore, the host cell is very different to the disease-vulnerable cell background.

Model organisms such as fruit flies, zebra fish and mice are commonly used for *in vivo* disease investigation due to the relative ease of manipulating their genomes. To understand disease aetiology, animal models have been pivotal and complement *in vitro* studies of disease-associated molecular events. They represent the only method to undertake mechanistic investigation of the pathophysiological process *in vivo*, particularly complex interactions of different systems. Animal models are also a valuable source of primary cells relevant to a disease process. Nevertheless, it is important to recognise their limitations. According to one published review, as few as 10% of gene knockouts in animals demonstrate phenotypes that may be relevant to target validation in drug discovery.[2] While several potential treatments have looked promising in rodents, many proved disappointing in the clinic. Inter-species differences in cellular processes, as well as pharmacokinetics and pharmacodynamics, can limit translation to man and the deeply embedded dependency on animal model-generated preclinical data can increase failure in clinical trials. For example, human cardiomyocytes are not easily available for efficacy and toxicity testing and the use of rodent-isolated cardiomyocytes or the Langendorff heart model have less translatability, as they have different excitation patterns to those in human cardiomyocytes. Penicillin methyl ester is hydrolysed in mice into its active form, while it is not hydrolysed in humans. Notoriously, thalidomide is teratogenic in humans but not in mice. Currently employed models are thus limited by availability, reliability and translatability towards major clinical outcome.

8.3 Advances in Human Stem Cell Technology

The term 'stem cell' is broadly used when referring to cells with the capacity to 'self-renew' to give rise to more copies of themselves and have the potential of 'differentiation' into functional cell types. Adult stem cells or tissue-resident somatic stem cells, present in most organs, contribute to tissue homeostasis in adulthood. They are multipotent and can give rise only to cell types of the organ in which they reside, but clearly have potential in regenerative medicine. Successful bone marrow transplantation using human adult stem cells was reported for the first time in the 1950s, and is now a routine practice in blood or bone marrow cancer patients. Although they are an attractive source of tissue repair and regenerative medicine, they have limited application in disease modelling and drug discovery because of their restricted differentiation ability towards certain lineages. Conversely, embryonic stem cells (ESCs), which are derived from the inner cell mass of the blastocyst stage of the embryo, are truly pluripotent, and can give rise to any existing cell type in the body (see Table 8.1 for stem cell classifications and their properties).

8.3.1 Embryonic Stem Cells

Human development had been considered a one-way street, where cells become more specialised through differentiation to give rise to a tissue or organ with a predefined function. Conrad Hal Waddington presented his model of epigenetic landscape to visualise the process of differentiation, where the peaks represent pluripotent or multipotent cells, and the base of valleys represent the end-differentiated state. Grooves along the valley guide the direction and represent developmental pathways. There has been significant progress over the last several decades to unravel these developmental pathways, identifying the key cellular decision-making points and the molecular

Table 8.1 Classification of stem cells based on their derivation

Types of stem cells	*Properties*
Totipotent stem cells	Ability to differentiate into embryonic and extra-embryonic cell types; can give rise to a viable organism
Embryonic stem cells	Ability to differentiate into any of the three germ layer cells (pluripotent); derived from the blastocyst stage of the embryo
Induced pluripotent stem cells	Ability to differentiate into any of the three germ layer cells (pluripotent); derived by the forced expression of pluripotent genes into somatic cells
Adult or tissue resident or somatic stem cells	Ability to differentiate into cell types of the organ in which they reside; multipotent; for example, neural stem cells or haematopoietic stem cells

determinants. This has been greatly facilitated by the use of ESCs. In 1981 Martin Evans successfully isolated mouse ESCs[3] and in 1998 James Thomson and colleagues obtained human ESCs (hESCs) from *in vitro* fertilised blastocysts,[4] heralding a new era of human stem cell experimentation and application. hESCs have proved to be an excellent tool to study human development and differentiation towards particular lineage, to model genetic diseases and for cell replacement therapies. Derivatives of hESCs also offer the potential for drug screens and toxicology testing. However, ethical concerns about their derivation from human embryos have presented a significant hurdle to wider adoption in research and their clinical application.

8.3.2 Reprogramming Somatic Cells

In an attempt to overcome ethical and technical barriers, the conversion of somatic cells into pluripotent stem cells similar to ESCs, through a process called reprogramming, has been a priority in modern biology. Natural reprogramming takes place when highly differentiated cells, sperm and oocyte, de-differentiate after fertilisation and give rise to the totipotent embryo. More than 40 years ago, King *et al.* pioneered artificial somatic cell nuclear reprogramming in frogs,[5] and in 1997 Wilmut and collaborators cloned the first sheep 'Dolly' by transplanting the nucleus from a mammary gland cell into an enucleated sheep egg.[6] The advent of somatic cell nuclear transfer (SCNT)[7] opened up the possibility of reprogramming somatic cells to produce patient-specific pluripotent cells. Along with SCNT, two other methods, cell fusion of somatic cells with ES cells[8] and induction of pluripotency by cell culture, were reported.[9] None of these methods, however, was shown to be efficient and highly reproducible. A major milestone was achieved in the field in 2006, when Shinya Yamanaka successfully overcame conceptual and technical barriers to the generation of pluripotent stem cells. He proposed an approach to convert differentiated somatic cells, mouse fibroblasts, into pluripotent cells by the delivery of just four transcription factors responsible for the maintenance of the pluripotent state.[10] The four transcription factors used were Oct3/4, Sox2, Klf4 and c-Myc, and the derived cells were termed induced pluripotent stem (iPS) cells. Most importantly these iPS cells demonstrated properties very similar to those of ES cells, such as pluripotency marker expression, teratoma formation, chimeras contribution and germline transmission. The same approach was later reproduced to derive iPS cells from human skin fibroblasts.[11] James Thomson's group replaced Klf4 and c-Myc with Nanog and LIN28 to successfully reprogram human fibroblasts to iPS cells.[12] Pioneering protocols have used a retroviral/lentiviral mode of gene delivery for the generation of iPS cells. As there is a clear risk that the use of integrating viral systems for the generation of iPS cells may lead to insertional mutagenesis or exogenous transgene reactivation, alternative methods using excisable cre/lox vector,[13] episomal plasmids,[14] piggyBac transposon system[15] and non-integrating approaches such as adenoviral vectors,[16] sendai viruses,[17] repeated transfections of mRNA[18] and cell penetrating proteins[19] have been developed. Use of small molecules, like valproic acid, and hypoxic conditions have also been reported to enhance the generation of iPS cells.[20] Furthermore, to avoid invasive procedures, cells derived from blood[21] and urine[22] were tested, and demonstrated successful reprogramming to iPSCs (see the review by Okano and Yamanaka on recent updates in iPSC technology).[23]

8.4 iPSC-Based Disease Models

Recapitulating the disease process *in vitro* is a major challenge for the development of effective therapeutics. Due to the invasiveness of biopsy procedures, there is limited access to diseased human tissue, and postmortem samples may show significant artefacts because of lengthy time intervals between death and tissue sampling. Primary animal cell lines and immortalised cell lines are widely used thanks to their robustness for experimentation. Clearly human-based cellular models from patients are advantageous in studying human diseases due to often marked differences in anatomical and physiological characteristics between species. Cells that are more accessible, such as skin fibroblasts and peripheral blood mononuclear cells, may not be affected significantly in many disorders. Pluripotent cell lines from pre-implantation genetic diagnosis (PGD) embryos have been derived to obtain disease cell lines for myotonic dystrophy type 1, cystic fibrosis and Huntington disease.[24] However, obtaining cell lines from PGD embryos is limited to rare genetic disorders, and may not be a widely accessible resource for the broader academic and industrial research community. In contrast, somatic cell reprogramming is now an easily accessible technology, and iPS cells can provide a continuous source of disease-specific cell types.

8.4.1 iPSC-Based Neurological and Psychiatric Disease Models

The archetypically difficult tissue to access is that from the human nervous system. iPSC-derived neurons and glia have not surprisingly generated huge interest among disease biologists and the pharmaceutical industry. The most immediate application for iPSC-based disease modelling is the study of hereditary disorders in which the disease process is strongly driven by a genetic mutation. In such circumstances, patient-derived iPSC lines should recapitulate the disease with little need for environmental manipulations. Furthermore, the role of the gene in the identified cellular disease phenotypes can be directly demonstrated through genome engineering by correcting the mutation.

iPS cell lines and transdifferentiated neurons have been generated from patients suffering from various neurological diseases and neurodevelopmental disorders to establish disease phenotypes in neurons.

Neurodevelopmental defects, such as autism spectrum disorders, have been studied using iPSCs derived from Rett syndrome (RTT) patients harbouring MeCP-2 gene mutation.[25] They are an attractive group of disorders to address with iPSC lines, because in many cases the cell types generated resemble the foetal rather than adult stage cells. iPSC-derived neurons from RTT patients manifest decreased glutamatergic synapses, correspondingly affected children display impaired neural development after one year of age. In another study, RTT-iPSC mutant astrocytes were shown to have an adverse effect on the morphology and functionality of wild-type neurons.[26] Down syndrome (trisomy 21) is the most common viable chromosomal disorder, which causes intellectual impairment and other developmental abnormalities. iPSCs derived from monozygotic twins discordant for trisomy 21 and harbouring the chromosomal abnormality resulted in changes in the architecture and density of neuronal and glial cultures, accompanied by altered expression of genes involved in neurogenesis.[27]

Adult onset disorders such as schizophrenia increasingly show evidence of neurodevelopmental abnormalities. Brennand et al.[28] demonstrated that schizophrenia iPSC-derived (SZ-iPSC) neurons have reduced neuronal connectivity along with decreased neurite number, PSD95-protein levels and glutamate receptor expression. SZ-iPSC NPCs have abnormal gene expression and protein levels related to cytoskeletal remodelling and oxidative stress, which subsequently demonstrated aberrant migration and increased oxidative stress. In another investigation on schizophrenia, SZ-iPSC were tested for hippocampal neurogenesis, and were found to have defects in the generation of dentate gyrus (DG) granule neurons, and generated DG neurons had immature electrophysiological properties.[29] Bundo et al.[30] reported that prefrontal cortical neurons of SZ patients and iPSC neurons had increased retrotransposition of L1 (long interspersed nuclear element-1), and they suggested that it could contribute to the susceptibility and pathophysiology of schizophrenia.

In the case of late-onset neurodegenerative diseases, individuals are born overtly healthy, and during the course of life, perhaps due to environmental exposures, neuronal cells become dysfunctional and/or degenerate. This gradual shift from the physiological to pathophysiological state may still be driven largely by genetic factors such as observed in hereditary Alzheimer's. iPSC disease models here offer the additional advantage of cells from the earliest stage of developmental maturity potentially allowing dissection of cause and consequence. For example, in Parkinson's disease (PD), almost 70% of dopaminergic neurons are lost before motor symptoms appear. Patient iPSC-derived dopaminergic neurons hence offer the unique opportunity to investigate the molecular mechanism of the events leading to targeted cell death, tracking events between an asymptomatic phase of the disease until the pathology is prominent. iPSC-derived mid-brain dopaminergic neurons generated from PD patients carrying leucine-rich repeat kinase-2 (LRKK2) mutation have shown higher expression of oxidative stress genes by microarray and susceptibility to hydrogen peroxide treatment even prior to overt degenerative cellular phenotypes.[31] The HD (Huntington's disease) consortium reported generation of HD iPSCs with CAG-repeat expansion-associated phenotypes.[32] Derived neurons showed increased risk of death over time in culture and after trophic factor withdrawal, and have increased susceptibility to stress and toxicity. iPSCs from Friedreich's ataxia (FRDA) demonstrate triplet repeat instability,[33] considered to underlie anticipation in families with FRDA. Furthermore physiological and ultra-structural abnormalities were detected in disease-relevant cell types, such as neurons and cardiomyocytes.[34] In SCA3, a hereditary cerebellar ataxia, iPSC patient neurons, mutant ataxin-3 inclusion bodies were evident. The authors also reported that neurotransmitter L-glutamate induced the formation of ataxin-3 aggregates, a phenotype that was abolished after inhibiting protease calpain.[35] Combining stem cell differentiation, gene editing and RNA sequencing, Kiskinis *et al.*[36] have identified several pathways perturbed in ALS motor neurons. ALS motor neurons were hyperexcitable due to abnormal protein folding[37] (see Table 8.2 for disease-specific iPS cell lines reported as *in vitro* models).

While clear progress has been made in understanding monogenic disorders, modelling of late-onset disorders is more challenging, because this may require exposure to environmental stressors and aging of cell lines. A recent approach to the latter problem was attempted by Miller *et al.*[38] by forced expression of progerin, a truncated form of lamin A associated with premature aging. The authors demonstrated that progerin over-expressing PD iPSC-derived dopamine neurons exhibited disease phenotypes, such as pronounced dendrite degeneration, progressive loss of tyrosine hydroxylase expression, and enlarged mitochondria or Lewy body precursor inclusions, which normally require both aging and genetic susceptibility.[38]

8.4.2 iPSC-Based Cardiovascular Disease Models

Cardiovascular disease is the leading cause of mortality worldwide, with 30% of global deaths caused by cardiovascular complications. Since the first demonstration of mutations in genes encoding cardiac ion channel subunits, there has been significant progress in our understanding of the genetic basis of these disorders. Genotype-(cellular) phenotype relations have been studied using patients' native cardiomyocytes isolated from biopsy tissue. As the procedure is not routinely performed, cardiac tissue is not easily available, and animal models do not faithfully recapitulate the disease. For example, mouse models of human long QT genes do not fully reproduce the human phenotype, especially for the K^+ channel-associated syndromes, as they have higher heart rates than humans, shorter action potentials and different repolarising K^+ currents.[39] Cardiological disease modelling has proved fertile ground, with numerous studies demonstrating relevant

Table 8.2 Reported disease-specific iPSC lines as *in vitro* models

Disease	Gene	Drug tested/drug screens	Reference(s)
Neurodevelopmental and psychiatric disorders			
Angelman syndrome	UBE3A	—	Chamberlain et al.[102]
Autism	Multifactorial	—	DeRosa et al.[103]
Down syndrome	APP	γ-secretase inhibitor	Shi et al.[104]
Fragile X syndrome	FMR1	—	Urbach et al.[105]
Prader Willi syndrome	N-K	—	Chamberlain et al.[102]
Rett's syndrome	CDKL5, MECP2	—	Amenduni et al.,[106] Marchetto et al., Kim et al.
Schizophrenia	DISC1, N-K	Loxapine	Chiang et al.,[107] Brennand et al.
Timothy syndrome	CACNA1C	Roscovitine, a cdk inhibitor	Pasca et al.,[108] Krey et al.[109]
Neurodegenerative disorders			
Adrenoleukodystrophy	ABCD1	Lovastatin, 4-phenylbutyrate	Jang et al.[110]
Alzheimer's disease	PS1, PS2	β- and γ-secretase inhibitors, NSAID	Israel et al.,[111] Yagi et al.,[112] Yahata et al.,[113] Qiang et al.[114]
Amyotrophic lateral sclerosis	SOD1	—	Dimos et al.[115]
	TDP-43	1757 bioactive compounds	Burkhardt et al.
Ataxia telangiectasia	ATM	—	Nayler et al.[116]
Duchenne muscular dystrophy	Dystropin,	—	Park et al.[117]
Familial dysautonomia	IKBKAP	Kinetin	Lee et al.
Friedreich's ataxia	FXN	—	Ku et al.
Huntington's disease	HTT	—	HD IPSC consortium,[118] Zhang et al.[119] Camnasio et al.,
Multiple sclerosis	MHC	—	Song et al.[120]
Olivopontocerebellar atrophy	SCA7	—	Luo et al.[121]
Parkinson's disease	SNCA, PINK1, LRRK2, Idiopathic, a-synuclein	—	Devine et al.,[122] Seibler et al.,[123] Sánchez-Danés et al.,[124] Soldner et al.
Spinal muscular atrophy	SMN	—	Ebert et al.,[125] Chang[126]
Spinal and bulbar muscular atrophy	Androgen receptor	17-allylaminogeldanamycin	Nihei et al.[127]
Spinocerebellar ataxia type 3	SCA3	Calpain inhibitor	Koch et al.
Taupathy	TAU-A152T	—	Fong et al.[128]
Cardiovascular disorders			
LQTS-1	KCNQ1	E4031, chrom. 293B, propran.	Egashira et al.,[129] Moretti et al.

Disease	Gene	Drug	Reference
LQTS-2	KCNH2	Nifedip., pinacidil, ranolazine Isoprenal., propran, nadolol, nicorandil, PD118057	Itzhaki et al.[130], Matsa et al.[131]
LQTS-3; 1	SCN5A	Sotalol, cisapride, erythromyc.	Lahti et al.[132]
LQTS-8/Timothy syndrome	CACNA1C	Mexiletine	Terrenoire et al.[133]
LEOPARD syndrome	PTPN11	Rescovitine	Yazawa et al.
CPVT-1	RYR2	—	Carvajal-Vergara et al.[134]
		Dantrolene	Jung et al.[135]
CPVT-2	CASQ2	Isoproterenol	Novak et al.[136]
Haematological			
Pancreatic ductal adenocarcinoma	—	—	Kim et al.
Chronic myelogenous leukemia	—	Imatinib	Kumano et al., Hu et al.[137]
Myeloproliferative disorder	JAK2-V617F	—	Ye et al.
Juvenile myelomonocytic leukemia	PTPN11	—	Gandre Babbe et al.[138]
Sickle cell anaemia	b-globin	—	Sebastiano et al.,[139] Zou et al.[140]
Fanconi anaemia	FA-A, FA-D2	—	Raya et al.[141]
Beta Thalassemia	b-globin	—	Ye et al.[142]
ADA-SCID	ADA	—	Park et al.
Metabolic			
Type 1 diabetes	—		Maehr et al.,[143] Thatava et al.
Type 1 diabetes	—		Kudva et al.
Maturity-onset diabetes of the young type 2	GCK		Hua et al.[144]
Lesch-Nyhan syndrome	HPRT1		Park et al.
Gaucher disease (GD) type III	GBA		Park et al.
A1ATD	AAT	3131 clinical Comp.	Rashid et al.,[146] Choi et al.
Familial hypercholesterolemia	LDLR		Rashid et al.[146]
GSD1a	Hepatic glucose-6-phosph. Defi.		Rashid et al.[146]
Crigler Najjar syndrome	UGT1A1		Rashid et al.[146]
Hereditary Tyrosinaemia type 1	FAH		Rashid et. al.[146]
Eye			
Retinitis Pigmentosa	RPO		Jin et al.[145]
Beat disease	BESTROPHIN1		Singh et al.[147]

Note: This table outlines published studies that generated patient-specific iPS cell lines and are included if these lines are utilised for drug testing and/or screening. LQTS = long QT syndrome; CPVT = catecholaminergic polymorphic ventricular tachycardia; ADA-SCID = adenosine deaminase deficiency-related severe combined immunodeficiency; A1ATD = alpha 1 Antitrypsin deficiency; GSD1a = glycogen storage disease type 1a.

cellular and electrophysiological phenotypes. In a seminal study, Moretti et al.[40] reprogrammed fibroblasts from a long QT1 syndrome patient with a KCNQ variant, and discovered that KCNQ1 protein was trapped in the endoplasmic reticulum of iPSC-CMs, suggesting that the variant resulted in a trafficking defect rather than an ion channel function defect.[40] Timothy syndrome patients are at risk of life-threatening ventricular arrhythmia due to missense mutation in the L-type calcium channel Ca(v)1.2. Calcium imaging and electrophysiological recording on iPSC-CMs from these patients revealed irregular contraction, excess Ca(2+) influx, prolonged action potentials, irregular electrical activity and abnormal calcium transients.[41] iPSC-CMs generated from familial hypertrophic cardiomyopathy recapitulated numerous aspects of the cellular phenotype, including cellular enlargement and contractile arrhythmia.[42]

8.4.3 Other Examples of iPSC-Based Disease Models

iPSC disease modelling has been applied to a very wide range of disorders, and nearly 70 hiPSC models of human diseases have been published (Table 8.2).

Zaret and co-workers reprogrammed pancreatic ductal adenocarcinoma cells to iPSCs, and upon differentiation, they underwent early developmental stages of the human cancer, thus providing a human cellular model for experimental access to early stages of the disease.[43] Furthermore, myeloproliferative disorder iPSC[44] and chronic myelogenous leukaemia patient-derived iPSC[45] models recapitulated the disease pathophysiology. These iPSC-based cancer models allow the study of both initiating and progressive phases of cancer arising from somatic genetic mutations, offering the possibility of dissecting the mechanism of tissue abnormal growth and biomarkers.

Diabetes is a chronic debilitating disease, affecting 382 million people worldwide, primarily arising from dysfunction of pancreatic β-cells.[46] Transgene-free iPS cells from type 1 and type 2 diabetes patients have been derived,[47] and their successful differentiation into glucose-responsive and insulin-producing cells has been reported.[48] These advances have raised hope of transplantation of exogenous functional pancreatic β-cells and design strategies for their endogenous repair/regeneration. A defective epidermal barrier is associated with a number of clinically diverse skin disorders, such as ichthyosis or atopic dermatitis. Functional human epidermal equivalents have been derived from PSCs, and can be used to model skin diseases with defective epidermal permeability. Children with defects of toll-like receptor 3 (TLR3) immunity are prone to HSV-1 (herpes simplex virus 1) encephalitis (HSE). iPSC neurons and oligodendrocytes derived from TLR3-and UNC-93B-deficient patients have increased susceptibility to HSV-1 infection, supporting the key role of these proteins for anti-HSV-1 immunity in the CNS.[49]

8.4.4 Genome Editing Tools

A causal relationship between genotype and phenotype can be difficult to establish. This is now being revolutionised through genome editing technologies. Rescuing disease-associated genes in patient cell types or introducing disease-causing genes in control iPS cells provides a mechanistic demonstration of the genotype–phenotype link. Zinc-finger nucleases,[50] transcription activator-like effector nucleases[51] and more recently, the CRISPR-Cas9 system,[52] are becoming widely adopted gene editing tools.[53]

Genome engineering in control iPSC can also be used to generate allelic series of mutations associated with hereditary disorders where patient recruitment may be difficult. Nevertheless, the challenges of single base pair gene editing should not be underestimated, with selection of clones successfully incorporating the base pair change being labour-intensive. Off-target effects also remain a concern, although recent modifications of nucleases have led to much higher specificity.[54] It is important to consider the possible contribution of genomic background even in the case of highly penetrant Mendelian disorders, and phenotype may be significantly modified by the choice of the host control line. Finally, while the aim is to generate isogenic lines, the process of selection may introduce genomic aberration resulting in clonal lines with very different behaviours.

8.5 Use of iPSCs in Drug Efficacy Assessment

Patient-derived iPSCs enable for the first time the development of an *in vitro* patient model, *i.e.* 'a patient in a dish'. Such an *in vitro* patient model makes it possible to study the mechanism of a disease *in vitro*, to search for biomarkers, and most important, to test the efficacy of possible drug candidates. If the identified phenotype is robust enough, it further enables full high-throughput screening in a phenotypic assay.

8.5.1 Target-Based Screening *versus* Phenotypic Screening

Many small molecule drugs were discovered by the pharmaceutical industries in the 1990s and 2000s using target-based drug discovery. Pharmaceutical researchers generally select the 'target' for a potential new medicine once they have enough understanding of the disease. Since the completion of the human genome project, the main focus of drug discovery efforts has been on targets, with strong evidence for genetic association. Phenotypic assays were generally employed for drug discovery in disorders with limited knowledge of the molecular basis of the disease process.[55] Paul Ehrlich's 'magic bullet' idea of targeting individual chemoreceptors with high efficacy and low toxicity has had some success, but the notion of 'one drug, one target, and one disease' is perhaps unrealistic for the majority of common diseases. Furthermore, current

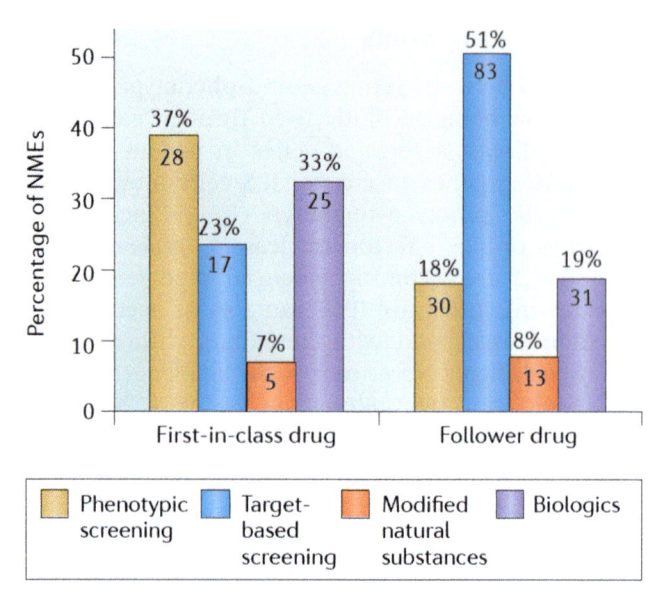

Figure 8.2 The different approaches resulting in the discovery of new medicines between 1998 and 2008. The graph represents the number of new molecular entities (NMEs) in each category. Phenotypic screening was the most successful approach for first-in-class drugs whereas target-based screening was the most successful approach for the follower drug category. The total number of new drugs discovered was similar in both categories using phenotypic screening, whereas the total number of drugs in the follower drug category was five times higher than the first-in-class drugs category using target-based screening. This figure is reproduced from Swinney et al.[57]

attrition rates challenge current target-based drug discovery paradigms. It has been suggested that 35% of biologically active compounds bind to more than one target.[56] Thus, alternative strategies are required to overcome the limitation of target-based screening, which should include the effect of a disease process on biological networks.

There is a now a resurgence of phenotypic screening which, over the last decade, has yielded more first-in-class new medicines than target-based screening (see Figure 8.2).[57] Complementary to target-based screening, phenotypic screening offers a more comprehensive view of drug discovery by integrating genetic, biochemical pathway and functional information into an integrated view of disease. Patient-derived iPSC lines offer an accessible cellular model to recapitulate disease-relevant phenotypes, and to screen compounds, with the advantages of being a human physiological model containing the genetic risk factors of the relevant disease. However, the critical issue that remains is to have strong evidence that a particular phenotype is relevant both to the disease and to drug efficacy. The robustness of the phenotype could be established by reproducing the phenotype in iPSC lines from multiple patients with the same disease. Further support could come from analogous phenotypes in animal models or, where available, from primary cells derived from patients. Human genetic studies can also be critical through establishing gene networks and pathways associated with common disease. After a phenotypic screen, there is the need to identify the target of a compound class that shows a desirable effect, and this target identification can involve considerable effort.

8.5.2 Examples of Drug Testing to Validate iPSC-Based Models

Whether or not they have been approved for human use, various drugs that have shown promising efficacy data in other models have been tested on some of the iPSC models to further validate their use in compound screening. For example, insulin-like growth factor 1 (IGF1) and gentamicin were successfully used on a Rett syndrome model, as they rescued glutamatergic synapses.[25] The plant hormone, kinetin, has been found to reverse aberrant splicing, and ameliorated neuronal differentiation and migration in a familial dysautonomia model.[58] Loxapine, an antipsychotic drug in clinical practice for schizophrenia, increased neuronal connectivity in a schizophrenia-iPSC model.[59] Furthermore, IGF1 corrected the excitatory synaptic transmission defect in Phelan–McDermid syndrome neurons, and restored mutated protein (SHANK3) expression.[60] In an Alzheimer's disease (AD) model, a γ-secretase inhibitor blocked production of Aβ peptides. hiPSC-derived neurons were used to determine the mechanism of toxicity of amyloid-β 42 accumulation, and it was observed that specific Cdk2 (cyclin-dependent kinase) inhibitors attenuated toxicity. Docosahexaenoic acid, a drug that failed in some clinical trials, might actually be beneficial to some patients, as shown by Kondo et al.[61] using an AD-iPSC model. An iPSC-based model was useful in identifying a successful hit for familial dysautonomia.[58] An anticonvulsive drug, retigabine, normalised ALS hyperexcitable motor neurons in an iPSC model.[37] Hibaoui et al.[27] reported that dual-specificity tyrosine-(Y)-phosphorylation regulated kinase 1A (DYRK1A) on chromosome 21 contributes to neuronal and glial defects in Down syndrome, and targeting it pharmacologically or with shRNA considerably corrected these defects. Using ALS motor neurons, Egawa et al.[62] screened four chemical compounds and found that a histone acetyltransferase inhibitor, anacardic acid, rescued the abnormal ALS motor neuron phenotype.

An innovative use of iPSC was demonstrated in a recent study examining why a drug with a promising preclinical profile was not effective in clinic: A subset of non-steroidal anti-inflammatory drugs (NSAIDs) were identified as γ-secretase modulators (GSMs) that lower the production of Aβ42, convincingly demonstrated preclinically using data from AD animal models and transgenic cell lines. Thus, using AD patient-derived neurons, Mertens et al.[63] demonstrated that pharmaceutically relevant concentrations of these GSMs that were clearly efficacious in other conventional AD cell models failed to evoke any effect on Aβ42/Aβ40 ratios in human neurons. These and other studies validate patient-derived iPSC models and support their potential use for larger scale drug screening examinations.

8.5.3 Drug Screens on iPSC-Based Models

For large-scale drug screens, it is crucial to develop scalable assays taking into consideration the characteristics of iPSC-derived cells. Challenges facing the development of a large-scale drug screen with iPSC include the typically long differentiation protocols, the heterogeneity of cells in iPSC cultures, the higher demand for stringent tissue culture maintenance, and the multiple sources of variability that may affect assay quality. The development of a stem cell high-throughput assay therefore requires careful consideration of which line is chosen to represent the disease, the target or phenotype to screen against, the stage of differentiation and the assay readout. Perhaps not surprisingly, high-content imaging screens have generated much interest, as they allow phenotype-based screens, and through careful design, they can circumvent some of the challenges discussed above.

The following are some examples of drug screens performed on iPSC-based assays. Lee et al.[64] tested thousands of compounds on FD-iPSC neurons, and found eight of them rescued IKBKAP, the gene responsible for FD. Characterisation of these compounds demonstrated that one of the hits, SKF-86466, was found to induce IKBKAP transcription through modulation of intracellular cAMP levels and PKA-dependent CREB phosphorylation. Xu et al.[65] screened a chemical library containing several hundred compounds on iPSC neurons, and discovered several small molecules as effective blockers of Aβ1-42 toxicity, including a Cdk2 inhibitor. Using high-throughput screening assays, Desbordes et al.[66] screened a library of 2880 small molecules that drive hESC self-renewal and differentiation, and identified several marketed drugs and natural compounds promoting short-term hESC maintenance and directing early lineage choice during differentiation. The Rho-kinase inhibitor Y-27632, which is routinely used to prevent dissociation-induced apoptosis of hESCs and iPSCs, was discovered in high-throughput chemical screens. Using a high-content assay, Burkhardt et al.[67] screened 1757 bioactive compounds on iPSC motor neurons from one ALS patient, and in this primary screen identified 38 hits that reduced the percentage of cells with mutant TDP-43 aggregates. Of 44 compounds screened, 16 showed a neuroprotective effect in a low-throughput assay with a small number of compounds using hiPS-dopaminergic neurons.[68] Using iPSC lines from patients with alpha-1 antitrypsin (AAT) deficiency, Choi et al.[69] conducted drug screening using their established library of 3131 clinical compounds with extensive safety profiles to reduce AAT accumulation in diverse patient iPSC-derived hepatocytes, and identified five clinical drugs.

8.6 Toxicity Testing Using iPSC-Based Models

Despite the use of various animal and human models, many drugs entering clinical development fail due to the appearance of unexpected, severe adverse effects in human trials. A review of FDA-approved drugs released from 1975 to 1999 estimated that 2.9% of marketed drugs were withdrawn from the market due to severe adverse effects.[70] Another report cites that approximately one in seven US FDA-approved NMEs were discontinued from the market in the period 1980–2009.[71] *In vitro* toxicology studies have again relied heavily upon animal cell lines and immortalised cell lines. The poor predictive power is in part due to inherent species differences in drug metabolising enzyme activities and cell type-specific susceptibility to toxicants.[72] For example thalidomide causes birth defects in humans but has little effect in rats.[73] Only 59% of 289 compounds known to be teratogenic in mouse are teratogenic in humans.[73] Development of highly predictive human-based *in vitro* assays is critical if we are to reduce drug attrition due to false-negative data interpretation arising from inter-species variations. Considering the diverse cell types that iPS cells can generate, this technology can be applied to determine tissue-specific toxicity of any given compound on a range of cell types. We particularly emphasised heart, liver and brain, three organs that are frequently adversely affected by novel medicament.

8.6.1 Cardiotoxicity

Cardiovascular (CV) and liver toxicity and are the most cited reasons for both market withdrawal and drugs failure during late-stage clinical trials.[74] Cardiovascular toxicity contributed to one-third of such withdrawals, emphasising the urgent need for reliable human preclinical models for candidate safety assessment. The protocols for deriving human iPSC-derived cardiomyocytes (hiPSC-CMs) are now well established, retain cardiac-specific functionality (such as spontaneous rhythmic beating) and can be maintained in culture for longer. Furthermore, electrophysiological approaches, such as patch clamp and MEA, fast kinetic fluorescence imaging of calcium-sensitive dyes and assays for mechanical contraction are established technologies that can be easily applied to iPSC-CMs for toxicity testing. For example, a study used multielectrode arrays on a set of reference compounds (E4031, nifedipine, verapamil, cisapride, terfenadine, flecainide, mexiletine and quinidine), demonstrating the utility of this assay for drug-induced cardiovascular electrophysiological risk.[75] Furthermore, the contribution that different genetic backgrounds make to an individual's susceptibility to cardiac arrhythmia was scrutinised by Liang *et al.*,[76] using a panel of patient-specific iPSC-CMs. And by performing single-cell PCR, the authors demonstrated that susceptibilities to cardiotoxic drugs and the use of disease-specific hiPSC-CMs may predict adverse drug responses more accurately. Guo *et al.*[77] reported a high-throughput functional assay employing a monolayer of beating iPSC-CMs on a 96-well plate with interdigitated electrode arrays, and tested 28 compounds with known cardiac effects. Wu and co-workers tested iPSC-CMs cultured on low-impedance MEA to identify underlying risk factors for drug-induced arrhythmia. Both these studies demonstrated that responses of iPSC-CMs were qualitatively and quantitatively consistent with reported drug effects in the literature, and concluded that the MEA/hiPSC-cardiomyocyte assay was a sensitive, robust and efficient platform for arrhythmia screening.[78]

8.6.2 Hepatotoxicity

The mechanism of liver toxicity is not entirely understood, and not always detected in preclinical studies. Furthermore, clinical trials with a selected cohort of patients may not reveal toxicity, and instead, toxicity may only be detected after marketing approval and more widespread use, when financial investment is high. Primary cultured hepatocyte-based assays are routinely used to assess generic cellular toxicity, metabolic drug activation, P450 induction signals, and formation of toxic drug metabolites, but are limited in supply, and cannot be readily used to investigate the differential susceptibility of subjects of different genetic background.[79] An important requirement for the use of hPSC-derived hepatocytes in toxicology is that the cells need to be functionally mature in order to metabolise drugs *via* the CYP450 family of phase I enzymes.[80] While a stem cell-based assay to test hepatotoxicity is work in progress, proof-of-concept studies are encouraging. For example, Kang *et al.*[81] evaluated the toxicity of chemicals at specific developmental stages of mouse ESC-derived hepatic differentiation, and demonstrated compound and cell maturation-specific toxicity. Using a high-throughput approach, Sirenko *et al.*[82] examined a number of assays and phenotypic markers, and developed automated screening methods to evaluate a diverse hepatotoxicity library of 240 known hepatotoxicants using hiPSCs. The establishment of robust protocols to derive metabolically active hepatocytes, and the co-culturing of these hepatocytes with other non-parenchymal cells, such as endothelial cells, stellate cells and Kupffer cells in 3D microfluidic liver platform are foreseen to better model liver physiology for effective toxicity assessment.[74]

8.6.3 Neurotoxicity

Drug-induced adverse effects in the CNS commonly include mood changes, dizziness, anxiety, sleep disturbances and headache. These disorders are very challenging to study *in vitro* considering the anatomical and functional networks of neuronal and glial cells that underlie human behaviours, and assays for CNS toxicity have been more challenging to establish than those for toxicity of liver or heart. Human iPSCs can provide subtype-specific neurons and glial cells, and their vulnerability can be tested, for example, by assays of neurite outgrowth and synapse activity. Measuring electrical excitability using a standard patch clamp or MEA platform can assess the effects of neurotoxicants on an established neuronal network. For example, Outinen et al.[83] used an MEA to demonstrate subtle perturbations in electrical activity by methyl mercury chloride in hESC-derived neurons, whereas no effect was detected with qRT-PCR, immunostainings or proliferation measurements. In a further example, hPSC-derived neurons were used to detect the mitochondria-dependent mechanism of neurotoxicity of anaesthetic ketamine.[75] In addition, metabolomics analysis on an 11 chemical subset of the ToxCast chemical library using hESC secretome proved 83% accurate in providing information valuable for predictive modelling and mechanistic understanding of mammalian developmental neurotoxicity.[84]

8.7 Integration of iPSCs in Drug Discovery

8.7.1 Challenges

Reprogramming efficiency of iPSC generation is currently very low (<1%), and their derivation is highly time-consuming. Stringent assays of pluripotency, such as chimera formation and germline transmission, are not possible with human iPSCs, and not required for applications of disease modelling and drug screening. Nevertheless, it is important to establish gold standard criteria for quality assurance, which would include propensity for differentiation into all three germ layers. A further important consideration is the use of media with defined conditions for both reprogramming and differentiation, as this will improve reproducibility.

Similarities and differences between hESCs and hiPSCs have been a subject of much debate, as investigators sought to establish whether reprogramming resulted in cells with all the features of embryonic pluripotency. Deng et al.[85] reported significant differences in iPS and ESC by targeted bisulfite sequencing in DNA methylation of a number of CpG sites across selected chromosomal regions. Chin et al.[86] subsequently identified hundreds of genes that were differentially expressed by microarrays. Three reports separately demonstrated epigenetic memories of donor cells in human iPS cells.[87-89] In addition to the copy number variations,[90] immunogenicity[91] and somatic mutation[92] were also accounted. While these differences and clonal variation in iPSC lines remain an important concern, other groups have reported that genetic differences found in iPSCs subsisted in starting somatic cells, and are independent of the reprogramming process.[93,94] Other evidence suggests that variations may arise from the culture conditions rather than the reprogramming factors.[95,96] It is also noteworthy that differences between human ES and iPS cells, apparent when comparing a few cell lines, are much less evident with larger sample sizes. Pragmatically, the advantages from induced pluripotency from patient tissue outweigh any possible differences between iPS and ESC, but they should nevertheless be considered when interpreting experiments. Progress has been made towards achieving ground state pluripotency,[97] and this may be an important means of avoiding clonal and inter-subject variation arising from reprogramming.

One of the most important considerations of iPSCs in drug discovery is that iPSC culture and differentiation protocols to various lineages needs to be robust, controlled, reproducible and consistent between laboratories. Many differentiated cell types with current differentiation protocols present immature stages of development, and lack functional correlation with their adult counterparts. Several disease-specific iPSCs have had no observable phenotype, perhaps due to immaturity of cell types generated or lack of environmental exposures or the subtlety of the disease-relevant phenotype.

8.7.2 Future Directions

In vivo, cells are born and interact with the extracellular milieu and the neighbouring cells in a 3D environment, supported by secreted growth factors and cytokines that are optimal for their normal growth and function. Culturing cells in 2D may change their shape, influencing their cytoskeleton, which in turn regulates gene and protein expression.[98] Hence there is increasing effort to develop enhanced physiological models where different cell types are cultured together in a 3D environment. Hydrogels, natural scaffold substrates and biodegradable substrate are some of the proposed material for 3D cell culture. Taking this into consideration and that conditions required for PSCs to self-renew and differentiate are distinct, Dixon *et al.*[99] have combined two hydrogels (alginate and collagen) for hPSC self-renewal, and the chemical microenvironmental switch was used to direct differentiation to allow dense tissue structure to be produced. This technology is still premature, and a number of issues need to be addressed before it is incorporated into drug screens and toxicity assessment, but it holds great potential. However, it seems likely that 3D culture technology will have a place in secondary screens and physiological validation rather than primary high-throughput screens.

8.7.2.1 Personalised Medicine

Patients have a wide spectrum of responses to the same drug for various reasons, including variation in the patient's genomic background. Personalised medicine and stratified medicine tailor treatment to the individual patient or a patient group, respectively. As discussed, iPSC capture the complex genome of the patient in the cellular model and are therefore the perfect substrate to develop personalised medicine approaches. The cost of reprogramming is such that at present, iPSC for personalised medicine require a strong starting hypothesis to differentiate treatment responders from non-responders or predefined patient subgroups based upon established pharmacogenomic markers. Stem cell models can then be used to validate the starting hypotheses, and serve as screening assays to identify compounds likely to be efficacious in the non-responder groups.

Wang et al.[100] combined 'organ-on-a-chip' and gene editing technologies to model Barth syndrome, a rare X-linked cardiac disorder. They demonstrated that iPSC-derived cardiomyocytes contracted very weakly and recapitulated patient-specific heart muscle electrophysiological abnormality on a chip. Gene replacement and genome editing confirmed that the underlying mutation was necessary and sufficient for relevant phenotypes and demonstrated that by quenching the excessive ROS production, contractile function could be restored.[100] Stem cell technology offers unprecedented opportunities to identify the repertoire of potential genetic and epigenetic factors that contribute to variable drug response in different populations. Heterogeneous banks of iPS cells, derived from individuals of different ages, sex and ethnic origin for a common disease, would help select the target population for subsequent clinical trials to realise the goal of precision or personalised medicine.

8.8 Emerging Resources of Diseased iPS Cell Lines

Disease modelling, high-throughput discovery and toxicology platforms within academia and pharmaceutical companies require large numbers of high-quality patient-specific and disease-relevant cells. The potential and realised value of such resources has led to high demand and commercial opportunities for biotechnology enterprises. Since the discovery by Shinya Yamanaka, iPS cells are being created by an increasing number of academic and industrial institutions, and housed in cell repositories. However, there is a general under-resourcing of infrastructure to distribute them to the interested community of researchers. In the last 2–3 years there have been significant endeavours to combine efforts towards a common goal to better furnish resources and overcome bottlenecks.

8.8.1 StemBANCC (Stem Cells for Biological Assays of Novel Drugs and Predictive Toxicology)

The Innovative Medicines Initiative (IMI), Europe's largest public–private partnership with the aim of improving the drug development process, announced in December 2012 the launch of the StemBANCC project. This project unites 35 academic and industrial partners sharing a common interest in using stem cells for drug discovery. StemBANCC aims to provide well-characterised patient-derived iPS cell lines and associated biomaterials in an accessible and sustainable biobank. Five hundred patients suffering from diseases such as Alzheimer's disease, Parkinson's disease, migraine, peripheral neuropathy, autism, schizophrenia and diabetes are currently being recruited, and 1500 high-quality iPSC lines will be derived. The project

Table 8.3 Disease and patient-specific human iPS cell line repositories

Sponsor	Disease categories	Approximate donor numbers
StemBANCC	Alzheimer's disease (AD), Parkinson's disease, bipolar disorder, autism, migraine, peripheral neuropathy, diabetes	500
EBiSC	Unknown	3000
UK Human iPSC Initiative (HipSci)	Unknown	1000
California Institute for Regenerative Medicine (CIRM)	Alzheimer's disease, autism, cerebral palsy, idiopathic	3000
US National Institute of Mental Health	Pulmonary fibrosis, idiopathic familial dilated	500
Michael Fox Foundation	Cardiomyopathy, blinding disease, viral hepatitis	
Personal Genome Project	Schizophrenia, bipolar disorder, autism	700
Farmingham Heart Study and	Parkinson's disease (PD)	300–400
Harvard Stem Cell Institute	Unknown	400
US National Institute of Neurological Disorder and Stroke	Heart attack, stroke, diabetes	
US NIH	Huntington's disease, PD, amyotrophic lateral Sclerosis	10–20 each 10–20000
NIH Undiagnosed Disease Program and	Wide-ranging	100
NY Stem Cell Foundation	Rare and undiagnosed diseases	
Guangzhou Institute of Biomedicine and Health	Wide-ranging	10–100 000

Note: This information is taken and adapted from Novak et al.[101]

is coordinated by F. Hoffmann-La Roche Ltd, Basel, and managed by the University of Oxford.

8.8.2 EBiSC (European Bank for Induced Pluripotent Stem Cells)

EBISC is a consortium of 26 partners supported by IMI, aiming to establish a sustainable repository of high-quality human iPS cell lines. The project, announced in February 2014, is coordinated by Pfizer Ltd in Cambridge, UK, and managed by Roslin Cells Ltd in Edinburgh. Both of the above-mentioned projects aim to generate and standardise genetically-defined iPS cell lines and protocols for their use as research tools.

8.8.3 HipSci (Human Induced Pluripotent Stem Cells Initiative)

The Wellcome Trust and Medical Research Council (MRC), UK, are supporting the establishment of a national iPS cell resource, and using it to carry out cellular genetic studies. The project aims to generate iPS cells from over 500 healthy individuals and 500 individuals with genetic disease, and is led by King's College London and the Welcome Trust Sanger Institute (see Table 8.3 for details on iPS cell lines global resources).

8.9 Summary

Stem cell research is one of the most rapidly developing areas of biomedicine. The application of this field in regenerative medicine is widely understood, but its potential role in drug discovery may be a nearer term achievement. Initial advances in SCNT and human embryonic stem cell derivation led to the discovery of the somatic cell reprogramming technology, and opened up many avenues for translational research. hESC-derived cells have recently been produced in large quantities for safety and toxicity assessment, are versatile but are limited in genetic diversity. Conversely, hiPSC offers the possibility of obtaining disease-affected human cells from different populations and races. iPSC derivatives from different genetic backgrounds can be employed to test potential drugs before clinical trials, so-called 'preclinical trials in a dish' that may allow the identification of more targeted cohorts of patients, increasing the chances of success in the identification of safe and effective new medicines.

References

1. F. Pammolli, L. Magazzini and M. Riccaboni, *Nat. Rev. Drug Discovery,* 2011, **10**, 428.
2. A. L. Hopkins, *Nat. Chem. Biol.,* 2008, **4**, 682.
3. M. J. Evans and M. H. Kaufman, *Nature,* 1981, **292**, 154.
4. J. A. Thomson, J. Itskovitz-Eldor, S. S. Shapiro, M. A. Waknitz, J. J. Swiergiel, V. S. Marshall and J. M. Jones, *Science,* 1998, **282**, 1145.
5. R. Briggs and T. J. King, *Proc. Natl. Acad. Sci. U. S. A.,* 1952, **38**, 455.
6. K. H. Campbell, J. McWhir, W. A. Ritchie and I. Wilmut, *Nature,* 1996, **380**, 64.
7. I. Wilmut, A. E. Schnieke, J. McWhir, A. J. Kind and K. H. Campbell, *Nature,* 1997, **385**, 810.
8. C. A. Cowan, J. Atienza, D. A. Melton and K. Eggan, *Science,* 2005, **309**, 1369.
9. K. Guan, K. Nayernia, L. S. Maier, S. Wagner, R. Dressel, J. H. Lee, J. Nolte, F. Wolf, M. Li, W. Engel and G. Hasenfuss, *Nature,* 2006, **440**, 1199.

10. K. Takahashi and S. Yamanaka, *Cell*, 2006, **126**, 663.
11. K. Takahashi, K. Tanabe, M. Ohnuki, M. Narita, T. Ichisaka, K. Tomoda and S. I. Yamanaka, *Cell*, 2007, **131**, 861.
12. J. Yu, M. A. Vodyanik, K. Smuga-Otto, J. Antosiewicz-Bourget, J. L. Frane, S. Tian, J. Nie, G. A. Jonsdottir, V. Ruotti, R. Stewart, I. I. Slukvin and J. A. Thomson, *Science*, 2007, **318**, 1917.
13. F. Soldner, D. Hockemeyer, C. Beard, Q. Gao, G. W. Bell, E. G. Cook, G. Hargus, A. Blak, O. Cooper, M. Mitalipova, O. Isacson and R. Jaenisch, *Cell*, 2009, **136**, 964.
14. K. Okita, M. Nakagawa, H. Hyenjong, T. Ichisaka and S. Yamanaka, *Science*, 2008, **322**, 949.
15. K. Woltjen, I. P. Michael, P. Mohseni, R. Desai, M. Mileikovsky, R. Hämäläinen, R. Cowling, W. Wang, P. Liu, M. Gertsenstein, K. Kaji, H. K. Sung and A. Nagy, *Nature*, 2009, **458**, 766.
16. M. Stadtfeld, M. Nagaya, J. Utikal, G. Weir and K. Hochedlinger, *Science*, 2008, **322**, 945.
17. N. Fusaki, H. Ban, A. Nishiyama, K. Saeki and M. Hasegawa, *Proc. Jpn. Acad. Ser. B Phys. Biol. Sci.*, 2009, **85**, 348.
18. L. Warren, P. D. Manos, T. Ahfeldt, Y. H. Loh, H. Li, F. Lau, W. Ebina, P. K. Mandal, Z. D. Smith, A. Meissner, G. Q. Daley, A. S. Brack, J. J. Collins, C. Cowan, T. M. Schlaeger and D. J. Rossi, *Cell Stem Cell*, 2010, **7**, 618.
19. H. Zhou, S. Wu, J. Y. Joo, S. Zhu, D. W. Han, T. Lin, S. Trauger, G. Bien, S. Yao, Y. Zhu, G. Siuzdak, H. R. Schöler, L. Duan and S. Ding, *Cell Stem Cell*, 2009, **4**, 381.
20. Y. Yoshida, K. Takahashi, K. Okita, T. Ichisaka and S. Yamanaka, *Cell Stem Cell*, 2009, **5**, 237.
21. Y.-H. Loh, O. Hartung, H. Li, C. Guo, J. M. Sahalie, P. D. Manos, A. Urbach, G. C. Heffner, M. Grskovic, F. Vigneault, M. W. Lensch, I. H. Park, S. Agarwal, G. M. Church, J. J. Collins, S. Irion and G. Q. Daley, *Cell Stem Cell*, 2010, **7**, 15.
22. T. Zhou, C. Benda, S. Dunzinger, Y. Huang, J. C. Ho, J. Yang, Y. Wang, Y. Zhang, Q. Zhuang, Y. Li, X. Bao, H. F. Tse, J. Grillari, R. Grillari-Voglauer, D. Pei and M. A. Esteban, *Nat. Protoc.*, 2012, **7**, 2080.
23. H. Okano and S. Yamanaka, *Mol. Brain*, 2014, **7**, 22.
24. I. Mateizel, N. De Temmerman, U. Ullmann, G. Cauffman, K. Sermon, H. Van de Velde, M. De Rycke, E. Degreef, P. Devroey, I. Liebaers and A. Van Steirteghem, *Hum. Reprod. Oxf. Engl.*, 2006, **21**, 503.
25. M. C. N. Marchetto, C. Carromeu, A. Acab, D. Yu, G. W. Yeo, Y. Mu, G. Chen, F. H. Gage and A. R. Muotri, *Cell*, 2010, **143**, 527.
26. E. C. Williams, X. Zhong, A. Mohamed, R. Li, Y. Liu, Q. Dong, G. E. Ananiev, J. C. Mok, B. R. Lin, J. Lu, C. Chiao, R. Cherney, H. Li, S. C. Zhang and Q. Chang, *Hum. Mol. Genet.*, 2014, **23**, 2968.
27. Y. Hibaoui, I. Grad, A. Letourneau, M. R. Sailani, S. Dahoun, F. A. Santoni, S. Gimelli, M. Guipponi, M. F. Pelte, F. Béna, S. E. Antonarakis and A. Feki, *EMBO Mol. Med.*, 2014, **6**, 259.

28. K. Brennand, J. N. Savas, Y. Kim, N. Tran, A. Simone, K. Hashimoto-Torii, K. G. Beaumont, H. J. Kim, A. Topol, I. Ladran, M. Abdelrahim, B. Matikainen-Ankney, S. H. Chao, M. Mrksich, P. Rakic, G. Fang, B. Zhang, J. R. 3rd Yates and F. H. Gage, *Mol. Psychiatry*, 2014, doi: 10.1038/mp.2014.22.
29. D. X. Yu, F. P. Di Giorgio, J. Yao, M. C. Marchetto, K. Brennand, R. Wright, A. Mei, L. McHenry, D. Lisuk, J. M. Grasmick, P. Silberman, G. Silberman, R. Jappelli and F. H. Gage, *Stem Cell Rep.*, 2014, 2, 295–310.
30. M. Bundo, M. Toyoshima, Y. Okada, W. Akamatsu, J. Ueda, T. Nemoto-Miyauchi, F. Sunaga, M. Toritsuka, D. Ikawa, A. Kakita, M. Kato, K. Kasai, T. Kishimoto, H. Nawa, H. Okano, T. Yoshikawa, T. Kato and K. Iwamoto, *Neuron*, 2014, 81, 306.
31. H. N. Nguyen, B. Byers, B. Cord, A. Shcheglovitov, J. Byrne, P. Gujar, K. Kee, B. Schüle, R. E. Dolmetsch, W. Langston, T. D. Palmer and R. R. Pera, *Cell Stem Cell*, 2011, 8, 267.
32. HD iPSC Consortium, Induced pluripotent stem cells from patients with Huntington's disease show CAG-repeat-expansion-associated phenotypes, *Cell Stem Cell*, 2012, 11, 264.
33. S. Ku, E. Soragni, E. Campau, E. A. Thomas, G. Altun, L. C. Laurent, J. F. Loring, M. Napierala and J. M. Gottesfeld, *Cell Stem Cell*, 2010, 7, 631.
34. A. Hick, M. Wattenhofer-Donzé, S. Chintawar, P. Tropel, J. P. Simard, N. Vaucamps, D. Gall, L. Lambot, C. André, L. Reutenauer, M. Rai, M. Teletin, N. Messaddeq, S. N. Schiffmann, S. Viville, C. E. Pearson, M. Pandolfo and H. Puccio, *Dis. Model. Mech.*, 2013, 6, 608.
35. P. Koch, Breuer P, M. Peitz, J. Jungverdorben, J. Kesavan, D. Poppe, J. Doerr, J. Ladewig, J. Mertens, T. Tüting, P. Hoffmann, T. Klockgether, B. O. Evert, U. Wüllner and O. Brüstle, *Nature*, 2011, 480, 543.
36. E. Kiskinis, J. Sandoe, L. A. Williams, G. L. Boulting, R. Moccia, B. J. Wainger, S. Han, T. Peng, S. Thams, S. Mikkilineni, C. Mellin, F. T. Merkle, B. N. Davis-Dusenbery, M. Ziller, D. Oakley, J. Ichida, S. Di Costanzo, N. Atwater, M. L. Maeder, M. J. Goodwin, J. Nemesh, R. E. Handsaker, D. Paull, S. Noggle, S. A. McCarroll, J. K. Joung, C. J. Woolf, R. H. Brown and K. Eggan, *Cell Stem Cell*, 2014, doi: 10.1016/j.stem.2014.03.004.
37. B. J. Wainger, E. Kiskinis, C. Mellin, O. Wiskow, S. S. Han, J. Sandoe, N. P. Perez, L. A. Williams, S. Lee, G. Boulting, J. D. Berry, R. H. J. Brown, M. E. Cudkowicz, B. P. Bean, K. Eggan and C. J. Woolf, *Cell Rep.*, 2014, 7, 1.
38. J. D. Miller, Y. M. Ganat, S. Kishinevsky, R. L. Bowman, B. Liu, E. Y. Tu, P. K. Mandal, E. Vera, J. W. Shim, S. Kriks, T. Taldone, N. Fusaki, M. J. Tomishima, D. Krainc, T. A. Milner, D. J. Rossi and L. Studer, *Cell Stem Cell*, 2013, 13, 691.
39. G. Salama and B. London, *J. Physiol.*, 2007, 578, 43.

40. A. Moretti, M. Bellin, A. Welling, C. B. Jung, J. T. Lam, L. Bott-Flügel, T. Dorn, A. Goedel, C. Höhnke, F. Hofmann, M. Seyfarth, D. Sinnecker, A. Schömig and K. L. Laugwitz, *N. Engl. J. Med.*, 2010, **363**, 1397.
41. M. Yazawa, B. Hsueh, X. Jia, A. M. Pasca, J. A. Bernstein, J. Hallmayer and R. E. Dolmetsch, *Nature*, 2011, **471**, 230.
42. F. Lan, A. S. Lee, P. Liang, V. Sanchez-Freire, P. K. Nguyen, L. Wang, L. Han, M. Yen, Y. Wang, N. Sun, O. J. Abilez, S. Hu, A. D. Ebert, E. G. Navarrete, C. S. Simmons, M. Wheeler, B. Pruitt, R. Lewis, Y. Yamaguchi, E. A. Ashley, D. M. Bers, R. C. Robbins, M. T. Longaker and J. C. Wu, *Cell Stem Cell*, 2013, **12**, 101.
43. J. Kim, J. P. Hoffman, R. K. Alpaugh, A. D. Rhim, M. Reichert, B. Z. Stanger, E. E. Furth, A. R. Sepulveda, C. X. Yuan, K. J. Won, G. Donahue, J. Sands, A. A. Gumbs and K. S. Zaret, *Cell Rep.*, 2013, **3**, 2088.
44. Z. Ye, H. Zhan, P. Mali, S. Dowey, D. M. Williams, Y. Y. Jang, C. V. Dang, J. L. Spivak, A. R. Moliterno and L. Cheng, *Blood*, 2009, **114**, 5473.
45. K. Kumano, S. Arai, M. Hosoi, K. Taoka, N. Takayama, M. Otsu, G. Nagae, K. Ueda, K. Nakazaki, Y. Kamikubo, K. Eto, H. Aburatani, H. Nakauchi and M. Kurokawa, *Blood*, 2012, **119**, 6234.
46. A. Rezania, J. E. Bruin, M. J. Riedel, M. Mojibian, A. Asadi, J. Xu, R. Gauvin, K. Narayan, F. Karanu, J. J. O'Neil, Z. Ao, G. L. Warnock and T. J. Kieffer, *Diabetes*, 2012, **61**, 2016.
47. Y. C. Kudva, S. Ohmine, L. V. Greder, J. R. Dutton, A. Armstrong, J. G. De Lamo, Y. K. Khan, T. Thatava, M. Hasegawa, N. Fusaki, J. M. Slack and Y. Ikeda, *Stem Cells Transl. Med.*, 2012, **1**, 451.
48. T. Thatava, Y. C. Kudva, R. Edukulla, K. Squillace, J. G. De Lamo, Y. K. Khan, T. Sakuma, S. Ohmine, A. Terzic and Y. Ikeda, *Mol. Ther. J. Am. Soc. Gene Ther.*, 2013, **21**, 228.
49. F. G. Lafaille, I. M. Pessach, S. Y. Zhang, M. J. Ciancanelli, M. Herman, A. Abhyankar, S. W. Ying, S. Keros, P. A. Goldstein, G. Mostoslavsky, J. Ordovas-Montanes, E. Jouanguy, S. Plancoulaine, E. Tu, Y. Elkabetz, S. Al-Muhsen, M. Tardieu, T. M. Schlaeger, G. Q. Daley, L. Abel, J. L. Casanova, L. Studer and L. D. Notarangelo, *Nature*, 2012, **491**, 769.
50. F. D. Urnov, E. J. Rebar, M. C. Holmes, H. S. Zhang and P. D. Gregory, *Nat. Rev. Genet.*, 2010, **11**, 636.
51. D. Hockemeyer, H. Wang, S. Kiani, C. S. Lai, Q. Gao, J. P. Cassady, G. J. Cost, L. Zhang, Y. Santiago, J. C. Miller, B. Zeitler, J. M. Cherone, X. Meng, S. J. Hinkley, E. J. Rebar, P. D. Gregory and F. D. Urnov, *Jaenisch R. Nat. Biotechnol.*, 2011, **29**, 731.
52. P. Mali, L. Yang, K. M. Esvelt, J. Aach, M. Guell, J. E. DiCarlo, J. E. Norville and G. M. Church, *Science*, 2013, **339**, 823.
53. K. Musunuru, *Dis. Model. Mech.*, 2013, **6**, 896.
54. F. A. Ran, P. D. Hsu, C. Y. Lin, J. S. Gootenberg, S. Konermann, A. E. Trevino, D. A. Scott, A. Inoue, S. Matoba, Y. Zhang and F. Zhang, *Cell*, 2013, **154**, 1380.
55. M. A. Lindsay, *Nat. Rev. Drug Discov.*, 2003, **2**, 831.

56. G. V. Paolini, R. H. B. Shapland, W. P. van Hoorn, J. S. Mason and A. L. Hopkins, *Nat. Biotechnol.*, 2006, **24**, 805.
57. D. C. Swinney and J. Anthony, *Nat. Rev. Drug Discov.*, 2011, **10**, 507.
58. G. Lee, E. P. Papapetrou, H. Kim, S. M. Chambers, M. J. Tomishima, C. A. Fasano, Y. M. Ganat, J. Menon, F. Shimizu, A. Viale, V. Tabar, M. Sadelain and L. Studer, *Nature*, 2009, **461**, 402.
59. K. J. Brennand, A. Simone, J. Jou, C. Gelboin-Burkhart, N. Tran, S. Sangar, Y. Li, Y. Mu, G. Chen, D. Yu, S. McCarthy, J. Sebat and F. H. Gage, *Nature*, 2011, **473**, 221.
60. A. Shcheglovitov, O. Shcheglovitova, M. Yazawa, T. Portmann, R. Shu, V. Sebastiano, A. Krawisz, W. Froehlich, J. A. Bernstein, J. F. Hallmayer and R. E. Dolmetsch, *Nature*, 2013, **503**, 267.
61. T. Kondo, M. Asai, K. Tsukita, Y. Kutoku, Y. Ohsawa, Y. Sunada, K. Imamura, N. Egawa, N. Yahata, K. Okita, K. Takahashi, I. Asaka, T. Aoi, A. Watanabe, K. Watanabe, C. Kadoya, R. Nakano, D. Watanabe, K. Maruyama, O. Hori, S. Hibino, T. Choshi, T. Nakahata, H. Hioki, T. Kaneko, M. Naitoh, K. Yoshikawa, S. Yamawaki, S. Suzuki, R. Hata, S. Ueno, T. Seki, K. Kobayashi, T. Toda, K. Murakami, K. Irie, W. L. Klein, H. Mori, T. Asada, R. Takahashi, N. Iwata, S. Yamanaka and H. Inoue, *Cell Stem Cell*, 2013, **12**, 487.
62. N. Egawa, S. Kitaoka, K. Tsukita, M. Naitoh, K. Takahashi, T. Yamamoto, F. Adachi, T. Kondo, K. Okita, I. Asaka, T. Aoi, A. Watanabe, Y. Yamada, A. Morizane, J. Takahashi, T. Ayaki, H. Ito, K. Yoshikawa, S. Yamawaki, S. Suzuki, D. Watanabe, H. Hioki, T. Kaneko, K. Makioka, K. Okamoto, H. Takuma, A. Tamaoka, K. Hasegawa, T. Nonaka, M. Hasegawa, A. Kawata, M. Yoshida, T. Nakahata, R. Takahashi, M. C. Marchetto, F. H. Gage, S. Yamanaka and H. Inoue, *Sci. Transl. Med.*, 2012, **4**, 145ra104.
63. J. Mertens, K. Stüber, P. Wunderlich, J. Ladewig, J. C. Kesavan, R. Vandenberghe, M. Vandenbulcke, P. van Damme, J. Walter, O. Brüstle and P. Koch, *Stem Cell Rep.*, 2013, **1**, 491.
64. G. Lee, C. N. Ramirez, H. Kim, N. Zeltner, B. Liu, C. Radu, B. Bhinder, Y. J. Kim, I. Y. Choi, B. Mukherjee-Clavin, H. Djaballah and L. Studer, *Nat. Biotechnol.*, 2012, **30**, 1244.
65. X. Xu, Y. Lei, J. Luo, J. Wang, S. Zhang, X. J. Yang, M. Sun, E. Nuwaysir, G. Fan, J. Zhao, L. Lei and Z. Zhong, *Stem Cell Res.*, 2013, **10**, 213.
66. S. C. Desbordes, D. G. Placantonakis, A. Ciro, N. D. Socci, G. Lee, H. Djaballah and L. Studer, *Cell Stem Cell*, 2008, **2**, 602.
67. M. F. Burkhardt, F. J. Martinez, S. Wright, C. Ramos, D. Volfson, M. Mason, J. Garnes, V. Dang, J. Lievers, U. Shoukat-Mumtaz, R. Martinez, H. Gai, R. Blake, E. Vaisberg, M. Grskovic, C. Johnson, S. Irion, J. Bright, B. Cooper, L. Nguyen, I. Griswold-Prenner and A. Javaherian, *Mol. Cell. Neurosci.*, 2013, **56**, 355.
68. J. Peng, Q. Liu, M. S. Rao and X. Zeng, *J. Biomol. Screen.*, 2013, **18**, 522.
69. S. M. Choi, Y. Kim, J. S. Shim, J. T. Park, R. H. Wang, S. D. Leach, J. O. Liu, C. Deng, Z. Ye and Y. Y. Jang, *Hepatol. Baltim. Md*, 2013, **57**, 2458.

70. D. A. Volpe, J. E. Tomaszewski, R. E. Parchment, A. Garg, K. P. Flora, M. J. Murphy and C. K. Grieshaber, *Cancer Chemother. Pharmacol.*, 1996, **39**, 143.
71. Z. P. Qureshi, E. Seoane-Vazquez, R. Rodriguez-Monguio, K. B. Stevenson and S. L. Szeinbach, *Pharmacoepidemiol. Drug Saf.*, 2011, **20**, 772.
72. M. L. Sutherland, K. M. Fabre and D. A. Tagle, *Stem Cell Res. Ther.*, 2013, **4**(Suppl 1), I1.
73. F. Homburger, S. Chaube, M. Eppenberger, P. D. Bogdonoff and C. W. Nixon, *Toxicol. Appl. Pharmacol.*, 1965, **7**, 686.
74. A. Bhushan, N. Senutovitch, S. S. Bale, W. J. McCarty, M. Hegde, R. Jindal, I. Golberg, O. Berk Usta, M. L. Yarmush, L. Vernetti, A. Gough, A. Bakan, T. Y. Shun, R. DeBiasio and D. Lansing Taylor, *Stem Cell Res. Ther.*, 2013, **4**(Suppl 1), S16.
75. K. Harris, M. Aylott, Y. Cui, J. B. Louttit, N. C. McMahon and A. Sridhar, *Toxicol. Sci. Off. J. Soc. Toxicol.*, 2013, **134**, 412.
76. P. Liang, F. Lan, A. S. Lee, T. Gong, V. Sanchez-Freire, Y. Wang, S. Diecke, K. Sallam, J. W. Knowles, P. J. Wang, P. K. Nguyen, D. M. Bers, R. C. Robbins and J. C. Wu, *Circulation*, 2013, **127**, 1677.
77. L. Guo, R. M. Abrams, J. E. Babiarz, J. D. Cohen, S. Kameoka, M. J. Sanders, E. Chiao and K. L. Kolaja, *Toxicol. Sci. Off. J. Soc. Toxicol.*, 2011, **123**, 281.
78. E. G. Navarrete, P. Liang, F. Lan, V. Sanchez-Freire, C. Simmons, T. Gong, A. Sharma, P. W. Burridge, B. Patlolla, A. S. Lee, H. Wu, R. E. Beygui, S. M. Wu, R. C. Robbins, D. M. Bers and J. C. Wu, *Circulation*, 2013, **128**, S3.
79. K. Kolaja, *J. Biol. Chem.*, 2014, **289**, 4555.
80. G. Brolén, L. Sivertsson, P. Björquist, G. Eriksson, M. Ek, H. Semb, I. Johansson, T. B. Andersson, M. Ingelman-Sundberg and N. J. Heins, *Biotechnol.*, 2010, **145**, 284.
81. S.-J. Kang, S. H. Jeong, E. J. Kim, J. H. Cho, Y. I. Park, S. W. Park, H. S. Shin, S. W. Son and H. G. Kang, *Cell Biol. Toxicol.*, 2013, **29**, 1.
82. O. Sirenko, J. Hesley, I. Rusyn and E. F. Cromwell, *Assay Drug Dev. Technol.*, 2014, **12**, 43.
83. L. Ylä-Outinen, J. Heikkilä, H. Skottman, R. Suuronen, R. Aänismaa and S. Narkilahti, *Front. Neuroengineering*, 2010, **3**.
84. N. C. Kleinstreuer, A. M. Smith, P. R. West, K. R. Conard, B. R. Fontaine, A. M. Weir-Hauptman, J. A. Palmer, T. B. Knudsen, D. J. Dix, E. L. Donley and G. G. Cezar, *Toxicol. Appl. Pharmacol.*, 2011, **257**, 111.
85. J. Deng, R. Shoemaker, B. Xie, A. Gore, E. M. LeProust, J. Antosiewicz-Bourget, D. Egli, N. Maherali, I. H. Park, J. Yu, G. Q. Daley, K. Eggan, K. Hochedlinger, J. Thomson, W. Wang, Y. Gao and K. Zhang, *Nat. Biotechnol.*, 2009, **27**, 353.
86. M. H. Chin, M. J. Mason, W. Xie, S. Volinia, M. Singer, C. Peterson, G. Ambartsumyan, O. Aimiuwu, L. Richter, J. Zhang, I. Khvorostov, V. Ott, M. Grunstein, N. Lavon, N. Benvenisty, C. M. Croce, A. T. Clark,

T. Baxter, A. D. Pyle, M. A. Teitell, M. Pelegrini, K. Plath and W. E. Lowry, *Cell Stem Cell,* 2009, **5**, 111.
87. K. Kim, R. Zhao, A. Doi, K. Ng, J. Unternaehrer, P. Cahan, H. Huo, Y. H. Loh, M. J. Aryee, M. W. Lensch, H. Li, J. J. Collins, A. P. Feinberg and G. Q. Daley, *Nat. Biotechnol.,* 2011, **29**, 1117.
88. Y. Ohi, H. Qin, C. Hong, L. Blouin, J. M. Polo, T. Guo, Z. Qi, S. L. Downey, P. D. Manos, D. J. Rossi, J. Yu, M. Hebrok, K. Hochedlinger, J. F. Costello, J. S. Song and M. Ramalho-Santos, *Nat. Cell Biol.,* 2011, **13**, 541.
89. R. Lister, M. Pelizzola, Y. S. Kida, R. D. Hawkins, J. R. Nery, G. Hon, J. Antosiewicz-Bourget, R. O'Malley, R. Castanon, S. Klugman, M. Downes, R. Yu, R. Stewart, B. Ren, J. A. Thomson, R. M. Evans and J. R. Ecker, *Nature,* 2011, **471**, 68.
90. S. M. Hussein, N. N. Batada, S. Vuoristo, R. W. Ching, R. Autio, E. Närvä, S. Ng, M. Sourour, R. Hämäläinen, C. Olsson, K. Lundin, M. Mikkola, R. Trokovic, M. Peitz, O. Brüstle, D. P. Bazett-Jones, K. Alitalo, R. Lahesmaa, A. Nagy and T. Otonkoski, *Nature,* 2011, **471**, 58.
91. T. Zhao, Z.-N. Zhang, Z. Rong and Y. Xu, *Nature,* 2011, **474**, 212.
92. A. Gore, Z. Li, H. L. Fung, J. E. Young, S. Agarwal, J. Antosiewicz-Bourget, I. Canto, A. Giorgetti, M. A. Israel, E. Kiskinis, J. H. Lee, Y. H. Loh, P. D. Manos, N. Montserrat, A. D. Panopoulos, S. Ruiz, M. L. Wilbert, J. Yu, E. F. Kirkness, J. C. Izpisua Belmonte, D. J. Rossi, J. A. Thomson, K. Eggan, G. Q. Daley, L. S. Goldstein and K. Zhang, *Nature,* 2011, **471**, 63.
93. M. A. Young, D. E. Larson, C. W. Sun, D. R. George, L. Ding, C. A. Miller, L. Lin, K. M. Pawlik, K. Chen, X. Fan, H. Schmidt, J. Kalicki-Veizer, L. L. Cook, G. W. Swift, R. T. Demeter, M. C. Wendl, M. S. Sands, E. R. Mardis, R. K. Wilson, T. M. Townes and T. J. Ley, *Cell Stem Cell,* 2012, **10**, 570.
94. L. Cheng, N. F. Hansen, L. Zhao, Y. Du, C. Zou, F. X. Donovan, B. K. Chou, G. Zhou, S. Li, S. N. Dowey, Z. Ye, NISC Comparative Sequencing Program, S. C. Chandrasekharappa, H. Yang, J. C. Mullikin and P. P. Liu, *Cell Stem Cell,* 2012, **10**, 337.
95. A. M. Newman and J. B. Cooper, *Cell Stem Cell,* 2010, **7**, 258.
96. M. G. Guenther, G. M. Frampton, F. Soldner, D. Hockemeyer, M. Mitalipova, R. Jaenisch and R. A. Young, *Cell Stem Cell,* 2010, **7**, 249.
97. O. Gafni, L. Weinberger, A. A. Mansour, Y. S. Manor, E. Chomsky, D. Ben-Yosef, Y. Kalma, S. Viukov, I. Maza, A. Zviran, Y. Rais, Z. Shipony, Z. Mukamel, V. Krupalnik, M. Zerbib, S. Geula, I. Caspi, D. Schneir, T. Shwartz, S. Gilad, D. Amann-Zalcenstein, S. Benjamin, I. Amit, A. Tanay, R. Massarwa, N. Novershtern and J. H. Hanna, *Nature,* 2013, **504**, 282.
98. A. Ben-Ze'ev, *BioEssays News Rev. Mol. Cell. Dev. Biol.,* 1991, **13**, 207.
99. J. E. Dixon, D. A. Shah, C. Rogers, S. Hall, N. Weston, C. D. Parmenter, D. McNally, C. Denning and K. M. Shakesheff, *Proc. Natl. Acad. Sci. U. S. A.,* 2014, **111**, 5580.

100. G. Wang, M. L. McCain, L. Yang, A. He, F. S. Pasqualini, A. Agarwal, H. Yuan, D. Jiang, D. Zhang, L. Zangi, J. Geva, A. E. Roberts, Q. Ma, J. Ding, J. Chen, D. Z. Wang, K. Li, J. Wang, R. J. Wanders, W. Kulik, F. M. Vaz, M. A. Laflamme, C. E. Murry, K. R. Chien, R. I. Kelley, G. M. Church, K. K. Parker and W. T. Pu, *Nat. Med.*, 2014, doi: 10.1038/nm.3545.
101. T. J. Novak, U. Grieshammer, M. Yaffe, S. Madore. *Drug Discovery World*, Winter 2013/14 (www.ddw-online.com/media/32/22621/the-repositioning-revolution.pdf, last accessed 20-06-2014).
102. S. J. Chamberlain, P. F. Chen, K. Y. Ng, F. Bourgois-Rocha, F. Lemtiri-Chlieh, E. S. Levine and M. Lalande, *Proc. Natl. Acad. Sci. U. S. A.*, 2010, **107**, 17668–17673.
103. B. A. DeRosa, J. M. Van Baaren, G. K. Dubey, J. M. Lee, M. L. Cuccaro, J. M. Vance, M. A. Pericak-Vance and D. M. Dykxhoorn, *Neurosci. Lett.*, 2012, **516**, 9–14.
104. Y. Shi, P. Kirwan, J. Smith, G. MacLean, S. H. Orkin and F. J. Livesey, *Sci. Transl. Med.*, 2012, **4**, 124–129.
105. A. Urbach, O. Bar-Nur, G. Q. Daley and N. Benvenisty, *Cell Stem Cell*, 2010, **6**, 407–411.
106. M. Amenduni, R. De Filippis, A. Y. L. Cheung, V. Disciglio, M. C. Epistolato, F. Ariani, F. Mari, M. A. Mencarelli, Y. Hayek, A. Renieri, J. Ellis and I. Meloni, *Eur. J. Hum. Genet.*, 2011, **19**, 1246–1255.
107. C. H. Chiang, Y. Su, Z. Wen, N. Yoritomo, C. A. Ross, R. L. Margolis, H. Song and G. L. Ming, *Mol. Psychiatry*, 2011, **16**, 358–360.
108. S. P. Paşca, T. Portmann, I. Voineagu, M. Yazawa, A. Shcheglovitov, A. M. Paşca, B. Cord, T. D. Palmer, S. Chikahisa, S. Nishino, J. A. Bernstein, J. Hallmayer, D. H. Geschwind and R. E. Dolmetsch, *Nat. Med.*, 2011, **17**, 1657–1662.
109. J. F. Kre, S. P. Paşca, A. Shcheglovitov, M. Yazawa, R. Schwemberger, R. Rasmusson and R. E. Dolmetsch, *Nat. Neurosci.*, 2013, **16**, 201–209.
110. J. Jang, H. C. Kang, H. S. Kim, J. Y. Kim, Y. J. Huh, D. S. Kim, J. E. Yoo, J. A. Lee, B. Lim, J. Lee, T. M. Yoon, I. H. Park, D. Y. Hwang, G. Q. Daley and D. W. Kim, *Ann. Neurol.*, 2011, **70**, 402–409.
111. M. A. Israel, S. H. Yuan, C. Bardy, S. M. Reyna, Y. Mu, C. Herrera, M. P. Hefferan, S. Van Gorp, K. L. Nazor, F. S. Boscolo, C. T. Carson, L. C. Laurent, M. Marsala, G. H. Gage, A. M. Remes, E. H. Koo and L. S. Goldstein, *Nature*, 2012, **482**, 216–220.
112. T. Yagi, D. Ito, Y. Okada, W. Akamatsu, Y. Nihei, T. Yoshizaki, S. Yamanaka, H. Okano and N. Suzuki, *Hum. Mol. Genet.*, 2011, **20**, 4530–4539.
113. N. Yahata, M. Asai, S. Kitaoka, K. Takahashi, I. Asaka, H. Hioki, T. Kaneko, K. Maruyama, T. C. Saido, T. Nakahata, T. Asada, S. Yamanaka, N. Iwata and H. Inoue, *PloS One*, 2011, 6, e25788.

114. L. Qiang, R. Fujita, T. Yamashita, S. Angulo, H. Rhinn, D. Rhee, C. Doege, L. Chau, L. Aubry, W. B. Vanti, H. Moreno and A. Abeliovich, *Cell,* 2011, **146**, 359–371.
115. J. T. Dimos, K. T. Rodolfa, K. K. Niakan, L. M. Weisenthal, H. Mitsumoto, W. Chung, G. F. Croft, G. Saphier, R. Leibel, R. Goland, H. Wichterle, C. E. Henderson and K. Eggan, *Science,* 2008, **321**, 1218–1221.
116. S. Nayler, M. Gatei, S. Kozlov, R. Gatti, J. C. Mar, C. A. Wells, M. Lavin and E. Wolvetang, *Stem Cells Transl. Med.,* 2012, **1**, 523–535.
117. I. H. Park, N. Arora, H. Huo, N. Maherali, T. Ahfeldt, A. Shimamura, M. W. Lensch, C. Cowan, K. Hochedlinger and G. Q. Daley, *Cell,* 2008, **134**, 877–886.
118. S. Camnasio, A. Delli Carri, A. Lombardo, I. Grad, C. Mariotti, A. Castucci, B. Rozell, P. Lo Riso, V. Castiglioni, C. Zuccato, C. Rochon, Y. Takashima, G. Diaferia, I. Biunno, C. Gellera, M. Jaconi, A. Smith, O. Hovatta, L. Naldini, S. Di Donato, A. Feki and E. Cattaneo, *Neurobiol. Dis.,* 2012, **46**, 41–51.
119. N. Zhang, M. C. An, D. Montoro and L. M. Ellerby, *PLoS Curr.,* 2010, **2**, RRN1193.
120. B. Song, G. Sun, D. Herszfeld, A. Sylvain, N. V. Campanale, C. E. Hirst, S. Caine, H. C. Parkington, M. A. Tonta, H. A. Coleman, M. Short, S. D. Ricardo, B. Reubinoff and C. C. Bernard, *Stem Cell Res.,* 2012, **8**, 259–273.
121. Y. Luo, Y. Fan, B. Zhou, Z. Xu, Y. Chen and X. Sun, *Tohoku J. Exp. Med.,* 2012, **226**, 151–159.
122. M. J. Devine, M. Ryten, P. Vodicka, A. J. Thomson, T. Burdon, H. Houlden, F. Cavaleri, M. Nagano, N. J. Drummond, J. W. Taanman, A. H. Schapira, K. Gwinn, J. Hardy, P. A. Lewis and T. Kunath, *Nat. Commun.,* 2011, **2**, 440.
123. P. Seibler, J. Graziotto, H. Jeong, F. Simunovic, C. Klein and D. Krainc, *J. Neurosci. Off. J. Soc. Neurosci.,* 2011, **31**, 5970–5976.
124. A. Sánchez-Danés, Y. Richaud-Patin, I. Carballo-Carbajal, S. Jiménez-Delgado, C. Caig, S. Mora, C. Di Guglielmo, M. Ezquerra, B. Patel and A. Giralt, *et EMBO Mol. Med.,* 2012, **4**, 380–395.
125. A. D. Ebert, J. Yu, F. F. Rose, V. B. Mattis, C. L. Lorson, J. A. Thomson and C. N. Svendsen, *Nature,* 2009, **457**, 277–280.
126. T. Chang, W. Zheng, W. Tsark, S. Bates, H. Huang, R. J. Lin and J. K. Yee, *Stem Cells Dayt, Ohio,* 2011, **29**, 2090–2093.
127. Y. Nihei, D. Ito, Y. Okada, W. Akamatsu, T. Yagi, T. Yoshizaki, H. Okano and N. Suzuki, *Biol. Chem.,* 2013, **288**, 8043–8052.
128. H. Fong, C. Wang, J. Knoferle, D. Walker, M. E. Balestra, L. M. Tong, L. Leung, K. L. Ring, W. W. Seeley, A. Karydas, M. A. Kshirsagar, A. L. Boxer, K. S. Kosik, B. L. Miller and Y. Huang, *Stem Cell Rep.,* 2013, **1**, 226–234.

129. T. Egashira, S. Yuasa, T. Suzuki, Y. Aizawa, H. Yamakawa, T. Matsuhashi, Y. Ohno, S. Tohyama, S. Okata, T. Seki, Y. Kuroda, K. Yae, H. Hashimoto, T. Tanaka, F. Hattori, T. Sato, S. Miyoshi, S. Takatsuki, M. Murata, J. Kurokawa, T. Furukawa, N. Makita, T. Aiba, W. Shimizu, M. Horie, K. Kamiya, I. Kodama, S. Ogawa and K. Fukuda, *Cardiovasc. Res.*, 2012, **95**, 419–429.
130. I. Itzhaki, L. Maizels, I. Huber, L. Zwi-Dantsis, O. Caspi, A. Winterstern, O. Feldman, A. Gepstein, G. Arbel, H. Hammerman, M. Boulos and L. Gepstein, *Nature*, 2011, **471**, 225–229.
131. E. Matsa, D. Rajamohan, E. Dick, L. Young, I. Mellor, A. Staniforth and C. Denning, *Eur. Heart J.*, 2011, **32**, 952–962.
132. A. L. Lahti, V. J. Kujala, H. Chapman, A. P. Koivisto, M. Pekkanen-Mattila, E. Kerkelä, J. Hyttinen, K. Kontula, H. Swan, B. R. Conklin, S. Yamanaka, O. Silvennoinen and K. Aalto-Setala, *Dis. Model. Mech.*, 2012, **5**, 220–230.
133. C. Terrenoire, K. Wang, K. W. C. Tung, W. K. Chung, R. H. Pass, J. T. Lu, J. C. Jean, A. Omari, K. J. Sampson, D. N. Kotton, G. Keller and R. S. Kass, *J. Gen. Physiol.*, 2013, **141**, 61–72.
134. X. Carvajal-Vergara, A. Sevilla, S. L. D'Souza, Y. S. Ang, C. Schaniel, D. F. Lee, L. Yang, A. D. Kaplan, E. D. Adler, R. Rozov, Y. Ge, N. Cohen, L. J. Edelmann, B. Chang, A. Waghray, J. Su, S. Pardo, K. D. Lichtenbelt, M. Tartaglia, B. D. Gelb and I. R. Lemischka, *Nature*, 2010, **465**, 808–812.
135. C. B. Jung, A. Moretti, M. Mederos y Schnitzler, L. Iop, U. Storch, M. Bellin, T. Dorn, S. Ruppenthal, S. Pfeiffer, A. Goedel, R. J. Dirscinger, M. Seyfarth, J. T. Lam, D. Sinnecker, T. Gudermann, P. Lipp and K. L. Laugwitz, *EMBO Mol. Med.*, 2012, **4**, 180–191.
136. A. Novak, L. Barad, N. Zeevi-Levin, R. Shick, R. Shtrichman, A. Lorber, J. Itskovitz-Eldor and O. Binah, *J. Cell. Mol. Med.*, 2012, **16**, 468–482.
137. K. Hu, J. Yu, K. Suknuntha, S. Tian, K. Montgomery, K. D. Choi, R. Stewart, J. A. Thomson and I. I. Slukvin, *Blood*, 2011, **117**, e109–e119.
138. S. Gandre-Babbe, P. Paluru, C. Aribeana, S. T. Chou, S. Bresolin, L. Lu, S. K. Sullivan, S. K. Tasian, J. Weng, H. Favre, J. K. Choi, D. L. French, M. L. Loh and M. J. Weiss, *Blood*, 2013, **121**, 4925–4929.
139. V. Sebastiano, M. L. Maeder, J. F. Angstman, B. Haddad, C. Khayter, D. T. Yeo, M. J. Goodwin, J. S. Hawkins, C. L. Ramirez, L. F. Z. Batista, S. E. Artandi, M. Wernig and J. K. Joung, *Stem Cells Dayt. Ohio*, 2011, **29**, 1717–1726.
140. J. Zou, P. Mali, X. Huang, S. N. Dowey and L. Cheng, *Blood*, 2011, **118**, 4599–4608.
141. A. Raya, I. Rodríguez-Pizà, G. Guenechea, R. Vassena, S. Navarro, M. J. Barrero, A. Consiglio, M. Castellà, P. Río, E. Sleep, F. Gonzalez, G. Tiscornia, E. Garreta, T. Aasen, A. Veiga, I. M. Verma, J. Surralles, J. Bueren and J. C. Izpisua Belmonte, *Nature*, 2009, **460**, 53–59.
142. L. Ye, J. C. Chang, C. Lin, X. Sun, J. Yu and Y. W. Kan, *Proc. Natl. Acad. Sci. U. S. A.*, 2009, **106**, 9826–9830.

143. R. Maehr, S. Chen, M. Snitow, T. Ludwig, L. Yagasaki, R. Goland, R. L. Leibel and D. A. Melton, *Proc. Natl. Acad. Sci. U. S. A.*, 2009, **106**, 15768–15773.
144. H. Hua, L. Shang, H. Martinez, M. Freeby, M. P. Gallagher, T. Ludwig, L. Deng, E. Greenberg, C. Leduc, W. K. Chung, R. Goland, R. L. Liebel and D. Egli, *J. Clin. Invest.*, 2013, **123**, 3146–3153.
145. Z. B. Jin, S. Okamoto, P. Xiang and M. Takahashi, *Stem Cells Transl. Med.*, 2012, **1**, 503–509.
146. S. T. Rashid, S. Corbineau, N. Hannan, S. J. Marciniak, E. Miranda, G. Alexander, I. Huang-Doran, J. Griffin, L. Ahrlund-Richter, J. Skepper, R. Semple, A. Weber, D. A. Lomas and L. Vallier, *J. Clin. Invest.*, 2010, **120**, 3127–3136.
147. R. Singh, W. Shen, D. Kuai, J. M. Martin, X. Guo, M. A. Smith, E. T. Perez, M. J. Phillips, J. M. Simonett, K. A. Wallace, A. D. Verhoeven, E. E. Capowski, X. Zhang, Y. Yin, P. J. Halbach, G. A. Fishman, L. S. Wright, B. R. Pattnaik and D. M Gamm, *Hum. Mol. Genet.*, 2013, **22**, 593–607.

CHAPTER 9

In Silico *Solutions for Predicting Efficacy and Toxicity*

GLENN J. MYATT* AND KEVIN P. CROSS

Leadscope, Inc., 1393 Dublin Road, Columbus, Ohio 43215, USA
*E-mail: gmyatt@leadscope.com

9.1 Introduction

Making informed decisions concerning what chemicals to pursue is the most critical task in the drug discovery process. These decisions are based on biological data related to efficacy and toxicity, ideally generated using human cells, used in combination with information on the composition of the chemicals tested. Through an analysis of these combined data, it is possible to understand the complex relationships between chemical structures and biological activity or toxicity data and use this understanding to guide the research process. In early discovery, where limited information is known about the types of chemicals that might make good drug candidates, screening data are analysed to identify the types of chemicals or lead series to focus on. Once a lead series has been identified, multiple iterations of focused screening and synthesis of new chemical libraries is performed, again to understand what chemicals to pursue as promising drug leads. As part of secondary efficacy and *in vitro* ADMET-related screening (Absorption, Distribution, Metabolism, Excretion, Toxicity), it is the knowledge of the relationship between this biological data and the chemicals that again drives the research decisions. Finally, as the most promising leads enter the preclinical phase, understanding what types of chemicals might be

RSC Drug Discovery Series No. 41
Human-based Systems for Translational Research
Edited by Robert Coleman
© The Royal Society of Chemistry 2015
Published by the Royal Society of Chemistry, www.rsc.org

associated with any undesirable effects helps prioritise the necessary safety assessment testing strategy prior to entering the clinic. Understanding and encoding the complex relationships between chemicals and their desirable as well as adverse biological effects of human relevance is fundamental to the entire drug discovery process. *In silico* approaches capture this knowledge from these historical successes and failures in databases and as computational models that are used to make decisions about future research directions. These decisions will increasingly be based on biological data generated from cell lines of human origin.

The drug discovery process is long and extremely expensive with estimates of between US$500 million and US$2 billion to discover a single new molecular entity.[1] One of the driving factors in the use of these *in silico* methods is the need to effectively prioritise the most promising research directions and avoid pursuing drug candidates that are later discovered to cause adverse effects. In addition, modern drug discovery technologies, such as high-throughput screening, generate data on such large numbers of chemicals that it is impossible to analyse them without the use of computational methods. Another factor is the desire to reduce, refine or replace the use of animal experiments (3 Rs) through the adoption of these computational approaches.[2] Finally, the use of *in silico* approaches is now required by international regulatory authorities in the assessment of genotoxic impurities[3] and is included in other regulatory guidance documents.[4–6]

This chapter provides a practical overview and describes a variety of *in silico* methods that provide support for research decisions on efficacy and toxicity. The chapter reviews the use of two-dimensional chemical structures and their associated biological data in computational approaches. This includes a discussion on how the data are represented for import into the computational tools. Searching databases of historical information helps to answer precise research questions and common approaches to querying this research information are outlined. To further analyse the relationships between the biological and chemical data requires the generation of molecular descriptors. These descriptors are subsequently used in advanced data mining methods, such as clustering or decision trees, to rapidly sift through the data. Encoding this knowledge into mathematical models enables the application of this historical experience to support both current and future research directions. Two case studies are presented, describing how these approaches can be used to support regulatory decisions on impurities and how these approaches can be used to predict human adverse events. Finally, this approach is summarised, and future *in silico* directions are discussed.

Two data sets are used in this chapter to illustrate analysis of efficacy ('PTP1B') and toxicity ('Hansen') data. The 'PTP1B' data set contains 118 chemicals with pIC_{50} data ($-\log IC_{50}$) representing the inhibition by benzofuran and benzothiophene biphenyl analogues of protein tyrosine phosphatase 1B (PTP1B), a target for insulin-resistant disease states.[7] The 'Hansen' data set contains non-proprietary data on *Salmonella* mutagenicity collected by Hansen et al.[8] from the scientific literature and further processed, leaving a total of 3700 compounds in the final test set.[9] A positive result is represented as '1' and a negative result as '0'. These data sets are used to illustrate the *in silico* approaches described in this chapter; however, these methods will be particularly valuable when applied to biological data generated from cell lines of human origin.

9.2 Representing Chemical Structures and Associated Data

9.2.1 Overview

The first step in any *in silico* analysis is to import data for both the chemical structures and the associated biological data into the computational tools. *In silico* analysis of chemicals requires information on the composition of the chemical. This electronic representation of the atoms, bonds and connectivity is usually referred to as a connection table. There are several common formats for standardising connection table information, but the two most widely used formats are the MOL file/SD files[10,11] and SMILES strings.[12,13] In addition, the chemical structures are usually named or assigned a unique identifier or registry number, which provides a convenient way of linking the chemicals to their associated data. Endpoint data can be represented as a data table (such as a spreadsheet); however, where the data are more complex (for example, where they represent a full toxicological study, including the experimental design along with raw and processed data), a more complex representation may be required. The following section addresses the representation of chemical structures, approaches to naming these chemicals as well as the representation of the associated biological data.

9.2.2 Chemical Representation

A MOL file is an electronic representation of a connection table for a chemical structure.[10,11] An example of a MOL file for the chemical aspirin is shown in Figure 9.1. The file format consists of five parts: (i) a header block, (ii) a count block, (iii) an atoms block, (iv) a bonds block, and (v) a properties block. The header block is contained on the first three lines of the file, which is dedicated to header information. The first line may contain the name or id of the chemical structure (*e.g.* LS-143); the second line is an optional line and often contains miscellaneous programme or usage data; the third line is again optional and is available for adding comments. The count block is shown in the fourth line and contains information concerning the number of atoms and bonds in the chemical structure, as well as some additional information. The line is formatted such that characters 1–3 are the number of atoms, characters 4–6 are the

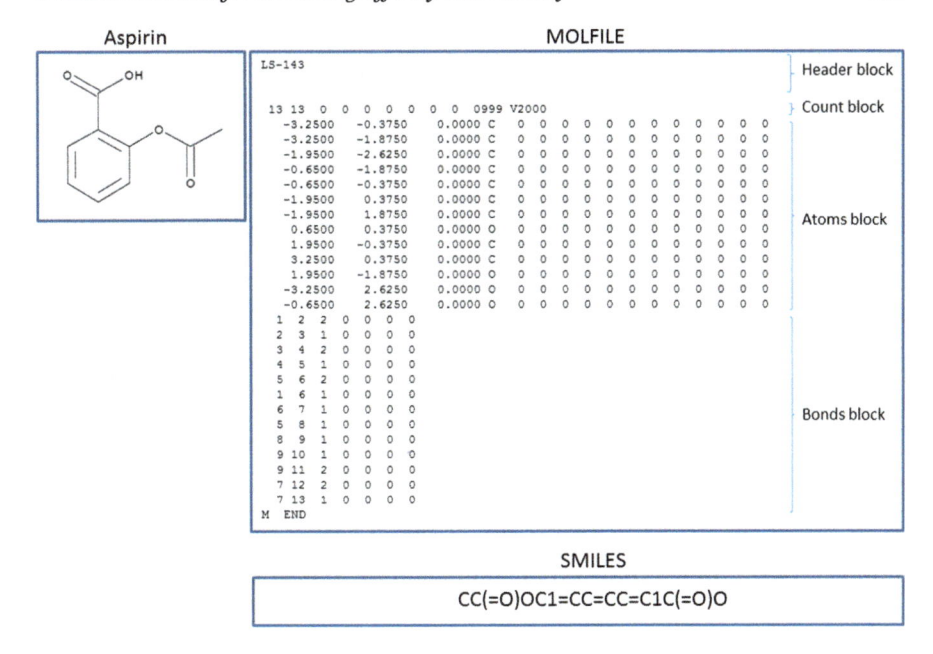

In Silico *Solutions for Predicting Efficacy and Toxicity* 197

Figure 9.1 MOLFILE and SMILES string for aspirin.

number of bonds, and characters 13–15 represent a flag as to whether the chemical is chiral or not (0 is not chiral and 1 is chiral). In the example in Figure 9.1 there are 13 atoms and 13 bonds in the chemical structure. Next is the atom block, where each line represents a single atom that is implicitly numbered from one to the number of atoms in the chemical. In the example in Figure 9.1, the first three numbers for atom number 1 are coordinates for displaying the chemical and the atom type is C or carbon. The bonds block again corresponds to the number of bonds identified in the count block, and each line describes a specific bond. In the example in Figure 9.1, the first bond is from atom number 1 to atom number 2 and is a double bond (2). Additional property information related to the atoms and bonds may be optionally found in the properties block (not shown in this example). This properties block includes information including charge, radical and isotope data. Each line contains the properties information and the properties block is terminated with the line 'M END'. An extension to the MOL file is an SD file, which represents multiple chemicals in a single file. It is simply a concatenation of a series of MOLFILEs, with a line containing the code '$$$$' separating each chemical. There are several extensions to the SD file format to account for large molecules and reactions, *etc.* While the standard SD file format allows for single property name/value pairs, an XML format is available for the representation of more complex relationships. For further information see http://download.accelrys.com/freeware/ctfile-formats/ctfile-formats.zip.[10]

SMILES is another format for representing chemical structures, and stands for Simplified Molecular Input Line Entry System.[12,13] Individual atoms in the connection table are represented by their atom symbol, such as C, Br, N, and so on. Additional information, such as charges, are represented by a square bracket around the atom symbol followed by the information. For example, [N+] represents a nitrogen with a single positive charge. Bonds are represented by a single character: '-' for single bonds, ' = ' for double bonds and '#' for triple bonds. For example, C-C represents a single bond between two carbons, and C#N represents a triple bond between a carbon and a nitrogen. It is not necessary to use single bonds explicitly, for example, C-C and CC both represent a single bond between two carbon atoms. Any time a path through a chemical structure splits, a branch is created. For example CC(=O)N represents the SMILES string for a four atom fragment with a branched carbonyl (=O). A cycle is formed where a continuous path can be traced through the chemical structure that starts and ends with the same atom. A cycle is represented in the SMILES format by designating an atom on which the cycle starts and ends, numbering the atom, and then closing the cycle using this designated atom number. For example, a six-membered ring with alternating single and double bonds can be represented by the SMILES string C1=CC=CC=C1. Stereochemistry, aromaticity and other atomic features can be represented in SMILES format. For further information see http://www.daylight.com/dayhtml/doc/theory/theory.smiles.html[12] and Weininger.[13] Figure 9.1 illustrates the SMILES representation for aspirin.

A unique identifier for a chemical structure provides a convenient link from the chemical to experimental data. It also provides a way of rapidly determining whether the data set has duplicate entries for the same compound by looking for repeated ids, which can be particularly important when combining data sets from different sources. In many data mining applications it is also important to represent both the tested form of the chemical as well as a form better suited for data analysis. For example, a chemical may be tested in a specific salt form, but the salt portion may be removed prior to its use in generating a computational model. Both the tested form and the model-ready structure should have a different id, because they are different chemical structures. However, the ids may be related syntactically (or *via* a named operator) to represent the relationship between the two. Organisations, such as pharmaceutical companies and vendors of chemical structure databases, rely on registration systems to ensure they have a systematic method of determining unique chemical structures, thus avoiding duplication in their databases. A registration system generates a unique identifier for each chemical structure that has not been previously seen. The systems generally operate by searching for the candidate chemical structure in the database of previously assigned chemicals. If the candidate is in the database, it is given the previously assigned id; otherwise a new id is assigned.

9.2.3 Biological Data Representation

Where there is a one-to-one correspondence between the chemical structure and the biological response, it is possible to represent such simple data as a table or in the data block of an SD file. Examples of simple end point data include IC50 (50% inhibition level) for biological activity or NOAEL (no observed adverse effect level) for toxicity data. Computational *in silico* analysis typically uses single-valued end point data for an entire compound for the representation of structure–activity signals. However, it may be necessary to include more complex data representations to represent the entire study design and results. This is especially important where the end point information is derived from one or more studies. ToxML[14,15] and SEND[16] are example formats that capture this type of information in a standardised manner.

9.2.4 Calculated Data

Properties of the whole molecule are often useful descriptors for chemicals. Some commonly calculated properties include molecular weight, hydrophobicity and polar surface area. Molecular weight is a calculation made by simply summing the individual isotopic masses for all atoms in the chemical structure. Hydrophobicity is related to the drug-likeness of a compound in terms of its ADME characteristics and its ability to bind to a target receptor. It is commonly estimated by the property logP or the partition coefficient between octanol and water. Generally, a score is assigned to a series of predefined atom types or molecular fragments, which are matched to the target molecule. The final logP estimate is generated by summing the individual scores for all atoms or fragments present. Polar surface area reflects the surface area over the polar atoms and is related to the drug's ability to permeate a cell. This property can also be estimated through the use of fragment scores. In addition, there are many properties of the whole molecule that can be calculated and potentially used as descriptors, such as topological indices, molar refractivity and Kappa shape indices (see Gasteiger and Engel,[17] Leach and Gillet[18] for more details).

9.3 Searching Chemical Databases

The ability to search the contents of a chemical structure database is critical to answer specific research questions. Did we test this compound before and what were the results? Do we have data on a similar compound with a specific effect and could we perform a read-across analysis[19] if the chemicals are sufficiently similar? For a single compound look-up where the name or id is known, a simple text search will return the chemical structure if it is in the database. However, the name may not be known and only the chemical structure may be available. In this situation, an exact structure search should be performed using the available structure as a query. For situations where classes of related chemical structures are required, search types such as a substructure search or a similarity search should be used.

In order to perform a structure search, a query describing the connection table of the chemical structure of interest should be defined. There are three primary ways of specifying a structure query. A structure query could be drawn in a chemical structure editor through a graphical user interface. The atoms and bonds are drawn graphically, as well as any necessary restrictions on the atoms or bonds. Alternatively, where a structure query has been committed to file, usually from a chemical structure editor, for example using the MOL file format, it can be uploaded to initiate a query. Finally, in the same way a text search is initiated, the connection table for the query can be typed in using a format such as a SMILES string.

Having specified a structure query, an exact structure search will compare the connection table of the query to each structure in the database to determine whether there is a match. Issues that should be addressed when searching an exact match include matching aromatic atoms and bonds, tautomers, stereochemical matching, and matching components within multi-component chemicals in the database (such as mixtures and salts).

A substructure search is another popular structure search where, as in the case of the exact match search, a structure query is initially specified. When a substructure search is then executed, a search of the database chemical structures is performed to identify those chemical structures containing the query chemical fragment. Again, like the exact match search, substructure searching needs to address issues concerning aromaticity, tautomers and stereochemistry to ensure appropriate recall. In addition to the connection table for the query, restrictions on the atoms and bonds can also be specified. Common restrictions include whether an atom or bond is part of a ring, the size of such a ring, whether particular atoms are open to further substitution, and so on. Atoms and bonds can also be defined as a member of a group, such as any one of the halogen atoms Cl, Br, I or F. Extensions to the MOL file and the SMILES (called SMARTS) formats accommodate these additional restrictions.

A third popular structure search is the similarity search. Like a substructure search, a similarity search returns multiple structures that are, for the most part, different from the chemical structure specified in the query. Unlike a substructure search, similarity searching may return structures with different atoms and bonds as well as chemical structures smaller than the query. A typical method for performing a similarity search is to use a fingerprint based on a set of predefined chemical fragments for both the query structure and the database chemical structures. A similarity score is generated for each database structure, based on a comparison of the fingerprints between the two structures (query and database).

In Figure 9.2 a query chemical structure is shown along with the results of an exact, substructures, and similarity search. Prior to performing any structure-based search the query and the database compounds go through

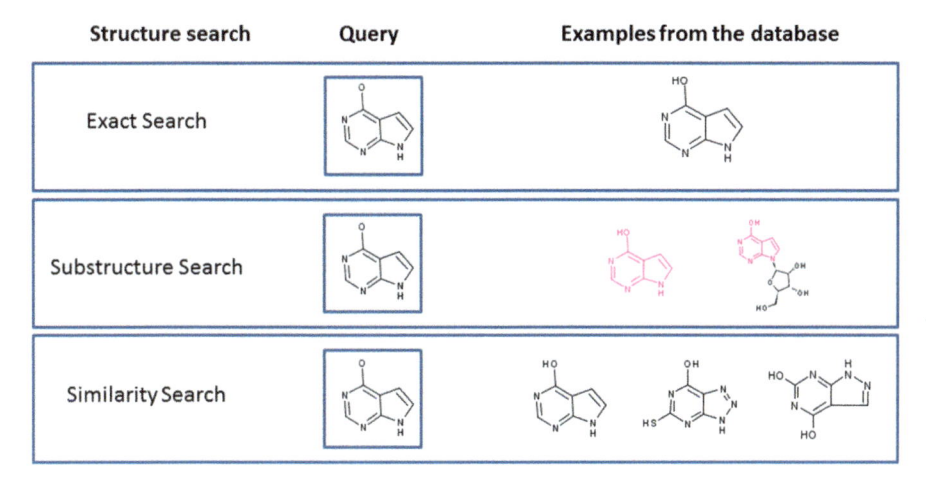

In Silico *Solutions for Predicting Efficacy and Toxicity* 201

Figure 9.2 Illustration of exact, substructure and similarity structure searching.

a normalisation process where atoms and bonds are annotated with information on whether they are aromatic or not, as well as bonds that could represent different tautomers. In addition, approaches to unambiguously represent the stereochemistry are performed. These automated processes ensure the structure searches return the correct results, taking into account the different alternative ways the same chemical could be described.

Most structure searchable databases also contain information on efficacy or toxicity. It is often possible to search these databases for chemicals with specific types and values of data, irrespective of the types of chemicals. Figure 9.3 illustrates an example search where chemicals are retrieved that have LD_{50} values of between 350 and 450 mg/kg for acute toxicity studies.

Chemical structure and data searches are often combined using Boolean logic to specify precise research questions. For example, it is possible to identify chemicals with positive Ames data in a specific strain with a particular substructure fragment.

9.4 Calculating Molecular Fragment Descriptors

9.4.1 Overview of Fragment Types

The basic premise of *in silico* methods is to understand the relationship between properties or attributes of a series of chemicals and associated biological data. Because many research decisions relate to understanding substructures of a molecule, generating fragment-based descriptors is an essential first step. The generation of fragments may be based on a dictionary of standard medicinal chemistry building blocks or on substructures

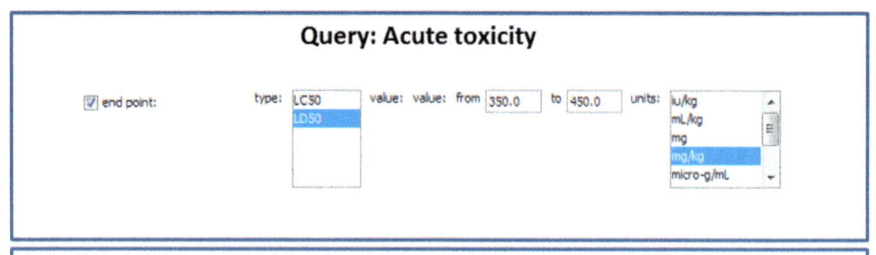

Figure 9.3 Searching for chemicals with specific toxicity data.

identified from the literature that have a known and documented relationship to specific effects. Alternatively, these fragments may be generated algorithmically from a set of chemicals, such as identifying large chemical substructures or scaffolds common to the chemicals of interest. These fragments may be generated to prioritise the generation of substructures that differentiate high or low biological response.

9.4.2 Predefined Features

This approach uses a dictionary of molecular fragments (substructure queries) constructed before any analysis. When a dictionary fragment matches a portion of the chemical structure a '1' is recorded for that feature, otherwise a '0' is entered. In Figure 9.4, eight predefined structural features are shown along the top of the table as an example: benzene, aldehyde, amine(NH2), carbonyl, carboxylic acid, halide, nitro and furan. For structure (a) aldehyde, carbonyl and furan fragments match shown as a '1', whereas benzene, amine(NH2), carboxylic acid, halide and nitro do not match. The presence or absence of the structural features for chemical structures (b) and (c) is also shown. There are a number of fragment collections that are used, including the MDL® MACCS public keys (166 fragments)[20] and the Leadscope® fingerprints (over 27 000 molecular fragments).[21]

In Silico *Solutions for Predicting Efficacy and Toxicity* 203

	benzene	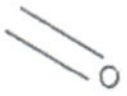 aldehyde	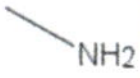 amine (NH2)	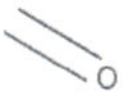 carbonyl	 carboxylic acid	 halide	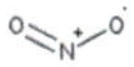 nitro	furan
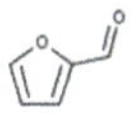 (a)	0	1	0	1	0	0	0	1
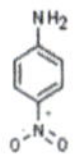 (b)	1	0	1	0	0	0	1	0
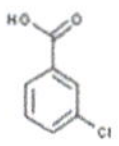 (c)	1	0	0	1	1	1	0	0

Figure 9.4 Use of predefined substructural fragments.

9.4.3 Structural Alerts

A structural alert is a specific type of predefined feature, where each alert is associated with an adverse biological outcome or an event potentially leading to this outcome. For example, a mutagenicity structural alert is based on a molecular substructure defining a DNA-reactive centre.[22–27] To illustrate, an aziridine substructure (three-membered ring containing two carbons and one nitrogen) has been cited in multiple publications as a structural alert for mutagenicity ('aromatic and aliphatic asiridinyl';[22] 'oxiranes and aziridines';[23] 'SA_7: epoxides and aziridines';[24,25]). For any mutagenicity structural alert, there should be a relationship between this reactive centre identified within the molecule and its ability to either directly or indirectly (through one or more metabolic steps) interact with DNA. For example, in Benigni *et al.*[25] aziridines have been described as "… extremely reactive alkylating agents that may react by ring-opening reactions … activity of these compounds depends on their ability to act as DNA cross-linking agents, *via* nucleophilic ring-opening of the aziridine moiety by N7 positions of purines."

In addition to the primary alert, it is also important to define any factors that would deactivate the alerts as a result of electronic or steric effects, solubility into an ion channel or cell, general ADME properties reflecting intercellular activity or by blocking an important metabolic step. For example, for the mutagenicity structural alert 'primary aromatic amine', an acidic group in the para position "…prevents proton abstraction from NH2 group of anilines … by the ferric peroxo intermediate of CYP1A2."[27]

Scaffolds generated to characterize different chemical classes

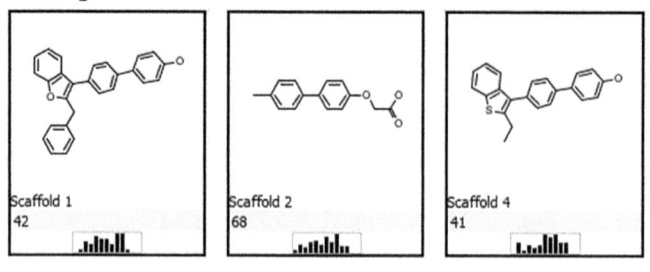

Scaffold 1
42

Scaffold 2
68

Scaffold 4
41

Scaffolds generated to differentiate the structural classes associated with high or low activity

Scaffold 2
17

Scaffold 3
15

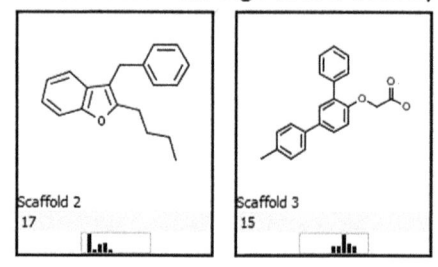

Figure 9.5 Commonly occurring and 'predictive' scaffolds generated from the PTB1B data set.

9.4.4 Common Chemical Scaffolds

Structural features can also be generated dynamically, for example paths of various lengths between pairs of atoms could be generated. Alternatively, substructures can be generated by combining building block fragments to generate large substructures representative of chemicals being analysed or scaffolds. For example, three large scaffolds generated for the PTB1B data set are shown in Figure 9.5 in the first row of scaffolds. Scaffold 1 is a large substructure contained in 42 chemicals from the data set, and these matching compounds have a frequency distribution of the biological activity data (pIC50) similar to the entire set (shown below the scaffold). These scaffolds could be used to start analysing the substitution pattern using an R-group analysis.

9.4.5 Scaffolds Associated with Data

Scaffolds can also be generated that characterise chemicals with either unusually high or low levels of biological activity or toxicity. For example, in the macrostructure assembly approach, a set of structural features are combined in different ways to form new fragments.[21] These new fragments are then matched against the underlying set to determine whether the resulting new fragment maps onto chemical structures with unusually high or low levels of biological activity, and the most differentiating fragments are retained. Figure 9.5 illustrates two generated scaffolds that were guided by the pIC$_{50}$ data in the PTB1B data set (shown on the second row of scaffolds). Scaffold 2 is contained in 17 chemicals, and these chemicals generally exhibit low pIC50 values (as shown in the frequency distribution below the scaffold), whereas with scaffold 3 (matching 15 chemicals), the biological activity data are generally higher (again illustrated in the frequency distribution shown below the scaffold).

9.5 Understanding Structure–Activity or Toxicity Relationships

9.5.1 Overview

The previous section described a number of approaches to generating molecular descriptors. This section describes ways in which these descriptors are used to help make sense of the complex relationships between the structural features of a molecule and the biological efficacy or toxicity.

9.5.2 Classification Based on Substructures

Once a series of chemical structural features has been identified as important, either through an analysis of the data (as in the case of the macrostructure assembly approach) or through a compilation of substructures from the literature, it is possible to use the information to classify a new set of chemicals. A new chemical may be assigned as potent or toxic where one or more of the identified substructures is present.

9.5.3 Clustering

Clustering is a popular approach to organising data and can be used to group chemical structures into sets of similar compounds. In order to perform this type of grouping it is important to measure the chemical similarity between pairs of chemicals. The most common approach is to use a set of sub-structural fragments as chemical descriptors and generate a similarity score based on a comparison of the presence and absence of the same set of descriptors between the two chemicals. Three counts are calculated: count$_{11}$ (the number of sub-structural fragments present in both chemicals); count$_{10}$ (the number of sub-structural fragments present in the first chemical but not in the second); and count01 (the number of sub-structural fragments present in the second chemical but not in the first). The similarity score most often calculated is the Tanimoto similarity coefficient (S_{Tanimoto}) shown in eqn (9.1):

$$S_{\text{Tanimoto}} = \frac{\text{count}_{11}}{\text{count}_{10} + \text{count}_{01} + \text{count}_{11}} \qquad (9.1)$$

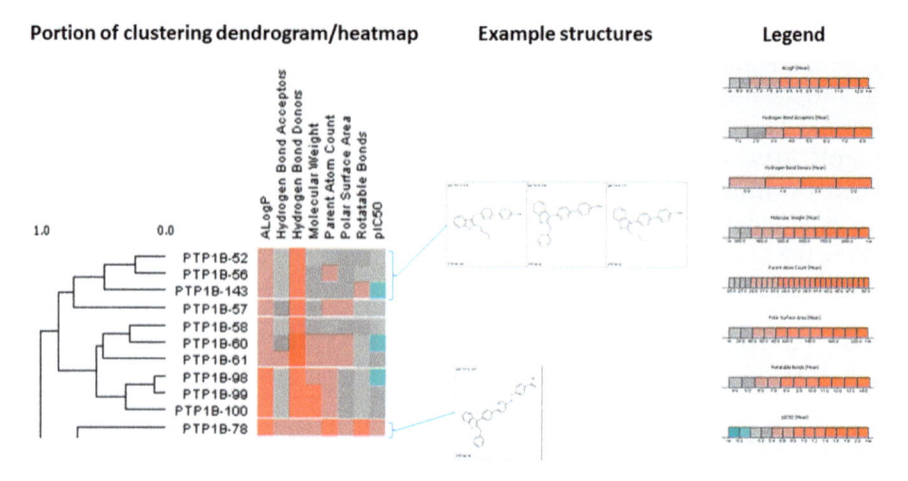

Figure 9.6 Structure-based clustering of the PTB1B data set with heat map and chemical subsets identified.

There are multiple approaches to clustering a set of chemicals. In the following example, the PTB1B data set will be clustered using the agglomerative hierarchical clustering approach.[28] Figure 9.6 will be used to illustrate how the clustering results coupled with a coloured heat map were constructed. Each row represents a single chemical in the data set (*e.g.* PTP1B-52, PTP1B-56, …). The ordering of the chemicals in the diagram is based on their structural similarity to their neighbours, as structures that are similar are adjacent to other similar structures. The complete hierarchical organisation is summarised using the clustering dendrogram shown on the left. The calculated similarity value (using the Tanimoto score) between pairs of compounds is shown as a vertical line between the two chemicals. The horizontal location of this line represents the degree of similarity between the two chemicals, with vertical lines to the right representing a higher degree of similarity than vertical lines to the left. These vertical lines also represent the similarity values between individual chemicals and other groups of chemicals, as well as between groups of chemicals. In addition to the clustering dendrogram, a coloured heat map is shown on the right representing both calculated physicochemical properties as well as the measured pIC_{50} data.

A portion of the clustering dendrogram and heat map is presented in Figure 9.6 for the chemicals in the PTB1B data set. Because chemicals that are adjacent to each other represent similar chemical structures, it is possible to identify groups of structurally related chemicals that represent interesting subsets (identified using the coloured heat map) as shown in Figure 9.6. Here a number of structurally related groups are identified that represent high and low pIC_{50} values.

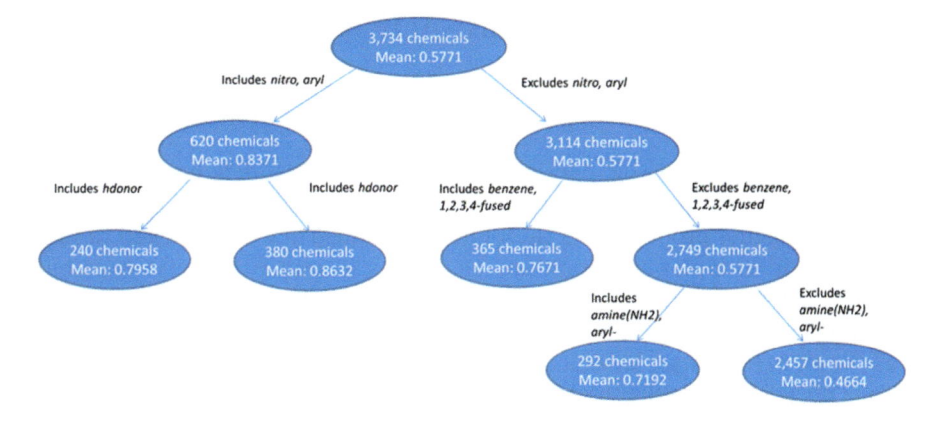

In Silico *Solutions for Predicting Efficacy and Toxicity* 207

Figure 9.7 Decision tree generated to analyse the Hansen data set.

9.5.4 Decision Trees

A decision tree is another analysis tools to help sift through large amounts of chemical structures and associated biological data. It again uses calculated chemical features to guide the analysis.[29] The analysis starts by looking at the whole set of chemicals being analysed where it inspects each individual feature in turn to identify which has the strongest positive or negative association with the biological data. Having identified this feature, the entire data set is split into two sets, one containing the feature, and the other the remaining chemicals. These two sets are then examined in a similar manner. All features are investigated, and again the feature with the strongest positive or negative association is identified. Again, this feature is used to divide the set into two – those compounds containing the feature and those that do not. This process creates a tree representing the analysis of the data set according to combinations of chemical features associated with the biological response. Figure 9.7 represents a decision tree generated from the Hansen data set. The initial set of 3734 chemicals is analysed, and the nitro, aryl-feature is most strongly associated with positive *Salmonella* mutagenicity data. The data set is initially segmented into a set of 620 chemicals containing a nitro, aryl- and 3114 chemicals containing no nitro, aryl- group. The 620 chemicals are examined further, and the most significant feature is the hdonor group and this set is now segmented into a group of 240 chemicals containing the hdonor group and 380 with no hdonor group. The set of 240 chemicals contains both a nitro-aryl- group and an hdonor group. Other significant features identified include a fused aromatic system (benzene, 1,2,3,4-fused) and an amine(NH2), aryl- group (both shown on the right-hand side of the tree).

9.6 Quantitative Structure–Activity Relationship (QSAR) Modelling

9.6.1 Overview

Quantitative structure–activity relationship (QSAR) models encode mathematically the relationships between important molecular substructures and whole-molecule properties to the biological data. They can be used to help understand complex relationships as well as to make predictions of biological activity or toxicity, before any experimental data have been generated. QSAR models have been built from a set of congeneric structures such as the PTB1B data set as well as diverse structures, such as the Hansen set. The OECD has specified a series of validation principles that QSAR models should adhere to: "1) a defined endpoint, 2) an unambiguous algorithm, 3) a defined domain of applicability, 4) appropriate measures of goodness-of-fit, robustness and predictivity, 5) a mechanistic interpretation, if possible."[30]

9.6.2 Building Models

The process of building a QSAR model often follows the following four steps: (i) create and characterise the training set, (ii) generate chemical features, (iii) prioritise features, and (iv) build and optimise the model. Creating the training set is often the most time-consuming part of the QSAR model building process, and care should be taken to generate a high-quality training set because the predictivity of the model will be dependent on both the quality and chemical diversity of these data. Once the training set is developed, the QSAR model building process continues by generating molecular descriptors such as whole-molecule properties or sub-structural fragments, as previously described. At this point many chemical descriptors across the set of chemicals may have been generated. This number is often reduced using a combination of automated approaches to prioritise descriptors with a strong positive or negative association with the biological data to be modelled (*i.e.* biological response data). Domain expertise may also be employed to select descriptors known to reflect particular response data. Once a suitable smaller set of chemical descriptors are selected, a mathematical model is built and optimised that encodes the potentially complex relationships between the chemical descriptors and the biological response data.

To illustrate this process we will use an example of building a QSAR model for *Salmonella* mutagenicity from publicly available data. The training set should contain a list of chemical structures and a numeric end point, such as 0 (for negative results) and 1 (for positive results). Each source of data (such as the National Toxicity Programme or a regulatory submission document) will contain information on the chemical being tested as well as the experimental design and test results. It will be necessary to thoroughly examine the

In Silico *Solutions for Predicting Efficacy and Toxicity* 209

data to understand which results are suitable to include. There may be conflicting data on the same compound from different sources. The chemicals may have been tested in different salt forms. For *Salmonella* different studies may have tested different strains of the bacteria. An overall *Salmonella* call for each compound in the training set will be used as a biological response for modelling.

Once a training set has been organised, it is helpful to characterise both the distribution of the biological response data as well as the chemical diversity of the compounds in the training set. A balanced number of positive and negative results can be helpful in distinguishing positive and negative signals when modelling. Also, if the biological response data are continuous, then a normal frequency distribution is desirable for modelling, and mathematic transformation of data are often performed to ensure this distribution (such as –log for IC_{50} data).

The next step in the process is to generate molecular descriptors from the chemical structures for both whole-molecule properties and sub-structural fragments (as discussed earlier in the chapter). To transform these molecular attributes for mathematical modelling, a data table is constructed from this information. To illustrate, Figure 9.8 shows how the presence and absence of sub-structural features can be represented as 0 (no feature present) and 1

$$y = g(f(x_i))$$

Statistical model

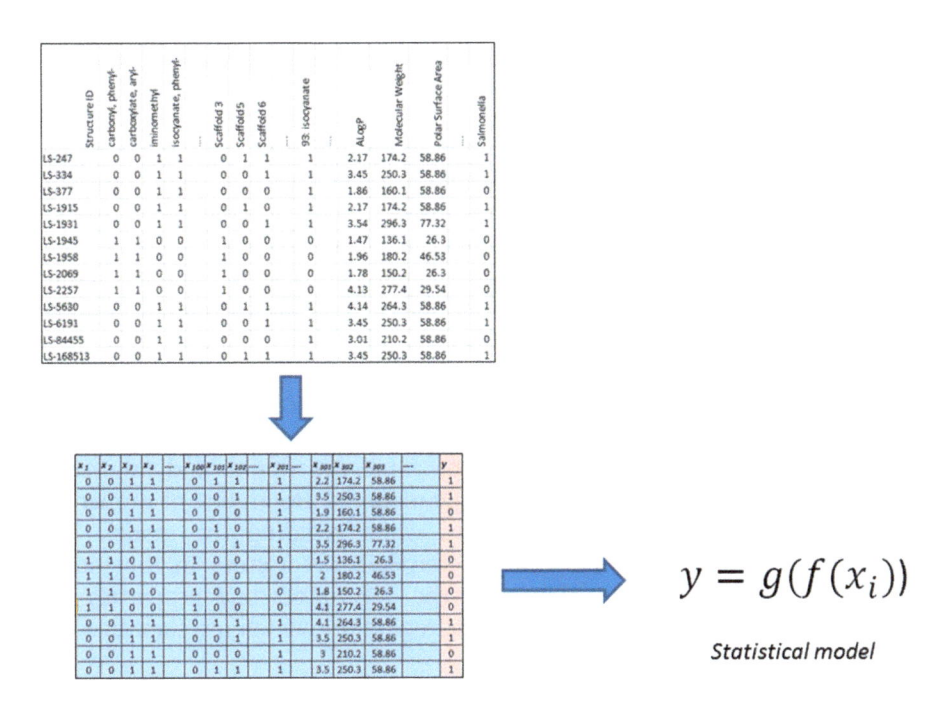

Figure 9.8 Build a QSAR model from the chemical descriptor data (x_i) to predict the biological response data (y).

(feature present). Calculated properties of the whole molecule are also included in the table. This table is now annotated with information on the biological response data, which can be used to generate a preliminary filter of the descriptors. Similar features (based on the correlation of structures that match or do not match) as well as low-frequency features can be removed. High-frequency features that are not discriminating can also be removed. Calculating associations between the descriptors and the biological response data can also help to prioritise descriptors for inclusion in the model. Finally, prior knowledge of the problem can also be used to help in identifying important descriptors to include in the model.

There are many approaches to generating mathematical models from a data table. Generally, non-linear modelling approaches that are capable of encoding the complex relationships in the data are best suited to this approach. Examples of modelling approaches include partial logistic regression (PLR),[7] partial least squares regression (PLS),[31,32] decision forests[33] and neural networks.[34] All modelling approaches describe the relationships mathematically between the chemical descriptors (x-variables) and the biological response data (y-variables). This is illustrated in Figure 9.8, where an equation is generated that is a function of a series of weighted x-variables used to predict the y-variable. The process of generating the statistical model often includes optimisation of the model parameters (such as number of factors to use in a PLS model) as well as further optimisation of the chemical features used in the model.

9.6.3 Applying Models

Once a QSAR model is built, it is possible to use it to make predictions for new chemicals. The model will calculate an estimate of the biological response specified when building the model, such as efficacy or toxicity using the calculated descriptors or properties. However, prior to making any prediction for a compound, an important first step is to determine whether it is appropriate for the test compound, *i.e.* it must be within the applicability domain of the QSAR model. This generally means that the training set used to build the model contains similar chemicals, and the descriptors present in the model should correspond to descriptors present in the test chemical. When the test chemical is within the applicability domain of the model, the next step is to generate the same descriptors and properties present in the model. Again a table of the matched descriptors or properties is calculated in a similar manner as when building the model. The descriptors calculated correspond to weighted descriptors in the model equation (x-variables). The values for the descriptors and properties are assigned to the x-variables in the model equation, which are then used to calculate estimated values.

QSAR models generate estimates for the biological response variable. If a QSAR model was built to predict pIC_{50} (for example, where the PTB1B data were used as a training set), then the QSAR model would predict a value for pIC$_{50}$. Alternatively, a QSAR model could be built to predict a categorical value (for example, if the Hansen set was used as a training set). These classification models may directly predict the positive or negative outcome. Alternatively, the model may predict a probability of a positive outcome, and this probability is subsequently converted into the desired category. As an example, probabilities above 0.6 may be assigned as positive, probabilities below 0.4 assigned as negative and probabilities between 0.4 and 0.6 assigned to an indeterminate category.

9.6.4 Model Validation

To understand the predictivity of the models generated, it is important to validate the models. This is ideally performed using a different data set than the one used to build the model. This external test set will present a less biased assessment of the performance of the model. When external validation data sets are not available, cross-validation is often used. Here the training set is divided into a series of paired (training and test) subsets. All chemicals in the original training set will be assigned to a single test set, and the training set (paired to the test set) will contain the remaining chemicals. For example, a 10% cross-validation will include 10 mutually exclusive test sets, each with a corresponding training set containing all the other remaining chemicals. Ten models will be independently built from the 10 training sets, and each one assessed using its corresponding test set. The final assessment is generated by concatenating the results from all the test sets. This approach limits the possible bias in assessing the performance, because each test compound is tested using a model built from a different set of compounds. This approach can be used to identify and prevent overtraining of a QSAR model.

The overall performance assessment for classification models is based upon the percentage of correctly predicted values. The number of true positives (TP) and true negatives (TN) are calculated, as well as the number of false positives (FP) and false negatives (FN). The overall concordance is (TN + TP)/(TN + TN + FP + FN). It is also important to understand the sensitivity (TP/(TP + FN)) and specificity (TN/(TN + FP)) of the model, as well as the negative predictive values (TP/(TP + FP)) and the positive predictive value (TN/(TN + FN)). The overall performance of a regression model is often based on how closely the predicted values in the test set match the actual experimental values in the test set (R-square is a typical calculation for this).

9.6.5 Explaining the Results

Inspecting the individual equation(s) to understand how the chemical descriptors were combined mathematically to arrive at a final prediction is one approach to explaining the results; however, these equations can be complex, making it challenging to understand how the prediction was performed. An effective approach to explaining the result is to examine the

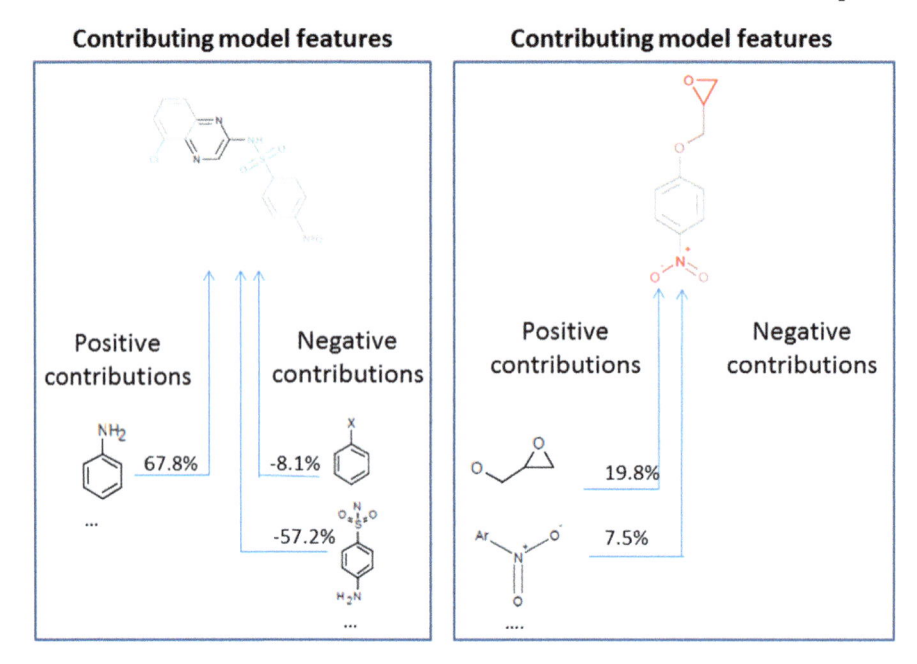

Figure 9.9 Explaining the results of a QSAR model.

contribution of each chemical feature to the overall prediction. Different chemical features may match different parts of the test compound, and some chemical features contribute positively whereas others contribute negatively to the overall prediction. The contribution (derived from the cumulative feature contributions) of each atom and bond can be calculated and colour coded. For example, darker shades of red indicate increasing levels of positive contribution, whereas lighter shades of green indicate increasing negative contribution. Shades of grey are used to represent little contribution to the overall prediction. This visualisation is illustrated in Figure 9.9. The example on the left would be predicted negative. The prediction is based on a number of chemical features (some shown in this example), which contribute positively and negatively to the overall prediction. Because some of the atoms and bonds in the positive and negative contributing features overlap in the test compound, the overall effect is little highlighting. The chemical on the left would be predicted positive, and the two features highlighted on the chemical in red show the atoms and bonds that contribute the most to the positive prediction.

9.7 Regulatory *in Silico* Case Study

The following case study presents a practical application of the *in silico* methods outlined in this chapter. Impurities are generated as part of the pharmaceutical manufacturing process. They result from using different reagents, catalysts and solvents in the synthesis of the drug substance and are also attributable to subsequent degradation. The International Committee on Harmonisation (ICH) has recently issued a draft guidance (currently out for public comment, termed Step 2) that covers the qualification of mutagenic impurities.[3] The purpose of the guidance is to ensure that any impurities present in the drug substance pose a low-level risk of causing cancer in humans. As such, the guideline's focus is on identifying DNA-reactive substances, usually detected by running the bacterial reverse mutation assay (Ames).[35] In the absence of carcinogenicity or bacterial mutagenicity data for actual or potential impurities, an *in silico* structure–activity analysis can be performed to help understand whether a substance can be classified as having no mutagenic concern. Where this negative classification is not possible, further *in vitro* or *in vivo* tests are required to support a negative classification or the impurity should be controlled below the limits established in the guideline.

To perform the computational structure–activity analysis, the guideline states that two complementary *in silico* methodologies should be used in the assessment. One should be expert rule-based and the second should utilise a statistical-based methodology. The methods used in the assessment should be compliant with the OECD validation principles,[30] and should be reviewed using expert analysis, especially in cases where conflicts result. If the results from the two methodologies indicate no predicted mutagenic potential, the impurity can be classified as having no mutagenic concern.

In this case study, eight impurities along with the main ingredient are analysed according to the *in silico* requirements to support ICH M7.[36] The first step is to import the chemical structures, which is commonly performed by generating a series of MOL files or a single SD file. The next step is to understand whether experimental test data exist for the compound. If no data exist, the two methodologies are used to predict the results of a bacterial mutagenesis test, one expert rule-based and one statistical-based. Figure 9.10 summarises the results from this analysis for three compounds. The input chemicals are shown in the fourth, seventh and eleventh columns, with the first column summarising the computational assessment based on all available information. This takes into account whether any data exist; however, in the absence of data, predictions are used. In the centre, two statistical models are presented that predict the results of a bacterial mutagenesis assay. These models cover both *E. coli*/*Salmonella* TA102 (A:T mutations) as well as *Salmonella* (G:C mutations). The second *in silico* methodology, the expert rule-based, is shown to the right. For the first example, API, it was possible to find historical experimental data, and so the overall M7 consensus is negative. For all other chemicals, there were no available data, and so the computational assessment is based solely on the *in silico* models. Two impurities are predicted positive – impurities C and D. Impurity C is predicted positive in all three models (as shown in Figure 9.10). The structural feature responsible for the negative prediction is highlighted in both the statistical models and the expert alerts, which consistently

214 Chapter 9

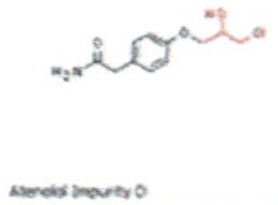

Figure 9.10 *In silico* assessment of the API and impurities C and D.

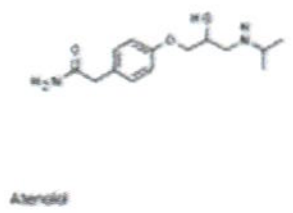

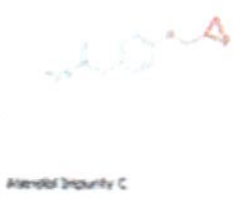

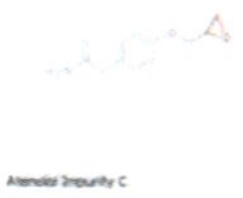

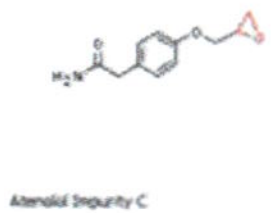

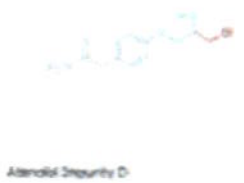

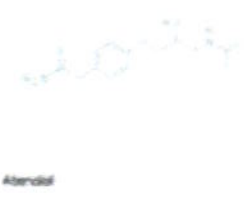

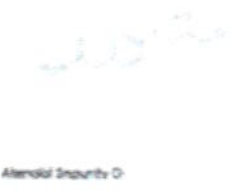

highlights the epoxide substructure. Because all three predictions were positive, the overall M7 consensus is assigned as positive.

It is essential that the results of an *in silico* assessment are accompanied by an expert opinion of both positive and negative calls. So it is important to look into why the results for this impurity were positive. In Figure 9.10 all atoms and bonds for impurity C are being evaluated because all are colour coded. The epoxide and its environment are highlighted as being responsible for the positive predictions (red colour). Looking in more detail at the QSAR results, *E. coli/Salmonella* TA102 prediction used 22 structural descriptors and seven calculated properties in the model to arrive at the high probability value of 0.974. Fourteen structural descriptors and seven calculated whole-molecule properties were used in the *Salmonella* model to arrive at the high probability of 0.985.

The expert alerts generated a positive prediction with a high precision value based on historical data containing this alert of 0.9053. The single alert '159: unhindered epoxide' was identified in impurity C and highlighted in the summary shown in Figure 9.10. Again, it is important to generate an expert opinion for this compound which could include the source of the alert as well as historical bacterial mutagenicity data for compounds that also matched this alert.

9.8 Prediction of Human Adverse Events Case Study

The *in silico* approaches outlined in this chapter can be applied to the prediction of human-based adverse events. The only difference in the biological response used to build the prediction model is whether a particular compound showed an adverse event in a clinical trial, or the event was reported some time after the drug was administered to a patient (outside of the clinical trial). For example, a biological response variable, *Torsades*, could be used to indicate whether a patient upon administration of a drug suffered the cardiac arrhythmia *Torsades de Pointe*. A value of 1 could be used to indicate the adverse event and 0 for compounds with no evidence of this event. A training set of chemicals could be used to build a QSAR model and used in priority setting.

There are a number of sources for these adverse event data, including the FDA's Adverse Events Reporting System (AERS).[37] These databases collect information from medical professionals, manufacturers and patients on millions of adverse events reported after the administration of marketed drugs. The major task in building such models is the processing of this raw information in order to generate a high-quality training set that can be used to build a QSAR model. This includes mapping the reported events to a controlled vocabulary, and processing the data to help understand where there is enough evidence to assign a specific compound as positive or negative for the event. This is especially challenging as patients may have taken a combination of drugs, or the drug may have been taken primarily by a patient population more susceptible to certain adverse events. In addition,

Structure	Prediction Call	Positive Prediction Probability
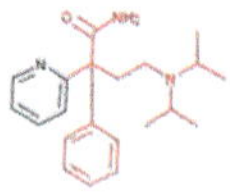 disopyramide	Positive	0.966
 dofetilide	Positive	0.995
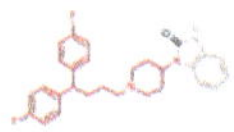 pimozide	Positive	0.973

Figure 9.11 Prediction of three compounds using the Torsade human adverse event QSAR model.

it is important to understand the number of incidents compared with background levels, which varies over time on the market. Only those events where significantly more incidents than you would expect to see should be identified. Finally, patient exposure is another factor to take into account when analysing the adverse event data. Once the training sets have been developed, QSAR models can be built to make predictions for these adverse events. A series of QSAR models have been built for adverse events in humans[38] and Figure 9.11 illustrates the application of one of these models (Torsades) to three compounds known to cause this adverse event.

9.9 Conclusions

This chapter has reviewed a number of practical considerations when using *in silico* methods for predicting efficacy and toxicity. This included an explanation of how chemical and biological data are represented in the most common *in silico* applications, as well as how *in silico* methodologies operate. Two practical case studies applying these approaches were also presented. These *in silico* approaches are currently widely used throughout the discovery process and with the ICH M7 guideline, as part of regulatory submissions. Although the examples used to illustrate the *in silico* methodologies outlined in this chapter are based on existing data, these approaches will be critical when analysing biological data generated from cell lines of human origin.

It is important to take a multidisciplinary approach to adopting *in silico* approaches – ideally including computational and medicinal chemists as well as the biologists or toxicologists in building and deploying these solutions, and using these approaches as part of an integrated testing strategy. These (Q)SAR models can be used to predict both biological activity and toxicity as well as to identify the chemical structural basis for any prediction. Highlighting important structural features is helpful in explaining the predictions as well as supporting the design of new molecules to optimise biological activity and avoid adverse effects. Additional biological data being generated including omics data will play an increasingly critical role as *in silico* methodologies begin to model more complex biological processes and understand mechanisms of action.

References

1. C. P. Adams and V. V. Brantner, *Health Affair,* 2006, **25**, 420.
2. F. Guhad, *J. Am. Assoc. Lab. Anim. Sci.,* 2005, **44**, 58.
3. *ICH M7 Mutagenic Impurities in Pharmaceuticals* http://www.ich.org/fileadmin/Public_Web_Site/ICH_Products/Guidelines/Multidisciplinary/M7/M7_Step_2.pdf accessed April 10th 2014.
4. *FDA Draft Genotoxic and Carcinogenic Impurities* http://www.fda.gov/downloads/Drugs/GuidanceComplianceRegulatoryInformation/Guidances/ucm079235.pdf accessed April 10th 2014.
5. *FDA ANDAs: Impurities in Drug Products* http://www.fda.gov/downloads/Drugs/GuidanceComplianceRegulatoryInformation/Guidances/UCM072861.pdf accessed April 10th 2014.
6. *FDA ANDAs: Impurities in Drug Substances* http://www.fda.gov/downloads/Drugs/GuidanceComplianceRegulatoryInformation/Guidances/UCM172002.pdf accessed April 10th 2014.
7. C. Yang, K. Cross, G. Myatt, P. Blower and J. Rathman, *J. Med. Chem.,* 2004, **47**, 5984.
8. K. Hansen, S. Mika, T. Schroeter, A. Sutter, A. Laak, S. T. Hartmann, N. Heinrich and K. P. Muller, *J. Chem. Inf. Model.,* 2009, **49**, 2077.
9. L. Stavitskaya, B. L. Minnier, R. D. Benz, N. L. Kruhlak, FDA Centre for Drug Evaluation and Research, SOT 2013 'Development of Improved Salmonella Mutagenicity QSAR Models Using Structural Fingerprints of Known Toxicophores'.
10. http://download.accelrys.com/freeware/ctfile-formats/ctfile-formats.zip accessed April 10th 2014.
11. A. Dalby, J. G. Nourse, W. G. Hounshell, A. K. I. Gushurst, D. L. Grier, B. A. Leland and J. Laufer, *J. Chem. Inf. Comput. Sci.,* 1992, **32**, 244.
12. http://www.daylight.com/dayhtml/doc/theory/theory.smiles.html accessed April 10th 2014.
13. D. Weininger, *J. Chem. Inf. Comput. Sci.,* 1988, **28**, 31.
14. http://toxml.org/ accessed April 10th 2014.

15. C. Yang, R. D. Benz and M. A. Cheeseman, *Curr. Opin. Drug Discovery Dev.*, 2006, **9**, 124.
16. http://www.cdisc.org/SEND accessed April 10th 2014.
17. J. Gasteiger and T. Engel. 2003. *Chemoinformatics: A Textbook*, Wiley.
18. A. R. Leach and V. J. Gillet., 2007. *An Introduction to Chemoinformatics*, Springer.
19. S. R. Vink, J. Mikkers, T. Bouwman, H. Marquart and E. D. Kroese, *Toxicol. Pharmacol.*, 2010, **58**, 64.
20. J. L. Durant, B. A. Leland, D. R. Henry and J. G. Nourse, *J. Chem. Inf. Comput. Sci.*, 2002, **42**, 1273.
21. P. E. Blower, K. P. Cross, M. A. Fligner, G. J. Myatt, J. S. Verducci and C. C. Yang, *Curr. Drug Discov. Technol.*, 2004, **42**, 93.
22. J. Ashby and R. W. Tennant, *Mutat. Res.*, 1988, **204**, 17.
23. A. B. Bailey, N. Chanderbhan, N. Collazo-Braier, M. A. Cheeseman and M. L. Twaroski, *Regul. Pharmacol. Toxicol.*, 2005, **42**, 225.
24. R. Benigni, C. Bossa, N. Jeliazkova, A. Worth. *The Benigni/Bossa Rulebase for Mutagenicity and Carcinogenicity – A Module of Toxtree* - EUR 23241 EN 2008.
25. R. Benigni and C. Bossa, *Chem Rev.*, 2011, **111**, 2507.
26. S. J. Enoch and M. T. D. Cronin, *Crit. Rev. Toxicol.*, 2010, **40**, 728.
27. I. Shamovsky, L. Ripa, L. Börjesson, C. Mee, B. Nordén, P. Hansen, C. Hasselgren, M. O'Donovan and P. Sjö, *J. Am. Chem. Soc.*, 2011, **133**, 16168.
28. J. M. Barnard and G. M. Downs, *J. Chem. Inf. Comput. Sci.*, 1992, **32**, 644.
29. D. M. Hawkins, S. Young and A. Rusinko, *Quant. Strct. Act. Relat.*, 1997, **16**, 296.
30. OECD (2007). *Guidance document on the validation of (Quantitative) structure activity relationships [(Q)SAR] models* (http://ihcp.jrc.ec.europa.eu/our_labs/predictive_toxicology/background/oecd-principles accessed April 10th 2014.
31. D. V. Nguyen and D. M. D. M. Rocke, *Bioinformatics*, 2002, **18**, 39.
32. P. Geladi and B. Kowalski, *Anal. Chim. Acta*, 1986, **185**, 1.
33. W. Tong, H. Hong, H. Fang, Q. Xie and R. Perkins, *J. Chem. Inf. Comput. Sci.*, 2003, **43**, 525.
34. L. Fausett. (1993). *Fundamentals of Neural Networks: Architecture, Algorithms, and Applications*. Pearson.
35. *OECD guideline for testing of chemicals: Bacterial Reverse Mutation Test 21st July 1997.* (http://www.oecd-ilibrary.org/environment/test-no-471-bacterial-reverse-mutation-test_9789264071247-en accessed April 10th 2014.
36. J. Siji., *Quantification of genotoxic 'Impurity D' in Atenolol by LC/ESI/MS/MS with Agilent 1200 Series RRLC and 6410B Triple Quadrupole LC/MS*, Agilent Application Note.
37. C. J. Ursem, E. J. Matthews, N. L. Kruhlak, J. F. Contrera and R. D. Benz, *Regul. Toxicol. Pharmacol.*, 2009, **54**, 1.
38. A. A. Frid and E. J. Matthews, Prediction of drug-related cardiac adverse effects in humans–B: use of QSAR programmes for early detection of drug-induced cardiac toxicities, *Regul. Toxicol. Pharmacol.*, 2010, **56**, 276.

CHAPTER 10

In Silico *Organ Modelling in Predicting Efficacy and Safety of New Medicines*

BLANCA RODRIGUEZ

Department of Computer Science, University of Oxford, Parks Road, Oxford OX1 3QD, UK
E-mail: blanca.rodriguez@cs.ox.ac.uk

10.1 Introduction

Over the last 50 years, a number of areas of physiology have experienced great progress in the development and integration of *in silico* models and simulation tools of different organs. Multiscale human organ computational models are available, for different organs including heart,[1,2] lungs[3-6] and cancer,[7,8] with fine-grained biophysical detail and a great level of integration from the genetic to the whole body level. Multiscale modelling and simulation have been crucial in improving our understanding of mechanisms of physiological and pathological phenomena, therapy and diagnosis. The advanced *in silico* technology is currently opening up promising and powerful avenues for the investigation and improvement of the efficiency of therapies and for the development of new medicines.

Perhaps the most advanced area of computational physiology is computational cardiology, concerned with the investigation of the physiology and pathology of the heart using computational modelling and simulating.

Computational heart models are multiscale both spatially and temporally, and integrate information across sub-cellular, cellular, tissue and organ levels, with spatial scales ranging from nanometres to metres and temporal scales from picoseconds to years (if processes such as ageing or disease remodelling are taken into account). Computational heart models have been demonstrated to be useful and also predictive, providing a strong basis for their credibility in a variety of settings, including the prediction of drug action.[9-16] This has led to an increase in interest by industry and regulators, as shown by the recent announcement by the Food and Drug Administration of their intentions to replace the Thorough QT study by a combined *in vitro/in silico* assay.[17] In the next sections, we describe the state-of-the-art in computational cardiac electrophysiology, as an example of an advanced area of *in silico* biology. We then describe how the technology is being developed and used to advance our understanding of the causes and modulators of variability in the response to medicines, through the application of computational population-based approaches. We finish the chapter with a discussion of the processes required for the validation and constant challenge of *in silico* methodologies in biology, as an iterative process that is required to build their credibility for applications such as the development of new medicines.

10.2 State-of-the-Art Computational Cardiac Electrophysiology

In this section, we describe the extensive computational technology available to investigate and predict the electrophysiology of the heart at the single cardiomyocyte and the whole organ level. Over 40 mathematical models of electrical excitation in single cardiac cells are now available. They include cellular models for a variety of species and cell types including ventricular, atrial and Purkinje cardiomyocytes from human, rabbit and dog, amongst others (see the CellML repository: www.cellml.org). Three-dimensional anatomical models of the cardiac ventricles and atria have also been constructed.[18-34] Simulations using sophisticated software and numerical techniques have been conducted to investigate the effect of drugs, disease or mutations at the ionic level on the electrical activity of the human heart.[13,24,27,30,34]

10.2.1 Single Cardiomyocyte Models

Cardiac cells are electrically excitable due to the presence in their membrane of proteins, which act as mechanisms of ionic transport across the cellular membrane, such as ion channels, pumps and exchangers. The transmembrane transport of ions such as potassium, sodium and calcium results in electrical and concentration gradients between the intracellular and the extracellular media.

In Silico *Organ Modelling in Predicting Efficacy and Safety of New Medicines* 221

In cardiac cells, the transmembrane potential at rest is negative due to the difference in ionic concentrations between the intracellular and the extracellular spaces. Following a sufficiently large electrical stimulus, the transmembrane potential experiences the action potential, *i.e.* a non-linear response which is a direct consequence of the voltage-dependent properties of the membrane ion channels, exchangers and pumps (Figure 10.1, middle panel). Firstly, the fast sodium ion channels become activated allowing sodium to enter the cell, and resulting in an increase or depolarisation of the transmembrane potential towards positive potentials. At depolarised potentials, calcium and potassium channels become activated, resulting in the plateau phase of the action potential, during which calcium enters the cell triggering the mechanical contraction of the cardiomyocytes through the excitation–contraction system. Then potassium channels open, allowing outward currents due to the higher potassium concentration in the intracellular than the extracellular space. Potassium currents repolarise the transmembrane potential towards its negative resting state. The shape of the

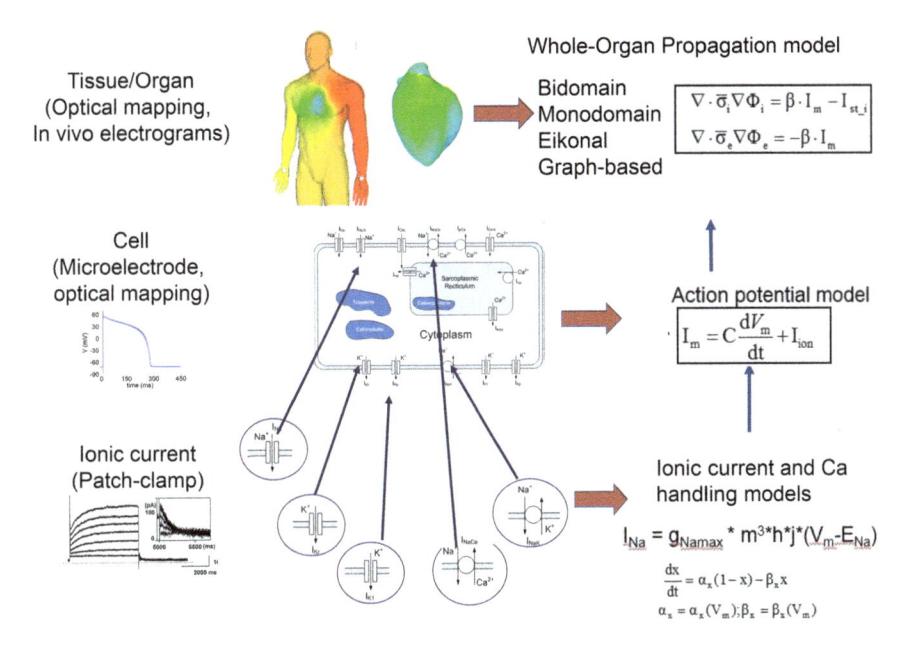

Figure 10.1 Construction of computational cardiac electrophysiology models from ionic current to whole organ level. For the last 50 years, ionic current models have been constructed mostly based on patch clamp data. The integration of ionic current models results in cellular level models for simulation of the action potential and concentration dynamics. Cardiac simulation at the whole organ level requires the construction of image-based anatomical models and a mathematical description of the processes underlying propagation of electrical excitation through cardiac tissue, which can take the form of bidomain, monodomain, eikonal or graph-based models.

action potential varies for different cell types in the heart, due to differences in the numbers and kinetics of the different voltage- and time-dependent ion channels in the heart.

Computational cardiac electrophysiology has progressed over the last 50 years, since in 1960 Denis Noble published the first mathematical model of the electrical activity of a cardiac cell.[35] Noble's doctoral work was based on the studies by Hodgkin and Huxley, who developed the first model of the electrical activity of a neuron (the axon of a giant squid)[36] and resulting in them being awarded the Noble Prize for Physiology in 1963. The initial models included only a single potassium and sodium current, based on their knowledge of electrophysiology at the time. Currently, most single-cell cardiac models include ionic current models based on the Hodgkin–Huxley formulation, and therefore they represent ionic currents as the product of maximum conductance, gates and driving force resulting from the concentration and electrical gradients acting on the ion across the membrane. The driving force is calculated as the difference between the transmembrane potential and Nernst or reversal potential for the ion. The Nernst potential is a function of the intracellular and extracellular concentrations for a specific ion, and it is equal to the transmembrane potential, resulting in no flux across the membrane due to the equilibration of the diffusion and electrical field acting on the specific ion.

The complexity of the cellular models has, however, increased owing to the integration of the new electrophysiology knowledge generated over the years on the types and modes of operation of cardiac ion channels, exchangers and pumps.[37–48] As illustrated in Figures 10.1 and 10.2, current single-cell cardiac electrophysiology models include detailed representation of the most important ionic transport mechanisms such as sodium and calcium and potassium ion channels, the sodium/calcium exchanger and the sodium/potassium pump, and also the biophysical processes underlying calcium dynamics across intracellular compartments of the sarcoplasmic reticulum. In some cases signalling pathways such as those underlying beta-adrenoceptor stimulation and the CaMKII cascade are also incorporated.[46,49]

Whereas the mathematical framework described above is generic for cardiac tissue, the models are tuned to be specific, for example to a particular animal species or spatial location within the heart (for example, dog *versus* rabbit, atria *versus* ventricles). Electrophysiological recordings obtained from laboratory experiments are used to estimate the parameter value for ionic current conductances and kinetics for specific cell types.[38–47,50–52] The parameter values in the models can sometimes be obtained directly from patch clamp recordings or indirectly through the use of pharmacological action to estimate the magnitude of specific properties such as ionic current conductances. Equations and parameters are determined by minimisation of the difference between simulated traces and mean experimental values. Therefore, models and parameters are linked to the specific conditions and experimental models in which the recordings were performed (including animal species and location of origin of the preparation, temperature,

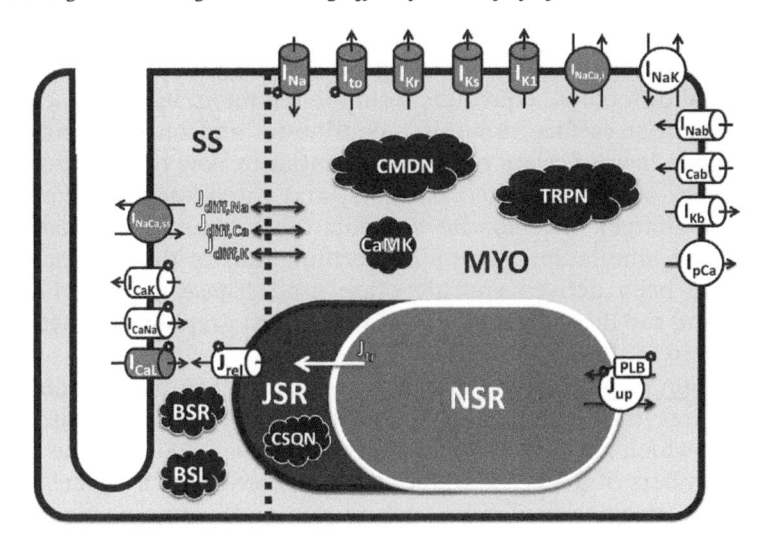

In Silico *Organ Modelling in Predicting Efficacy and Safety of New Medicines* 223

Figure 10.2 Structure of the human action potential model by O'Hara–Virag–Varro–Rudy. The schematic represents ion channels, exchangers and pumps acting as mechanisms of ionic transport across the surface and sarcoplasmic reticulum membranes to produce the action potential and calcium transients (with permission from Kudva *et al.*[46]).

solutions, recording technique, *etc.*). Novel experimental findings lead to model updates and also to new models with novel and improved features, but also aspects inherited from previous models.[35,45,53–55]

In view of the great complexity of measuring and modelling ionic currents, an alternative approach is also adopted to model cardiac action potential dynamics, consisting of simplified, minimal or phenomenological models.[56–59] The equations in this case aim at capturing cellular-level properties, such as morphology and rate dependence, without including a detailed description of specific ionic currents. The advantage is that the models have a reduced number of equations and parameters to be identified, matching the limited information obtained for specific cells and experimental recordings. The phenomenological cellular models can capture the effect of drugs on cellular properties such as the action potential shape and duration, but they do not allow representation of the effect of drug action on specific ionic currents.

10.2.2 Whole Organ Heart Models

Simulations of cardiac electrophysiology at the tissue and whole organ level require mathematical modelling of the processes underlying intercellular electrical coupling and propagation of electrical excitation from cell to cell through cardiac tissue. The cardiac bidomain model is currently the gold standard and arises by representing the cardiac tissue as a continuum with two domains: the intracellular and the extracellular spaces, electrically connected through the membrane of cardiomyocytes. Two partial differential equations form the bidomain model, based on the *assumption* that cardiac tissue is a continuum and on two fundamental laws: Ohm's law (to relate electrical potential to flow of transmembrane, intracellular and extracellular currents) and Kirchhoff's law (for conservation of charge). Due to the computational expense associated with whole-organ simulations, simplified formulations of the bidomain equations have been derived and are often used. These include the monodomain and the Eikonal equations as well as fast graph-based methods, as described in Wallman *et al.*[60]

Simulation of propagation of electrical excitation through cardiac tissue also requires the definition of spatial tissue characteristics such as its geometry, which are defined by a mesh. Anatomical models of the upper or lower chambers of the heart (whole-atria or whole-ventricle models) are obtained from imaging modalities such as MRI and histology using segmentation techniques to obtain a binary image that defines the boundaries of the cardiac tissue (Figure 10.3).[1,20,21,31,33,61,62] The processed image is then used to generate a volumetric mesh by applying a discretisation process in space, which depends on the numerical method used to solve the propagation model equations, which often is the finite element method. Thus the underlying tissue model is a discrete one composed of cells (the cardiac tissue), which is turned into a continuous model (the bidomain model), and finally turned back into a discrete anatomico-mathematical model to conduct the computer simulations.[2] The anatomically-based volumetric mesh comprises the elements or points in the cardiac domain over which the solution to the tissue model equation (*e.g.* the bidomain model) is calculated. In order to resolve the spatial and temporal scales of the electrophysiological dynamics, human ventricular meshes can contain millions of elements, making simulations computationally expensive.[13,24–30]

Simulations of whole organ level physiological activity are conducted using software, developed through the efficient implementation of the computational physiology models and techniques often run in high-performance computing platforms. In the case of the heart, this allows the simulation of the multiscale effects of drug action on the heart from the ionic level to the electrocardiogram (ECG). An example is illustrated in Figure 10.3. The simulation of the ECG requires an anatomical model of the human torso with biophysically detailed and anatomically-based descriptions of the human cardiac ventricles, as described in various reviews.[13,32,63] The simulations are computationally expensive as the computational mesh contains 14 336 528 tetrahedral elements and single-cell membrane kinetics are represented heterogeneously by the Ten Tusscher and Panfilov model[64] for epicardial, endocardial and mid-myocardial behaviour based on transmural location. The human cardiac ventricular model is coupled to the torso using the diffusion equation to account for the electrical propagation from heart to

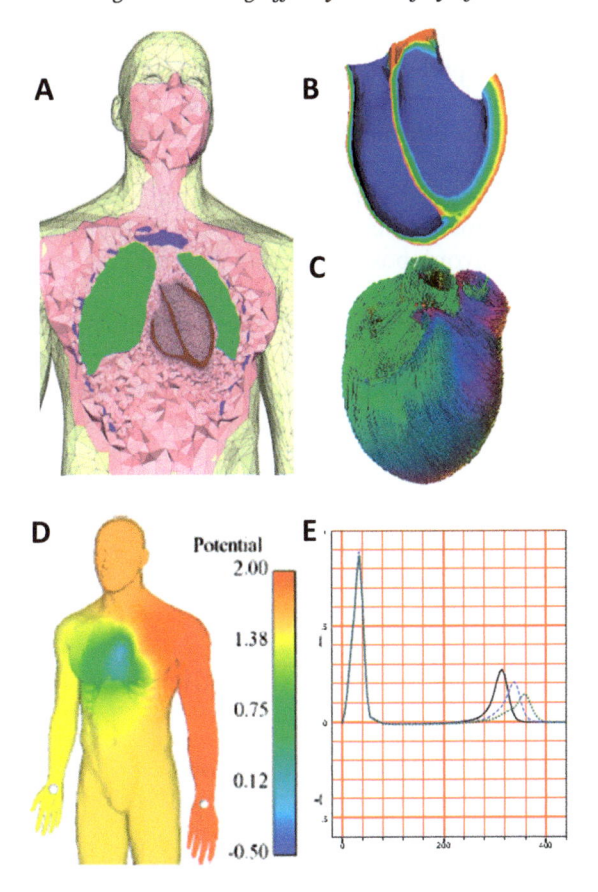

In Silico *Organ Modelling in Predicting Efficacy and Safety of New Medicines*   225

Figure 10.3 Computer simulation of the effect of drug action on the heart from ionic currents to body surface potentials using a human electrophysiological model of the heart and the body torso. Anatomical model of the human torso (A) with human ventricular model (B) including realistic fibre orientation (C). Simulated body surface potentials (D) and electrocardiogram (E) under control and following potassium channel block. Modified with permission from Soldner *et al.*[13]

torso. The ECG can then be measured on the body surface, as shown in Figure 10.3. The multiscale models described here provide the computational framework required to simulate the effects of medicines at the ionic level, as described in the next section, and how they propagate to cellular, tissue and whole organ activity.

Whole-organ simulations of the electrophysiological activity of the heart require finding accurate approximations of the bidomain equations using numerical techniques, such as the finite element method. This is because an exact solution cannot be obtained. The numerical method required to conduct the simulations determines how anatomical meshes are discretised into smaller spatial steps. The mesh spatial resolution is determined to achieve convergence of numerical algorithms and has no relationship to the size of the cells. Simulation complexity requires a trade-off within numerical routines between efficiency, robustness and accuracy, and subtle interplays between spatial and temporal resolutions. Verification of the numerical routines is key to ensuring the numerical stability of the simulations, and to avoid possible errors or artefacts that would affect simulation results.[65–68]

At the whole organ and tissue level, the intracellular and extracellular conductivity tensors include two important pieces of information: fibre architecture and conductivity (Figure 10.3C). Fibre architecture, *i.e.* the local direction of the conductivity tensor at all 'points' of the model, is usually incorporated into tissue models by extracting information from histological or DT-MRI images, using image processing algorithms or using a mathematical rule that relates fibre rotation at a particular location with distance to the surfaces of the ventricular wall.[27,33,69–71] Conductivity values along and across fibres are indirectly determined to yield conduction velocities, as measured experimentally in tissue.[72]

10.2.3 Simulating Interactions of Medicines with Ionic Channels

In order to simulate the action of medicines on the heart, the equations in the ionic models need to be altered to include terms representing the modulation of ion channels by drugs. Drug action is usually modelled as a function of concentration and mode of action of each drug. The complexity of the mathematical models of ionic current/medicine interaction can vary from a simple pore block model to more complicated Markovian models including kinetics states.[9,73]

Drug/ionic current models are based on receptor theory in pharmacology, based on the application of receptor models developed in physical chemistry to explain the mechanisms of drug molecules binding with cellular receptors (for review, Brennan *et al.*[73]). Clark was the first to apply the receptor model to describe the effects of drugs on ion channels based on a framework previously used for enzyme kinetics.[74]

Receptor theory provided the basis for simulating the effects of drugs on ion channels. It was however not until the late 1970s that Neher and Sakmann developed the patch clamp experimental technique to record single ion channel currents through a glass micropipette clamped to the cell membrane.[75] The innovation enabled experimentalists to investigate and characterise the interaction between drug compounds and single ion channels. A number of early experimental studies of drug action on cardiac ion channels by Hille[76] paved the way for a mathematical representation of the interaction drug molecule/ion channel receptor, based on classical receptor theory and laws of physical chemistry.

Models of the interaction of molecules and ion channels can be loosely categorised as:

1. Simple pore block: the flow of ions is restricted by the drug binding with a continuously accessible ion channel receptor.
2. State and voltage-dependent binding theories: the modulated and guarded receptor theories.
3. Allosteric effectors: drug binds with an allosteric receptor on the channel protein, affecting the channel kinetics.

The mechanisms of drug/ion channel binding are usually a combination of all three processes. Therefore, the identification of the dominant effect in any given interaction is key to constructing their mathematical models.

The simplest application of receptor theory to ion channel block is the simple pore block. The molecule is assumed to have continuous access to the receptor site and the affinity of the drug for the receptor is assumed to be time and voltage independent. There are two basic ways in which drug molecules binding with ion channel receptors can inhibit the flow of ions through the channel pore. The first one is if the receptor site is located in or near the channel pore. Then, the bound drug molecule physically restricts the flow of ions through the channel pore. Alternatively, drug binding with an allosteric receptor can cause a conformational change which leads to restriction of the ion flow.

In the simple pore block model, the degree of channel block is a function of drug concentration, the half maximal inhibitory concentration (IC_{50}) and the Hill coefficient for the targeted ionic channel. Thus, for a given drug dose $[D]$, IC_{50} value and Hill coefficient n with respect to the channel j the formulation of drug action on the ionic current conductance g_j is given by:

$$g_j([D]) = g_j \left(\frac{1}{1 + \left(\frac{[D]}{IC_{50}}\right)^n} \right)$$

If the law of mass action applies, binding of a ligand to one binding site should not affect the affinity of another binding site. However, in many instances binding experiments show cooperation between binding sites. In other words, a ligand binding or dissociating from a binding site alters the affinity of other binding sites. The Hill equation is often used in cooperative binding assays to assess the degree of cooperation between more than one ligand binding to the same receptor and was originally formulated to describe oxygen binding to haemoglobin by Hill in 1910.[77] The Hill equation is also used to describe the relationship between the binding ligand concentration and the fractional occupancy of binding sites and is used in cardiac pharmacology to describe the affinity between a particular drug molecule and its binding sites on ion channels. A Hill coefficient of

1 indicates independent ligand binding where the binding affinity is not affected by whether or not other ligand molecules are already bound. A Hill coefficient different than 1 means that a ligand binding to a receptor would affect the affinity of the unblocked receptors for remaining ligands.

The Hill coefficient n can be determined empirically from the slope of the straight line estimate of the curve. The IC_{50} value and the Hill coefficient are used in physical chemistry to define the affinity of an enzyme for a particular binding site. These values can similarly be used to represent the binding dynamics between drug molecules and ion channel binding sites for a simple pore block. However, the drug action on ion channels is complicated by the voltage and time dependencies of the ion channel proteins.

As described in previous sections, cardiac ion channels are dynamic protein structures, which activate and deactivate in response to changes in cellular transmembrane potential and concentrations. The voltage dependency of cardiac ion channels is fundamental to cardiac activity as it regulates both excitation and contraction of cardiomyocytes. Drug action on cardiac ion channels is often also voltage and time dependent. However, because ion channel state transitions are voltage dependent, it is difficult or even impossible to distinguish between state and voltage dependence of drug/ion channel interactions. Two main hypotheses have been proposed to describe state-dependent block, named the modulated and guarded receptor theories, as reviewed by Brennan and colleagues.[73]

The modulated receptor theory was proposed by Hille in 1977, based on his work on the effects of local anaesthetics on sodium channels in nerve cells.[76] The premise of the modulated receptor theory is that the affinity of the drug for the ion channel binding site is dependent on the state of the ion channel. This implies that the binding affinity is voltage and time dependent. Figure 10.4 illustrates the basic kinetic scheme of the modulated receptor theory proposed by Hille. In this case, as an example, the model of the ion channel includes three states, namely open (O), closed (C) and inactivated (I). The modulated receptor theory proposes that the drug can bind in all three states but the affinity of the drug for the ion channel binding sites is voltage and time dependent and is different for each state. Drug binding with each state causes the channel to enter modified bound states, all three non-conducting (indicated by asterisks). From the bound states, the channel can either transition to an unbound state or to one of the other bound states until it recovers from binding.

In contrast, the guarded receptor theory proposes that the drug molecule binds with a single bindable conformation with fixed affinity and that access to the binding site is restricted due to the conformational changes of the protein. Thus, the molecule action on the ion channel is dependent on the conformation of the channel, which is voltage and time dependent. The guarded receptor theory was pioneered by the work by Starmer and colleagues in the mid-1980s, who investigated the effects of local anaesthetics on cardiac sodium channels.[78] The main difference between the guarded *versus* modulated receptor theory, as depicted in Figure 10.4, is that

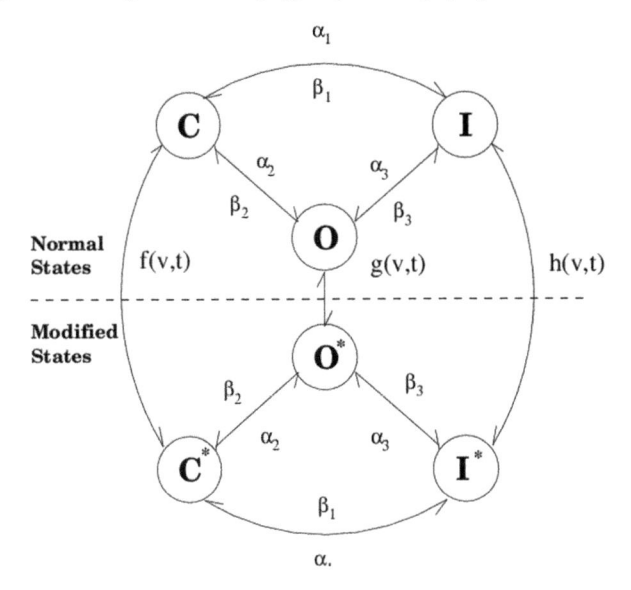

In Silico Organ Modelling in Predicting Efficacy and Safety of New Medicines

Figure 10.4 Receptor modelling theory for drug/ion channel interactions. The schematic represents the generic ion channel as having three conformational states: closed (C), open (O) and inactivated (I) in drug unbounded and bounded states (without and with an asterisk, respectively). The state transition rates between channel states are αn and βn, and the voltage and time-dependent affinity of the drug molecule for the channel states is represented by the functions f(v, t), g(v, t) and h(v, t). (Modified from Bhushan et al.[73] with permission.)

the drug molecule binds to the particular channel state conformation, and recovery from binding is often also state dependent. In the example of open-state block, this could be caused by the drug molecule being trapped by the closed or inactivated conformations of the channel and thus the channel is only able to recover from block when it returns to the activated state.

Both the modulated and the guarded receptor theories focus on voltage and time dependent association and dissociation processes, due to conformation of the channel protein, and do not consider the drug effects on ion channel conformation state transition dynamics. However, many drugs also have allosteric effects, causing changes in the dynamics between protein conformation states. For instance, lidocaine binding to sodium channels in rat skeletal muscle was shown to produce a functional change in the channel dynamics by slowing down its inactivation dynamics.[79] A consequence of this is that the sodium channel is preferentially in the late scattered mode when the drug is bound compared to the transient (fast) mode of the unbound sodium channel. Thus, the drug compound does not simply block the ion channel, but alters its dynamics by binding to its allosteric receptor. This implies that no explicit drug bound states need to be represented in the model, as the binding of the drug molecule is taken into account in the model as changes in the transition rates between conformation states. However, the binding process of the drug compound to the allosteric receptor can also abide by the modulated or guarded receptor theories. Therefore, a more comprehensive model of drug–ion channel interaction including allosteric regulation consists of both unbound and bound states, with a conducting bound open state and transitions between unbound states being different than for bound states.

10.3 Investigating Variability: Population of Models Approach

Biological variability is a common feature of all physiological systems, and a key challenge in the development of new medicines. Variability manifests itself at all levels in all organs of living organisms as differences in physiological function between individuals of the same species or as differences in time for the same individual. Variability becomes even more important when it involves significant differences in the outcome of exposure to pathological conditions or treatment. Thus, healthy cells of the same species and location usually exhibit a qualitatively similar behaviour. For example, cardiomyocytes of a similar location and species exhibit a similar action potential. However, significant quantitative inter-subject and intra-subject differences exist, which in cardiac cells are manifested as quantitative differences in action potential morphology and duration. The small differences under physiological conditions may, however, be important in explaining the different individual response of each of the cells and patients to disease and treatments.

To date, experimental and theoretical research has often ignored the variability underlying the physiological and pathological responses of different individuals to disease and medicines, ultimately hampering the extrapolation of results to a population level. Often an average of the recordings is performed to reduce experimental error, therefore also averaging out the effects of inter-subject variation, and resulting in an important loss of information. This averaging of experimental data is inherited by theoretical research, and consequently, models such as those described in the previous sections are often developed for a 'typical' behaviour within a particular sub-population. This is certainly the case in cardiac electrophysiology, and therefore, although all experimentally measured action potentials (APs) are different, even within a homogeneous population, a single AP model is developed and used in the simulations, again losing all information regarding inter-subject variability.

Understanding how inter-subject variability determines the efficacy and safety of medicines for specific patients or groups of patients is still a major challenge for both the practice of medicine and the development of new medicines. The importance of temporal and intra-subject variability has also been highlighted in recent years. For example, an experimental study showed evidence of remodelling of ionic currents in cardiac cells cultured for 24 hours in the presence of dofetilide, a drug blocking the rapid component of the delayed rectifier potassium current.[80] This means that cells change over time to adapt to the presence of drugs, potentially affecting the efficacy and safety of medicines. An increasing number of studies also demonstrate the strong effects circadian rhythms have on our bodies and the increasing importance given to chronotherapy.[81–83] A recent study conducted in mice demonstrated the effects of circadian rhythms on cardiac repolarisation and arrhythmias, providing measurements from the mRNA to the electrocardiogram level, and showing oscillations in ionic currents and the QT interval of the electrocardiogram during day and night.[83] Circadian rhythms may also influence the higher propensity of sudden cardiac death at specific times of the day.[81] Therefore, understanding the effect of circadian rhythms on the response to medicines could also be potentially important, and *in silico* methods have the potential to make an important contribution to this effect.[84]

The high degree of inter-species, inter-subject, temporal and spatial biological variability has become an increasingly important topic in computational physiology.[11,16,43,52,54,85,86] Computational population-based approaches have been proposed in recent studies to shed light on ionic determinants of intercellular variability to the response of individual cells to medicines and disease.

Davies *et al.*[16] adjusted model parameters in a canine cellular model to obtain 19 different models, which were fitted to specific action potential recordings. An alternative approach has recently been developed, based on the construction of populations of cell models, building on studies by Marder and colleagues in neuroscience.[85] Thus, Sarkar *et al.*[11] constructed populations of around 300 cardiac cell models by varying ionic properties in an arbitrary range, and demonstrated variability in cardiomyocyte response to ion channel block.

More recently, Britton *et al.*[12] have proposed a methodology to tightly couple experimental measurements and *in silico* models to construct and calibrate populations of cell models to investigate the causes of experimentally measured variability in physiological conditions and following drug response (Figure 10.5). The research builds on previous studies showing the importance of sensitivity analysis in investigating the ionic determinants of inter-subject variability in biological properties.[43,54]

The experimentally calibrated population of cell models proposed by Britton *et al.*[12] aims to capture the variability exhibited in specific experimental recordings under physiological conditions and to predict inter-subject variability in response to potassium channel block. The main assumption of the novel approach is that inter-subject variability in experimentally measured APs is primarily caused by quantitative differences in the properties of ionic currents, rather than by qualitative differences in the biophysical processes underlying the currents. All models in the population therefore share the same equations, which are used as the model structure to

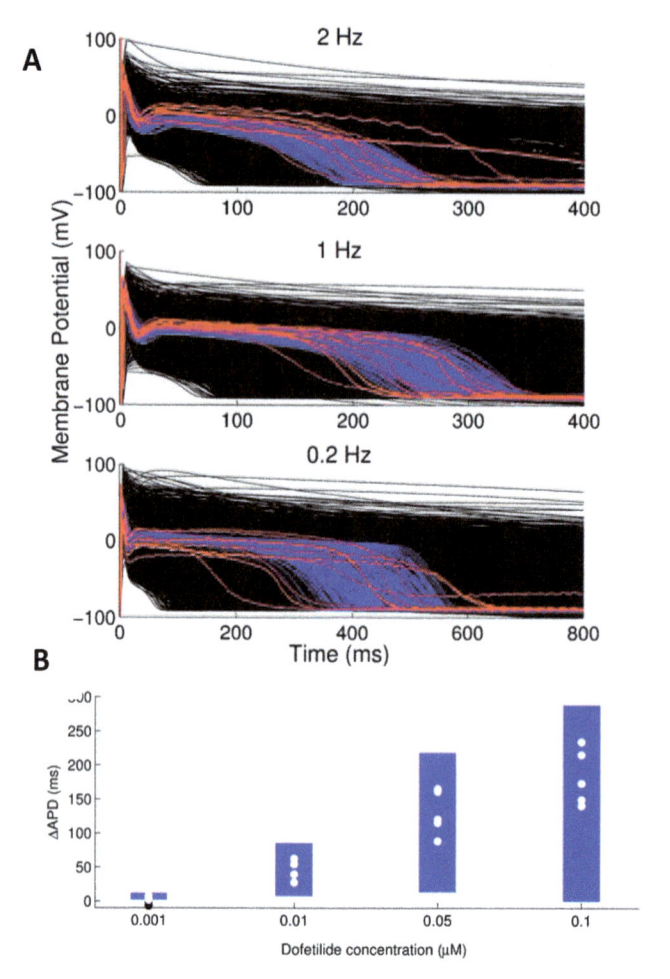

Figure 10.5 Inter-subject variability in cellular cardiac electrophysiology. A. Simulated action potentials obtained with an experimentally-calibrated population of rabbit Purkinje cell models (blue traces), in range with experimental action potentials (red traces). Black traces correspond to models excluded from the model population for being out of experimental range. B. Range of action potential duration prolongation caused by four doses of the HERG block dofetilide obtained using the experimentally-calibrated model population (bars) and in experiments (white dots). (Modified from Yu et al.[12] with permission.)

generate more than 10 000 candidate models with different sampled parameter values for the ionic properties within a wide range (Figure 10.5A). The cell model population is then calibrated using a set of cellular biomarkers extracted from experimental recordings at three pacing frequencies to capture key rate-dependent action potential properties. The experimentally calibrated model population is then used to identify the ionic mechanisms that may underlie inter-subject variability for each biomarker, yielding information on the relative importance of ionic currents in the generation of the AP at each pacing frequency.

As illustrated in Figure 10.5B, the same paper by Britton *et al.*[12] shows that the population of cell models can then be used to predict variability in the response of rabbit Purkinje fibres to drug action (using an independent data set). Figure 10.5B shows how the calibrated cell model population quantitatively predicts the prolongation of action potential duration caused by exposure to four concentrations of dofetilide, a blocker of the rapid component of the delayed rectifier potassium current (I_{Kr}). This intervention was considered because I_{Kr} block is the main assay required in safety pharmacology assessment due to its potential importance in drug-induced Torsades de pointes. It is therefore important in the development of all new medicines.

A computational population-based approach was also applied in Walmsley *et al.*[87] to investigate ionic causes of differences in the response of human ventricular myocytes to heart failure, a common disease that leads to ion channel remodelling. The ability of computational models to reflect and explain variability under disease conditions could be very powerful in the development of new medicines for cardiovascular disease treatment and also for safety pharmacology.

As well as investigations into inter-subject variability, a number of experimental and clinical studies highlight the importance of temporal or beat-to-beat cardiac variability in quantifying arrhythmic risk caused by disease or medicines. Indeed, changes in the magnitude of beat-to-beat cardiac variability in response to drug or disease have been linked to the pro-arrhythmic potential in experimental and clinical studies, also showing their predictive power to identify patients at high risk of arrhythmia.[88-92]

Recent studies hypothesised that the intrinsically stochastic opening and closing of cardiac ion channels contribute to beat-to-beat cardiac variability, and that pro-arrhythmic drugs might unmask and enhance their effect.[52,86,93] This is based on the idea that slight beat-to-beat differences in the number of channels opening and closing at specific times could manifest themselves as beat-to-beat differences at the cellular level (for example in the duration of the action potential). Several *in silico* studies[52,86,93] have investigated the mechanisms underlying these processes, and have linked stochasticity in ion channel behaviour to the variability of cellular outcomes following drug action.

In a combined *in vitro/in silico* study, Pueyo *et al.*[52] investigated the contribution of the stochastic behaviour of the slow delayed rectifier potassium current (I_{Ks}) to beat-to-beat cardiac variability. The same study also shows how this effect is enhanced following the application of an I_{Kr} blocker (Figure 10.6). In isolated cells, the stochasticity in I_{Ks} could also explain how I_{Kr} block results in potentially pro-arrhythmic instabilities in repolarisation (named early after depolarisations) in some cells and some beats.

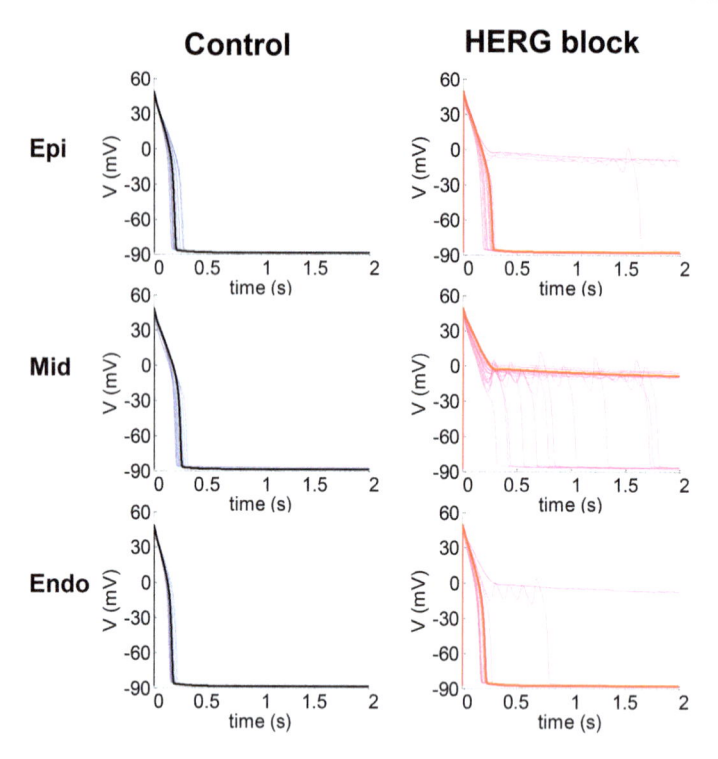

Figure 10.6 Temporal variability in simulated cardiac action potentials caused by the stochastic nature of the slow component of the delayed rectifier potassium current, under control conditions and following the application of HERG block. Pro-arrhythmic abnormalities in repolarisation (*i.e.* soon after depolarisations) are obtained in some simulations following a deceleration of the stimulation frequency. (Modified from Musunuru[52] with permission.)

Stochasticity in I_{Ks} may therefore explain, at least in part, some mechanisms of inter- and intra-subject variability in repolarisation in response to drug action. Interestingly, the repolarisation instabilities were suppressed in tissue simulations due to the electrotonic effects of intercellular coupling. These results highlight the importance of multiscale simulations in elucidating the complex interplay of mechanisms determining the response of biological systems such as the heart to pharmacological compounds.

10.4 Validation of *in Silico* Models and Simulations

As *in silico* models and simulation techniques develop both conceptually and technically, the question of their verification and validation becomes crucial for their uptake beyond scientific research into industrial and regulatory arenas.[2,17] The process of validation of *in silico* models and simulations is complex and still under investigation but it is essential to increase the credibility of *in silico* methods as powerful techniques to augment the information extracted from experimental and clinical investigations.

The paper by Carusi *et al.*[2] specifically explores the issue of validation of computational physiological models in the context of their construction and applications. It highlights the iterative nature of the process of validation and the fact that it needs to consider the ensemble of equations, parameters, simulation techniques, software and experiments used for a particular study or purpose, namely, the model–simulation–experiment system. This is of course similar to any other experimental science involved in the development of new medicines, and specifically experimental biology, where the interpretation of experimental recordings depends on knowledge with the entire history of the samples and techniques used to obtain the results.

In computational physiology, a necessary (although insufficient) condition for *in silico* methods validation is the verification of the robustness and convergence of numerical techniques and software used in the simulations, as they are implicated throughout model construction and simulation.[2,66] Once this aspect is demonstrated, the credibility of *in silico* models is based on physiological validation, which is the evaluation of the simulation results, through comparison to independent experimental data sets, not used in the model construction. It is important to re-emphasise the iterative nature of validation, which is a process rather than a final result.[2]

As discussed in previous sections and in Carusi *et al.*,[2] multiscale models often result in data used in construction and validation being acquired at different levels (*e.g.* ionic *versus* cellular or organ level), and therefore in different preparations and using different techniques. As a consequence, the multiscale models are often constructed from experiments using a variety of techniques and preparations. Each experimental technique involves a modification of the preparation with respect to its *in vivo* state, therefore introducing alterations in the measurements. Therefore, validation through comparison of experiments and simulations needs to take into account the potential distortion introduced by experimental recording techniques in order to be able to interpret the success or failure of simulations in matching/predicting experimental results.

Biological variability is of course one of the identified challenges in the evaluation of *in silico* models, as it can greatly influence experimental results and their comparison to simulated data. As discussed in the previous section, the sources of electrophysiological variability are multiscale, spanning a wide range of spatio-temporal scales and the more we understand them, the more we will be able to advance in the quantitative evaluation of *in silico* models for their biological validation. Given the dynamic interplay and the variability in models, experiments and simulations, standards and interoperability at the modelling, simulation and experimental levels are essential, as reflected in the development of specific languages such as CellML, SBML and FieldML for model encoding, repositories and databases.[94,95]

Physiological validation needs to be considered like a continuous and iterative process between experimental and computational research.[2] The aim is augmenting the information on the physiological system or process under investigation, such as for example the mode of action of a new medicine on specific organs. The dynamics of the iterative process are driven by advances in both experimental and computational techniques as well as investigations on their combined use. The process is driven both by matches and mismatches of simulations and experiments (or success and failure in validation).[2,96] The iterative nature of this process is clear from the evolution of single-cell electrophysiology models of different types and species, which has often resulted in families of several generations of models,[35,45,96] possibly capturing different aspects of cellular physiology given the large biological variability. As new experimental data and techniques become available, inconsistencies between computer simulation predictions and experimental recordings allow accelerating the refinement and improvement of our knowledge and its integration in quantitative and qualitative models.

With respect to cardiac electrophysiology at the whole organ level, the process of iteration between experiments and simulations has been crucial, for example, to improving the understanding of important human health phenomena such as ventricular fibrillation,[30,50] the most dangerous type of lethal arrhythmia, and the mechanisms underlying electrical defibrillation, the only effective therapy against sudden cardiac death.[1] In both cases, whole ventricular electrophysiology simulations were successfully combined with clinical and experimental methods to bring two important advantages:

1. High spatio-temporal resolution on the three-dimensional simulated electrophysiological activity of the ventricles overcome limitations of clinical or experimental recordings, which often exhibit low spatio-temporal resolution and/or just the surface of the heart.
2. Computer models and simulations provide the ability to dissect the relative importance of different factors, which result in the identification of key properties, which may determine the progress of disease and might inform the development of new therapies.

Similar approaches are likely to make a similar impact in the development of new medicines through the use of combined *in silico*, *in vitro* and *in vivo* studies, building on the great advances being achieved in computational, experimental and clinical medicine.

References

1. N. A. Trayanova, *Circ. Res.*, 2011, **108**, 113.
2. A. Carusi, K. Burrage and B. Rodriguez, *Am. J. Physiol. Heart Circ.*, 2012, **303**, H144.
3. K. S. Burrowes, A. R. Clark and M. H. Tawhai, *Pulm. Circ*, 2011, **1**, 365.

4. K. S. Burrowes, A. R. Clark, A. Marcinkowski, M. L. Wilsher, D. G. Milne and M. H. Tawhai, *Philos. Trans. A Math. Phys. Eng. Sci.*, 2011, **369**, 4255.
5. J. G. Venegas, T. Winkler, G. Musch, M. F. Vidal Melo, D. Layfield, N. Tgavalekos, A. J. Fischman, R. J. Callahan, G. Bellani and R. S. Harris, *Nature*, 2005, **434**, 777.
6. L. A. De Backer, W. G. Vos, R. Salgado, J. W. De Backer, A. Devolder, S. L. Verhulst, R. Claes, P. R. Germonpré and W. A. De Backer, *Int. J. Chron. Obstruct. Pulmon. Dis.*, 2011, **6**, 637.
7. M. R. Owen, I. J. Stamper, M. Muthana, G. W. Richardson, J. Dobson, C. E. Lewis and H. M. Byrne, *Cancer Res.*, 2011, **71**, 2826.
8. M. Kim, R. J. Gillies and K. A. Rejniak, *Front. Oncol.*, 2013, **18**, 278.
9. C. Clancy, Z. I. Zhu and Y. Rudy, *Amer.. J. Physiol. Heart. Circ. Physiol.*, 2007, **292**, H66.
10. S. Polak, B. Wiśniowska, K. Fijorek, A. Glinka and A. Mendyk, *Drug Discov. Today*, 2013, **19**, 275.
11. A. X. Sarkar and E. A. Sobie, *Heart Rhythm*, 2011, **8**, 1749.
12. O. Britton, A. Bueno-Orovio, K. Van Ammel, H. R. Lu, R. Towart, D. Gallacher and B. Rodriguez, *Proc. Nat. Acad. Sci. U. S. A.*, 2013, **110**, E2098.
13. N. Zemzemi, M. Bernabeu, J. Saiz, G. Mirams, J. Cooper, J. Pitt-Francis and B. Rodriguez, *Br. J. Pharm.*, 2013, **168**, 718.
14. M. Wilhelms, C. Rombach, E. P. Scholz, O. Dössel and G. Seemann, *Europace*, 2012, **14**(Suppl. 5), v90.
15. T. O'Hara and Y. Rudy, *Am. J. Physiol. Heart Circ. Physiol.*, 2012, **302**, H1023.
16. M. R. Davies, H. B. Mistry, L. Hussein, C. E. Pollard, J. P. Valentin, J. Swinton and N. Abi-Gerges, *Am.. J. Physiol Heart Circ. Physiol.*, 2012, **302**, 1466.
17. P. T. Sager, G. Gintant, J. R. Turner, P. Syril, *Am. Heart J.*, 2013. XXXXX.
18. A. Atkinson, S. Inada, J. Li, J. O. Tellez, J. Yanni, R. Sleiman, E. A. Allah, R. H. Anderson, H. Zhang, M. R. Boyett and H. Dobrzynski, *J. Mol. Cell. Cardiol.*, 2011, **51**, 689.
19. A. Baher, M. Uy, F. Xie, A. Garfinkel, Z. Qu and J. N. Weiss, *Heart Rhythm*, 2011, **8**, 599.
20. M. J. Bishop, G. Plank, R. A. Burton, J. E. Schneider, D. J. Gavaghan, V. Grau and P. Kohl, *Am. J. Physiol. Heart Circ. Physiol.*, 2010, **298**, H699.
21. R. Bordas, K. Gillow, Q. Lou, I. Efimov, P. Kohl, D. Gavaghan, V. Grau and B. Rodriguez, *Prog. Biophys. Mol. Biol.*, 2011, **107**, 90.
22. M. Deo, P. Boyle, G. Plank and E. Vigmond, *Heart Rhythm*, 2009, **6**, 1782.
23. D. U. Keller, D. L. Weiss, O. Dossel and G. Seemann, *IEEE Trans. Biomed. Eng.*, 2012, **59**, 311.
24. J. D. Moreno, Z. I. Zhu, P. C. Yang, J. R. Bankston, M. T. Jeng, C. Kang, L. Wang, J. D. Bayer, D. J. Christini, N. A. Trayanova, C. M. Ripplinger, R. S. Kass and C. E. Clancy, *Sci. Transl. Med.*, 2011, **3**, 98ra83.
25. S. Niederer, L. Mitchell, N. Smith and G. Plank, *Front. Physiol.*, 2011, **2**, 14.

26. J. Okada, T. Washio, A. Maehara, S. Momomura, S. Sugiura and T. Hisada, *Am. J. Physiol. Heart Circ. Physiol.*, 2011, **301**, H200.
27. M. Potse, B. Dubé, J. Richer, A. Vinet and R. M. Gulrajani, *IEEE Trans. Biomed. Eng.*, 2006, **53**, 2425.
28. D. Romero, R. Sebastian, B. H. Bijnens, V. Zimmerman, P. M. Boyle, E. J. Vigmond and A. F. Frangi, *Ann. Biomed. Eng.*, 2010, **38**, 1388.
29. G. Seemann, C. Höper, F. B. Sachse, O. Dössel, A. V. Holden and H. Zhang, *Philos. Transact. A. Math. Phys. Eng. Sci*, 2006, **364**, 1465.
30. K. H. ten Tusscher, R. Hren and A. V. Panfilov, *Circ. Res.*, 2007, **100**, 87.
31. F. Vadakkumpadan, H. Arevalo, A. Prassl, J. Chen, F. Kickinger, P. Kohl, G. Plank and N. Trayanova, *WIREs Syst. Biol. Med.*, 2010, **2**, 489.
32. N. Zemzemi, M. Bernabeu, J. Saiz and B. Rodriguez, *Lecture Notes Computer Sci.*, 2011, **6666/2011**, 259.
33. J. Zhao, T. D. Butters, H. Zhang, A. J. Pullan, I. J. Legrice, G. B. Sands and B. H. Smaill, *Circ Arrhythm. Electrophysiol.*, 2012, **5**, 361.
34. M. Wilhelms, O. Dössel and G. Seemann, *IEEE Trans. Biomed. Eng.*, 2011, **58**, 2961.
35. D. Noble, A. Garny and P. J. Noble, *J. Physiol.*, 2012, **590**, 2613.
36. A. L. Hodgkin and A. F. Huxley, *J. Physiol.*, 1952, **117**, 500.
37. J. Carro, J. F. Rodríguez, P. Laguna and E. Pueyo, *Philos. Transact. A. Math. Phys. Eng. Sci.*, 2011, **369**, 4205.
38. A. Corrias, W. Giles and B. Rodriguez, *Am. J. Physiol. Heart Circ. Physiol.*, 2011, **300**, H1806.
39. K. F. Decker, J. Heijman, J. R. Silva, T. J. Hund and Y. Rudy, *Am. J. Physiol. Heart Circ. Physiol.*, 2009, **29**(6), H1017.
40. E. Grandi, F. S. Pasqualini and D. M. Bers, *J. Mol. Cell. Cardiol.*, 2010, **48**, 112.
41. E. Grandi, S. V. Pandit, N. Voigt, A. J. Workman, D. Dobrev, J. Jalife and D. M. Bers, *Circ. Res.*, 2011, **109**, 1055.
42. L. Li, S. A. Niederer, W. Idigo, Y. H. Zhang, P. Swietach, B. Casadei and N. P. Smith, *Am. J. Physiol. Heart Circ. Physiol.*, 2010, **299**, H1045.
43. P. Li and Y. Rudy, *Circ. Res.*, 2011, **109**, 71.
44. A. Mahajan, Y. Shiferaw, D. Sato, A. Baher, R. Olcese, L. H. Xie, M. J. Yang, P. S. Chen, J. G. Restrepo, A. Karma, A. Garfinkel, Z. Qu and J. N. Weiss, *Biophys. J.*, 2008, **94**, 392.
45. D. Noble and Y. Rudy, *Philos. Transact. A. Math. Phys. Eng. Sci.*, 2001, **359**, 1127.
46. T. O'Hara, L. Virág, A. Varró and Y. Rudy, *PLoS Comput. Biol.*, 2011, **7**, e1002061.
47. T. W. Shannon, F. Wang, J. Publisi, C. Weber and D. M. Bers, *Biophys. J.*, 2004, **87**, 3351.
48. M. Wilhelms, H. Hettmann, M. M. Maleckar, J. T. Koivumäki, O. Dössel and G. Seemann, *Front. Physiol.*, 2013, **3**, 487.
49. J. Heijman, P. G. Volders, R. L. Westra and Y. Rudy, *J. Mol. Cell. Cardiol.*, 2011, **50**, 863.

50. O. V. Aslanidi, R. N. Sleiman, M. R. Boyett, J. C. Hancox and H. Zhang, *Biophys. J.*, 2010, **98**, 2420.
51. S. N. Flaim, W. R. Giles and A. D. McCulloch, *Heart Rhythm*, 2007, **4**, 768.
52. E. Pueyo, A. Corrias, L. Virag, N. Jost, T. Szel, A. Varro, N. Szentandrássy, P. P. Nánási, K. Burrage and B. Rodriguez, *Biophys. J.*, 2011, **101**, 2892.
53. S. A. Niederer, M. Fink, D. Noble and N. P. Smith, *Exp. Physiol.*, 2009, **94**, 486.
54. L. Romero, B. Carbonell, B. Trenor, B. Rodriguez, J. Saiz and J. M. Ferrero, *Prog. Biophys. Mol. Biol.*, 2011, **107**, 60.
55. A. Bueno-Orovio, A. Sanchez, E. Pueyo and B. Rodriguez, *Pflüg. Arch. Eur. J. Physiol.*, in press, **2013**.
56. F. Fenton and A. Karma, *Chaos*, 1998, **8**, 20.
57. A. Bueno-Orovio, E. M. Cherry and F. H. Fenton, *J. Theor. Biol.*, 2008, **253**, 544.
58. A. Bueno-Orovio, B. Hanson, J. Gill, P. Taggart and B. Rodriguez, *PLoS One*, 2012, 7, e52234.
59. J. Walmsley, G. Mirams, M. Bahoshy, C. Bollensdorff, B. Rodriguez and K. Burrage, *Conf. Proc. IEEE Eng. Med. Biol. Soc.*, 2010, **2010**, 1457.
60. M. Wallman, N. Smith and B. Rodriguez, *Trans. Biomed. Eng.*, 2012, **59**, 1739.
61. G. Plank, R. A. B. Burton, P. Hales, M. Bishop, T. Mansoori, M. O Bernabeu, A. Garny, A. J. Prassl, C. Bollensdorff, F. Mason, F. Mahmood, B. Rodriguez, V. Grau, J. Schneider, D. Gavaghan and P. Kohl, *Phil. Trans. Roy. Soc. A*, 2009, **367**, 2257.
62. M. Plotkowiak, B. Rodriguez, G. Plank, J. E. Schneider, D. P. Gavaghan, Kohl and V. Grau, *Lect. Notes Computer Sci.*, 2008, **5101**, 571.
63. D. Chapelle, M. A. Fernandez, J. F. Gerbeau, P. Moireau, J. Sainte-Marie, N. Zemzemi. *Functional Imaging and Modeling of the Heart*, Volume 5528 of Lecture Notes in Computer Science. Springer: Berlin, pp. 357–365, 2009.
64. K. H. ten Tussscher and A. V. Panfilov, *Am. J. Physiol. Heart Circ. Physiol.*, 2006, **291**, H1088.
65. E. M. Cherry and F. H. Fenton, *J. Theor. Biol.*, 2011, **285**, 164.
66. R. H. Clayton, O. Bernus, E. M. Cherry, H. Dierckx, F. H. Fenton, L. Mirabella, A. V. Panfilov, F. B. Sachse, G. Seemann and H. Zhang, *Prog. Biophys. Mol. Biol.*, 2011, **104**, 22.
67. P. Pathmanathan, G. R. Mirams, S. Southern and J. P. Whiteley, *Int. J. Numer. Method. Bi.omed. Eng.*, 2011, **27**, 1751.
68. P. Pathmanathan and R. A. Gray, *Int. J. Numer. Method. Biomed. Eng.*, in press, **2013**.
69. A. A. Holmes, D. F. Scollan and R. L. Winslow, *Magn. Reson. Med.*, 2000, **44**, 157.
70. I. J. LeGrice, B. H. Smaill, L. Z. Chai, S. G. Edgar, J. B. Gavin and P. J. Hunter, *Am. J. Physiol.*, 1995, **269**, 571.
71. F. J. Vetter and A. D. McCulloch, *Prog. Biophys. Mol. Biol.*, 1998, **69**, 157.
72. D. A. Hooks, K. A. Tomlinson, S. G. Marsden, I. J. LeGrice, B. H. Smaill, A. J. Pullan and P. J. Hunter, *Circ. Res.*, 2002, **91**, 331.
73. T. Brennan, M. Fink and B. Rodriguez, *Eur. J. Pharm. Sci.*, 2009, **36**, 62.
74. A. J. Clark, *The mode of action of drugs on cells*. Arnold, London, UK, 1933.

75. E. Neher and B. Sakmann, *Nature*, 1976, **260**, 799.
76. A. Hille, *J. Gen. Physiol.*, 1977, **69**, 497.
77. A. Hill, *J. Physiol.*, 1910, **40**, 4.
78. A. F. Starmer, A. O. Grant and H. C. Strauss, *Biophys. J.*, 1984, **46**, 15.
79. J. R. Balser, B. Nuss, D. W. Orias, D. C. Johns, E. Marban, G. F. Tomaselli and J. H. Lawrence, *J. Clin. Invest.*, 1996, **98**, 2874.
80. Xiao_Circulation.
81. J. E. Muller, P. L. Ludmer, S. N. Willich, G. H. Tofler, G. Aylmer, I. Klangos and P. H. Stone, *Circulation*, 1987, **75**, 131.
82. G. Gagandeep Kaur, C. Phillips, K. Wong and B. Saini, *Int. J. Clin. Pharm.*, 2013, **35**, 344.
83. A. Jeyaraj, S. M. Haldar, X. Wan, M. D. McCauley, J. Ripperger, K. Yuan Lu, B. L. Eapen, N. Sharma, E. Ficker, M. J. Cutler, J. Gulick, A. Sanbe, J. Robbins, S. Demolombe, R. V. Kondratov, S. Shea, U. Albrecht, X. Wehrens, D. Rosenbaum and M. K. Jain, *Nature*, 2012, **483**, 96.
84. K. Fijorek, M. Puskulluoglu and S. Polak, *Comput. Math. Methods Med.*, 2013, **2013**, 429037.
85. A. L. Marder, *Taylor. Nat. Neurosci.*, 2011, **14**, 133.
86. J. Heijman, A. Zaza, D. M. Johnson, Y. Rudy, R. L. Peeters, P. G. Volders and R. L. Westra, *PLoS Comput. Biol.*, 2013, **9**, e1003202.
87. J. Walmsley, J. F. Rodriguez, G. R. Mirams, K. Burrage, I. R. Efimov and B. Rodriguez, *PLoS One*, 2013, **8**, e56359.
88. L. M. Hondeghem, L. Carlsson and G. Duker, *Circulation*, 2001, **103**, 2004.
89. M. B. Thomsen, S. C. Verduyn and M. A. Vos, *Circulation*, 2004, **110**, 2453.
90. M. Johnson, J. Heijman and P. G. Volders, *J. Mol. Cell. Cardiol.*, 2010, **48**, 122.
91. M. Hinterseer, B. M. Beckmann and S. Kääb, *Am. J. Cardiol.*, 2010, **106**, 216.
92. R. C. Myles, F. L. Burton and G. L. Smith, *J. Mol. Cell. Cardiol.*, 2008, **45**, 1.
93. M. Lemay, E. de Lange and J. P. Kucera, *J. Theor. Biol.*, 2011, **281**, 84.
94. D. A. Beard, R. Britten, M. T. Cooling, A. Garny, M. D. Halstead, P. J. Hunter, J. Lawson, C. M. Lloyd, J. Marsh, A. Miller, D. P. Nickerson, P. M. Nielsen, T. Nomura, S. Subramanium, S. M. Wimalaratne and T. Yu, *Philos. Transact. A Math. Phys. Eng. Sci.*, 2009, **367**, 1845.
95. T. A. Quinn, S. Granite, M. A. Allessie, C. Antzelevitch, C. Bollensdorff, G. Bub, R. A. Burton, E. Cerbai, P. S. Chen, M. Delmar, D. Difrancesco, Y. E. Earm, I. R. Efimov, M. Egger, E. Entcheva, M. Fink, R. Fischmeister, M. R. Franz, A. Garny, W. R. Giles, T. Hannes, S. E. Harding, P. J. Hunter, G. ,Iribe, J. Jalife, C. R. Johnson, R. S. Kass, I. Kodama, G. Koren, P. Lord, V. S. Markhasin, S. Matsuoka, A. D. McCulloch, G. R. Mirams, G. E. Morley, S. Nattel, D. Noble, S. P. Olesen, A. V. Panfilov, N. A. Trayanova, U. Ravens, S. Richard, D. S. Rosenbaum, Y. Rudy, F. Sachs, F. B. Sachse, D. A. Saint, U. Schotten, O. Solovyova, P. Taggart, L. Tung, A. Varró, P. G. Volders, K. Wang, J. N. Weiss, E. Wettwer, E. White, R. Wilders, R. L. Winslow and P. Kohl, *Prog. Biophys. Mol. Biol.*, 2011, **107**, 4.
96. D. Noble, *Heart Rhythm*, 2011, **8**, 1798.

CHAPTER 11

Human Microdosing/Phase 0 Studies to Accelerate Drug Development

R. COLIN GARNER

Garner Consulting, 5 Hall Drive, Sand Hutton, York YO41 1LA, UK
E-mail: colin@consultgarner.com

11.1 Introduction

There is an urgent need to accelerate the development of new drugs to treat unmet medical needs both in the developed and underdeveloped parts of the world. There is also an urgent need to improve the efficiency of drug development through time and cost reduction because development costs hinder the introduction of new therapies. There are many life-threatening human diseases that still require better therapies including cancer, bacterial and viral infection and lifestyle-associated diseases. The costs of bringing a drug to market are so high that there are very few organisations able to afford to discover, develop and introduce drugs on their own. Estimates of drug development costs vary between US$1.2 and 12.0 billion dollars, making the costs of newly approved drugs very high in order that companies can recover the costs of all the failed drugs in their pipelines.[1,2] It has been estimated that only three out of 10 drugs make a return on their investment and that over 99% of molecules researched by a pharmaceutical company are likely to fail at some point during the development process.[3] Even when a drug enters into

---

RSC Drug Discovery Series No. 41
Human-based Systems for Translational Research
Edited by Robert Coleman
© The Royal Society of Chemistry 2015
Published by the Royal Society of Chemistry, www.rsc.org

Phase I clinical trials it has only a 10% chance of making it to market.[4] Recognising that there were major inefficiencies in the drug development process, in 2004 the US FDA launched the Critical Path Initiative to encourage pharmaceutical companies to re-examine their drug development programmes and to see if these could be conducted smarter.[5] The FDA argued that many drug failures could have been prevented if smarter development approaches had been adopted. Since the Critical Path Initiative document's publication, the FDA and the EMA have published a number of guidance documents steering the drug developer towards more human-based research.[6–8] While some of the FDA's recommendations have been adopted by the pharmaceutical industry, in far too many development programmes the old unthinking box-ticking mentality prevails. There appears to be a reluctance to adopt new drug development approaches despite the fact that many of our currently used *in vitro* and animal models are poorly predictive of human pharmacokinetics (PK).[9]

When considering a new development molecule, there are a number of properties that will determine whether the molecule will ever become a marketed drug. After knowing about a drug's pharmacology (proof of concept), in the author's opinion the next most important biological parameter to understand is how humans metabolise the drug. Figure 11.1 depicts the relationship between a drug's safety, efficacy and dose. The common feature determining safety and efficacy is the drug's pharmacokinetics (PK). If too low exposure of drug reaches the target for too short a time then the drug will lack efficacy. Conversely if too high exposure of drug reaches the target for too long a time then toxicity might arise through off-target effects.

Gaining an understanding of human PK as early as possible in a candidate drug's development will enable better informed go/no go decisions to be made and hence ultimately lead to an improvement in development efficiency. While this would appear to be a statement of the obvious, drug

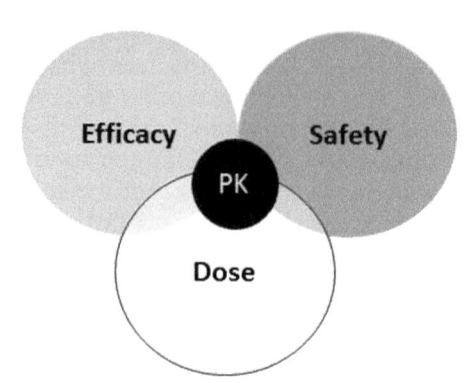

Figure 11.1 Links between a drug's PK, efficacy and safety.

developers still stick to their *in vitro*, *in silico* and preclinical models despite the evidence showing that these models are poorly predictive of human PK. In a major collaboration coordinated by US PhRMA, 12 pharmaceutical companies submitted blinded data on 108 compounds for which they had preclinical and clinical PK data. The intention was to establish with this large data set the predictivity of various methodologies to determine human PK parameters. When the Wajima allometry scaling method was used, the PK parameters clearance (CL) and volume of distribution at steady state (Vss) for oral administration was 58% predictive (within 3-fold of actual values); the curve shape prediction was also poor.[9] Using the alternative approach of *in silico* PBPK modelling for the same data set, human PK prediction of plasma concentration–time profiles of orally administered drugs was 23%.[10]

Our current human PK predictive tools are poor and so other approaches need to be considered. It has been stated that few drugs currently fail because of PK issues[11] and yet lack of clinical response could well be related to inadequate target exposure. Often target exposure data is not available and so it is overly simplistic to state that all PK prediction issues have been solved. In any event, knowing and understanding the human PK profile, including metabolite production, of a development drug early, before large sums of money are committed to development programmes, will provide confidence to drug developers in their decision making.

In the late 1990s, the author proposed that using the ultrasensitive bioanalytical tool, accelerator mass spectrometry (AMS), homeopathic doses of drug could be administered to humans to obtain essential ADME/PK information.[12] The concept of 'microdosing' arose from earlier patient studies in which trace doses of various human carcinogens were administered in a hospital setting such as aflatoxin B1, benzo(a)pyrene, MeIQx, *etc.* and DNA adducts measured in surgically obtained specimens.[13–17] Administered doses of these xenobiotics were in the ng kg^{-1} range. Using the same approach, in collaboration with Johnson and Johnson Pharmaceuticals, the first human drug development microdose study was undertaken with AMS bioanalysis.[18] Since that time many other microdose studies have been performed including validation studies such as CREAM[19] and EUMAPP,[20,21] as well as microdose studies of development drugs.[22,23] This chapter centres on studies of small organic molecules primarily and focuses on where AMS bioanalysis was used. Later in this chapter mention will be made of the use of microdosing with biologicals.

11.2 What is Microdosing (Human Phase 0 Studies) and Where Does it Add Value?

In 2004, the EMA published a position paper on the preclinical requirements to support a single microdose study.[24] A microdose was defined as 1/100th of the pharmacological dose based on primary pharmacodynamic data or a maximum dose of 100 µg, whichever is the lesser. Subsequently the US FDA in 2006 published its Exploratory IND Guidance, which widened the dose range for microdose studies.[25] These documents, together with their Japanese equivalent, were incorporated into ICH M3 (R2).[26]

Microdosing is not constrained by the analytical methodology but in practice most published microdose studies have utilised AMS bioanalysis. AMS is a nuclear physics instrument originally developed for radiocarbon dating that is increasingly being used in drug development.[27–30] AMS has the attraction of high sensitivity (attograms to zeptograms), thus permitting drug and metabolite blood concentrations to be measured after low-dose drug administrations. There have been isolated reports on the use of LC/MS to measure parent drug concentration after microdose administration,[31–33] but for an unknown development drug AMS will always be the superior analytical technique because of its large dynamic range, its ability to quantitate metabolites and not requiring internal standardisation. Blood concentrations of drugs with high volumes of distribution and low bioavailability will always be detected by AMS but not necessarily by LC/MS. Mass balance information as well as relative metabolite:parent concentrations are obtained with AMS bioanalysis – these cannot easily be obtained using LC/MS.

11.3 Scientific, Regulatory and Ethical Aspects of Microdosing

11.3.1 Scientific Considerations

When the microdosing concept was proposed in the late 1990s, a number of critical papers were published arguing that many drugs showed saturation PK and therefore PK information obtained at a microdose would not scale to a therapeutic dose (non-linear PK).[34] Since that time two large validation studies have been completed. The first study, known as CREAM,[19] examined five generic drugs (diazepam, midazolam, Schering ZK253, warfarin and erythromycin). For the latter drug, where no therapeutic correlation was obtained, it was determined after the study was completed that in all probability the drug microdose had broken down in the stomach. Erythromycin is acid labile with a degradation half-life of 15 minutes at pH 3 and an even shorter half-life at lower gastric pH values. All but warfarin scaled linearly between a microdose and therapeutic dose (75% predictivity). A follow-up study known as EUMAPP[20,21] examined a further six generic drugs (fexofenadine, clarithromycin, sumatriptan, propafenone, paracetamol and phenobarbital). All but sumatriptan and propafenone showed linear PK between a microdose and a therapeutic dose. A review of these studies was published in 2010.[35] Sugiyama and his colleagues have published a number of papers on the linearity question[32,33,36–38] and have endeavoured to predict using *in vitro* methods such as K_m measurement which drugs might deviate from linearity.[39] A commentary around the issues of linearity and the utility of microdosing has been published by Rowland.[40]

Table 11.1 Summary table for 37 new chemical entities (NCEs) that had linear PK which were microdosed and subsequently administered at therapeutic doses (unpublished observations)

	No. of NCEs
Microdosed	37
In Phase 1	5
In Phase 2/3	5
Marketed	1
No. of programmes	17
Linear PK	11 out of 11

Where intravenous (IV) and oral administration comparisons have been made, the IV route of administration has always shown linearity between a microdose and a therapeutic dose, suggesting that non-linearities are always associated with the gut wall. It is recommended that the IV dosing route always be part of a microdose study design for this reason because the role of the gut wall and hepatic metabolism in the overall PK can be established.

The overall predictivity of artificially selected drugs in the microdose validation studies, *i.e.* drugs in which animal PK failed to predict human PK was approximately 70%, a much better figure than was obtained in the US PhRMA prediction exercise. For real development drugs, which were first microdosed and then subsequently taken forward into Phases 1 and 2, the PK prediction was 100% (see Table 11.1) suggesting that overall prediction between a microdose and a therapeutic dose might be higher than 70%.[35]

11.3.2 Regulatory

Conducting a microdose study on a new development molecule requires regulatory authorisation. In the USA this would be from the FDA under the Exploratory IND Guidance. Concerns have been expressed that filing an Exploratory IND takes as much time as filing an IND and requires a similar amount of paperwork. On the other hand, because the amount of preclinical testing is much reduced for a microdose study compared to a Phase 1 study, time and cost savings can be made. The optimum use of microdosing is to make PK comparisons of several molecules. Provided these are structurally related a single eIND can be filed. In the UK permission to conduct a microdose study would need to be sought from the MHRA. Table 11.2 sets out the main components required to conduct a microdose study together with the author's comments on these requirements. Regulatory guidance is provided in ICHM3 (R2)[26] and manufacturing guidance in the US FDA's eIND guidance documents.[25]

Table 11.2 Requirements for the various components in a microdose study

Module	Requirements	Comments
Preclinical toxicology	Extended single-dose toxicity study in one species, usually rodent, by intended route of administration with toxicokinetic data, or *via* the IV route. A maximum dose of 1000-fold the clinical dose on a mg kg^{-1} basis for IV and mg m^{-2} for oral administration can be used. Genotoxicity studies are not recommended, but any studies or SAR assessments conducted should be included in the clinical trial application.	IV route may be preferable as no toxicokinetics required. Surprising recommendation of no genetox necessary. Author would always want to see at least 'Ames' test data. ICHM3(R2) states: "Generally, extended single dose toxicity studies should be designed to evaluate hematology, clinical chemistry, necropsy, and histopathology data (control and high dose only if no treatment-related pathology is seen at the high dose) after a single administration, with further evaluations conducted 2 weeks later to assess delayed toxicity and/or recovery. The usual design for rodents consists of 10 animals/sex/group to be assessed on the day following dosing, and 5 animals/sex at the dose level(s) selected to be assessed on day 14 post-dose. The usual design for non-rodents consists of 3/sex/group for all groups on day 2 and 2/sex for the dose level(s) assessed on day 14."
^{14}C-labelled drug	Manufactured to at least GLP standards and formulated in a GMP accredited facility.	Some companies insist that GMP radiosynthesised material be used. As the amount of radioactivity administered is generally 7 kBq/adult then this equates usually to more than 1–2 μg of drug substance at a maximum and generally much less. Stability studies of undiluted and formulated radiolabel need to be conducted. For IV doses, a check should be made for binding of radiolabel to the infusion apparatus.
Unlabelled drug	Up to 20 g drug substance required.	Manufacturing conditions determined by the proposed human dose (≤100 μg). GLP quality material may suffice provided the same batch is used for the preclinical testing. To save discussion with regulators GMP material preferred.
Clinical	No prescribed study design.	Can be conducted in volunteers or patients (including paediatric). Minimum of 5 subjects. Oral/IV crossover design. Solution formulations. Collect blood for at least $3 \times t_{1/2}$. Excreta can be collected. Appropriate biopsy samples may be taken.

The most challenging aspect from a regulatory perspective in conducting microdose studies is the radiosynthesis and formulation of the ^{14}C-labelled drug. Regulatory affairs personnel and pharmaceutical companies' in-house protocols are often not written for exploratory clinical studies and tend to request the same procedures be adopted for microdose studies as for Phase 1 studies. As a microdose study is not designed to test a molecule's safety or efficacy and the dose is no more than 100 µg, this requirement is excessive and means that many of the cost and time advantages of conducting microdose studies are lost.

11.3.3 Ethical Considerations in Conducting Microdose Studies

Following the traditional pathway of drug development, starting with a Phase 1 study to study a drug's safety and tolerability and then on to Phase 2 to evaluate proof of concept, there is at least a 50% chance that the molecule will be abandoned at this stage. Inappropriate human PK may be one of the reasons for abandonment, meaning that a significant investment has been made with no positive outcome. A large number of animals will have been used for preclinical testing of the failed drug as well as unnecessarily exposing humans to a drug that will not progress. A financial investment of between US$3 and 5 million will have been made into the failed drug over a period of 18–24 months. The judicious use of microdosing will allow the drug developer to 'fail fast, fail cheap'. The numbers of animals used (only rodents) will have been much reduced and volunteers subject only to trace drug doses. The ethics of administering a microdose to human subjects who gain no potential therapeutic benefit has been discussed;[41,42] in the final analysis a Phase 0 study is no different to a classical Phase 1 study in that for either scenario, the individual gains no benefit but society as a whole does. Interestingly, in very few of the microdose publications is there a discussion about the ethical advantages of the approach. It has taken animal welfare groups to raise awareness of the reduction in animal usage that should flow from general adoption of microdosing.[43,44] Is one reason for the slow uptake of microdosing by pharmaceutical companies that they do not want to be seen as being on the same side as the animal welfare lobby? Is there not an opportunity here for a joining of minds on this subject?

11.4 Analytical Technologies used for Microdosing Studies

Most of the studies discussed so far in this chapter have focused on AMS bioanalysis in human microdose studies because this is the technology most used to date in the published literature. On the other hand the pharmaceutical industry does not have ready access to AMS instruments except through specialist suppliers or CROs. A number of publications have

Table 11.3 AMS vs LC/MS/MS comparison for microdose study bioanalysis

Parameter	AMS	LC/MS/MS
Ionisation methods	Negative ions of carbon isotopes only	Many but usually generating positive ions
Transmission efficiency from ion source to detector	High (> 70% is typical)	Low (often < 1% and molecule dependent)
Molecule characterisation	Isotope measurement only	Molecular mass and fragmentation pattern for parent drug molecule and metabolite characterisation
Sensitivity	fg to ag/mL can be measured	Best is low pg/mL
Sample work-up	Same for all molecules	Molecule and matrix dependent
Amount of sample	Typically µg or µl	mg or mL
LC hyphenation	Under development but currently off-line	Yes
Internal standard	AMS is fully quantitative without the need for internal standards	Method is semi-quantitative unless stable isotope standards are available
Instrument availability	Only few centres	Widely available

reported on the use of LC/MS as an alternative analytical technology.[31,45,46] Many of these publications have used *in vitro* or spiked plasma samples to argue that LC/MS is an alternative technology to AMS and because of the wider availability of instruments should become the analytical method of choice for routine microdose studies. Two reports of the use of LC/MS/MS to support human microdose studies have been published.[47,48] In one study nicardipine was studied after microdose (100 µg) and therapeutic dose administration (20 mg) to six healthy volunteers. The LLOQ for parent drug was in the region of 1 pg/mL. Interestingly the parent AUCs in both the microdose and therapeutic dose arms were taken out to 8 hours only, *i.e.* less than one half-life for the drug. The metabolism of nicardipine is complex, with up to nine metabolites identified. When the authors analysed for these metabolites, the data was expressed as peak area ratios rather than absolute quantities. This study highlights the strengths and weaknesses of the LC/MS/MS approach. No stable isotope standards were available for the parent drug or any of its metabolites. A related drug, nifedipine, was used as an internal standard. Table 11.3 sets out a comparison of AMS *versus* LC/MS/MS bioanalysis for microdose studies. The choice of analytical method will be entirely determined by the questions to be answered. Drugs that have poor oral bioavailability and high volumes of distribution will be most easily measured by AMS. If parent drug and metabolite quantification are needed, then AMS provides these without any previous knowledge of metabolite structures. On the other hand, LC/MS/MS is much more widely available and therefore companies will naturally consider using this instrument in preference to AMS.

11.5 Applications of Microdosing

The arguments for the usefulness of microdosing fall into a number of broad headings:

- Obtaining early human PK data on one or more molecules to establish which had the best PK to take forward into Phase 1 studies.
- Clarification of conflicts in metabolism between *in vitro*, animal models and human PK.
- Assistance in determining the starting dose in a Phase 1 study.
- Validating animal pharmacology or toxicology models.
- Determination of likely cost of goods.

11.5.1 Example of a Microdosing Study with Several Molecules and Practical Implications

A US biotech company was interested in developing a new H1-antagonist taking advantage of the known sleep-inducing properties of the antihistamine diphenhydramine.[23] A number of analogues had been synthesised but when these were tested *in vitro* and in preclinical models conflicting PK results were found. Phase I PK data for the lead molecule NBI-1 had been obtained – the company wished to obtain PK data for follow-up molecules. An oral/intravenous crossover microdose study was performed and the standard PK parameters determined. Table 11.4 summarises the PK data for this study. Linear PK was seen for diphenhydramine and NBI-1, giving confidence that the other molecules in the series would also demonstrate linear PK. NBI-2 had a higher oral C_{max} than diphenhydramine and NBI-1, higher oral bioavailability and higher AUCs, indicating that NBI-2 had superior PK properties to diphenhydramine and NBI-1. The result was that the company nominated NBI-2 for further development.

Microdosing is optimally used for several drug candidates as in the example highlighted here. On the other hand, only one or two rather than several drug candidates might be available. In this situation it might still be worth conducting a microdose study to establish human PK but consideration should be given to continuing to conduct the necessary programmes for the molecule to enter Phase 1 in parallel. In this way time is not lost while the microdose study is conducted. If the microdose PK look favourable then the preclinical Phase I programme is continued to completion with no time losses. On the other hand if the microdose PK indicate poor human PK properties, then the Phase I programme can be terminated at this stage.

Table 11.4 Mean PK parameters for five H1 antagonists after IV and oral microdose administration to male volunteers. Taken from Madan et al.[23].[a]

(A) IV dosing

Drug	AUC_{0-t} (ngh ml^{-1})	$AUC_{0-\infty}$ (ngh ml^{-1})	$t_{1/2}$ (h)	CL (l h^{-1})	V (l)	V_{ss} (l)	$MRT(inf)$ (h)
DPH	4.37 (3.93–5.17)	4.43 (4.19–5.20)	9.3 (5.7–12)	24.7 (20.7–25.9)	302 (214–433)	158 (99.6–264)	6.4 (4.8–10)
NBI-1	4.80 (4.12–6.75)	4.95 (4.13–6.80)	9.1 (6.5–13)	22.3 (16.1–27.0)	308 (170–361)	128 (103–175)	6.8 (3.8–8.0)
NBI-2	11.2 (7.62–12.6)	11.3 (7.79–12.8)	9.0 (7.5–10)	9.79 (8.58–14.0)	134 (96.5–176)	77.6 (46.9–117)	7.9 (5.3–8.9)
NBI-3	6.97 (6.67–9.01)	7.43 (6.69–9.42)	11 (6.0–16)	14.3 (10.7–16.7)	198 (144–327)	130 (113–248)	10 (7.8–17)
NBI-4	2.47 (1.49–6.03)	2.66 (1.65–6.54)	15 (8.0–23)	43.7 (16.4–66.7)	685 (538–1060)	401 (201–565)	10 (4.7–17)

(B) oral dosing

Drug	T_{max} (h)	C_{max} (ng ml^{-1})	AUC_{0-t} (h ng ml^{-1})	$AUC_{0-\infty}$ (h ng ml^{-1})	$t_{1/2}$ (h)	$MRT(inf)$ (h)	% F
DPH	2.5 (1.0–3.0)	0.195 (0.153–0.223)	1.36 (1.11–1.60)	1.52 (1.39–1.89)	12 (7.2–26)	15.2 (11.9–16.5)	34 (29–42)
NBI-1	1.0 (0.75–2.0)	0.625 (0.532–1.20)	2.78 (1.90–4.23)	2.86 (1.94–4.28)	6.7 (3.6–10)	6.9 (4.5–7.5)	52 (41–88)
NBI-2	1.0 (0.75–3.0)	0.704 (0.525–0.887)	6.54 (5.05–7.53)	6.63 (5.17–7.71)	9.5 (5.7–9.8)	11 (8.7–12)	60 (50–75)
NBI-3	2.0 (1.0–2.0)	0.349 (0.279–0.482)	3.05 (2.31–5.17)	3.42 (2.57–5.56)	11 (7.4–17)	13 (9.3–20)	52 (33–53)
NBI-4	1.5 (1.0–2.0)	0.106 (0.040–0.203)	0.636 (0.210–1.86)	0.778 (0.229–1.87)	13 (7.6–15)	12 (9.6–25)	26 (11–48)

[a]Data are median values of four individual subjects. Data in parentheses are the range for the four subjects. Per cent bioavailability (% F) was calculated by multiplying by 100 the ratio of AUC_{0-t} after oral and IV administration, correcting for the actual dose received by each subject. MRT, mean residence time.

One of the key factors in estimating the value of microdose studies centres on whether any drug development time and cost savings result. Examinations of the potential value of microdosing using a cost-effectiveness model[49] and using data from Japanese pharma companies suggested that an incremental cost saving of US$159 million could be made per drug approved if microdosing was introduced into the go/no go decision-making process.

With the trend towards NME identification and early development occurring more and more in small biotech companies rather than big pharma, microdosing offers a cheap and rapid way to obtain human PK data and hence add value to a molecule. Furthermore big pharma companies are using microdose studies to rapidly screen molecules that are being offered to them by the small biotechs to help establish whether the molecule has the right PK characteristics to take further into development.

11.5.2 Microdosing Use for Clarification of Conflicts in Metabolism between *in Vitro* and Animal Models for Human PK Prediction

In the early stages of drug development, metabolism information is obtained using *in vitro*, *in silico* and preclinical studies in a variety of animal species (typically mouse, rat, dog and monkey) to predict the likely human PK.[50,51] Candidate selection is made based on this information. There is an ongoing argument in the scientific community about how predictive these models are. What is undeniable is that having early human PK data must help in the selection process. Our metabolism models are generally poor in predicting human oral bioavailability.[52] On occasion animal models provide different values for clearance and bioavailability, making it difficult to determine which animal model reflects the human situation. This could mean that an incorrect choice of drug candidate is made or that much time and effort is expended on formulation in order to improve exposure to a poorly bioavailable drug. While there are many successful drugs with low oral bioavailability, given a choice the drug developer would prefer high bioavailability as well as a once a day administration. Microdosing provides a means to address some of these issues, particularly if an oral/intravenous crossover study design is used.

11.5.3 Microdosing to Assist in Determining the Starting Dose in a Phase 1 Study

The current paradigm for establishing the starting dose in single ascending dose Phase 1 study is to give a dose (defined in mg per metres squared of body surface area) associated with 1/10th the no observed adverse effect level in the most sensitive animal species and calculating a human equivalent dose. This approach uses the concept of allometric scaling where toxicity as a function of body weight or surface area is assumed to be roughly constant across species. As the recent US PhRMA study showed, allometric scaling is not that predictive of human PK. Determination of human microdose PK allows the initial Phase 1 starting dose to be calculated based on an exposure assessment rather than by using species scaling factors. A starting dose might be administered that is closer to the presumed therapeutic dose exposure, *e.g.* the starting dose is the dose that results in 1/10th therapeutic dose exposure. This approach should save clinical time, reduce animal use for toxicity testing and reduce costs.

11.6 Microdosing and Drug Targeting

A major challenge in drug development is to establish whether the drug reaches its cellular and molecular target. As discussed earlier, a lack of clinical efficacy can arise from no or too little drug reaching its target or when that which reaches the target is present for too short a time. While microdose studies have focused primarily on obtaining human PK information, the analytical sensitivity of AMS permits one to study drug concentration in individual tissues and cells. Indeed the author's first clinical studies with AMS bioanalysis centred on measuring DNA adducts from surgically obtained tissue after ^{14}C-carcinogen administration. The administered doses of carcinogen were in the ng range and 1 adduct per 10^{10} nucleotides were capable of being detected (1 adduct per 10 cells). There have been few literature reports of measurements of drug concentrations in cells. Figure 11.2a and b shows drug concentrations in cellular targets. In Figure 11.2a volunteers were administered 100 µg/7 kBq ^{14}C-AFN1252, a potential antibiotic under development, either orally or intravenously. A suction blister was formed and sampled for drug concentration measurement by AMS. Blister fluid represents the extravascular space where bacterial infection resides. A 10 µL fluid sample was analysed, which was sufficient volume for AMS analysis. In Figure 11.2b, drug concentrations in bronchial mucosa biopsy samples and plasma are shown from a microdose study of drug being developed to treat respiratory infections.[53] In the same study bronchoalveolar lavage samples were obtained. Extracts of these were analysed by HPLC followed by AMS analysis. Figure 11.2c shows a chromatographic profile of the same drug obtained after extraction of bronchoalveolar lavage samples. The samples were obtained 12 hours after drug administration and a clear parent drug peak was seen at femtogram concentrations. The drug dose for this study was 100 µg, containing 7 kBq of ^{14}C-drug, *i.e.* 100 fg of drug is equivalent to 0.007 mBq (0.00042 dpm) demonstrating the sensitivity of the AMS technique.

Measurement of prodrug to active drug conversion intracellularly can be an analytical challenge. As an example take a drug used for HIV treatment, azidothymidine (AZT), which inhibits viral reverse transcriptase by undergoing intracellular phosphorylation in target CD4$^+$ T cells. The resulting AZT-triphosphate (AZT-TP) inhibits reverse transcriptase and hence viral replication. A comparison was made between measurement of AZT-TP in target white blood cells using LC/MS *versus* AMS.[54] Subjects were dosed with

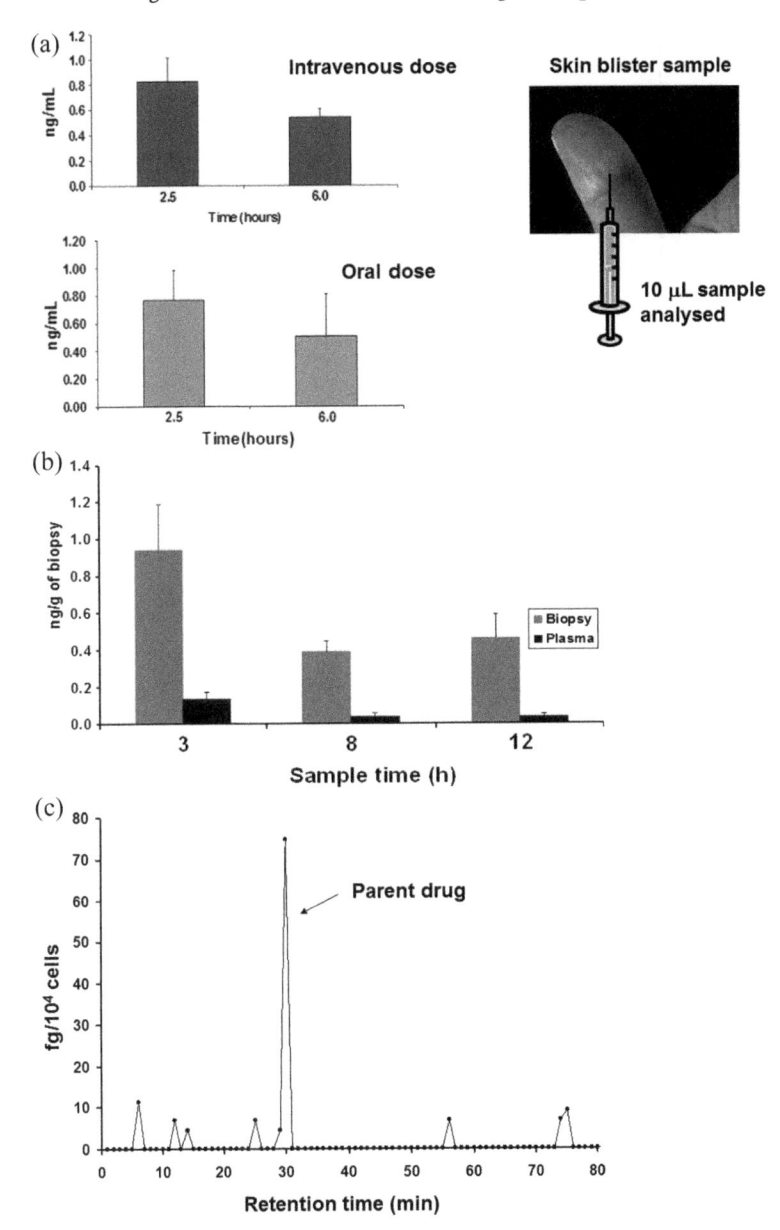

Figure 11.2 Measurement of human drug concentrations in (a) skin blister fluid after either an intravenous or oral 100 μg microdose of[14] C-AFN-1252, $n = 4$ subjects, mean ± SD and (b) 2 mg of bronchial biopsy material after an intravenous microdose of anti-infective development drug AR-709. Plasma concentrations = ng equivalents/mL. Data from $n = 5$ subjects. Mean ± SD. (c) HPLC AMS profile of a bronchoalveolar macrophage extract obtained 12 hours after intravenous drug administration showing the presence of parent drug (data from taken from Lappin et al.[53]).

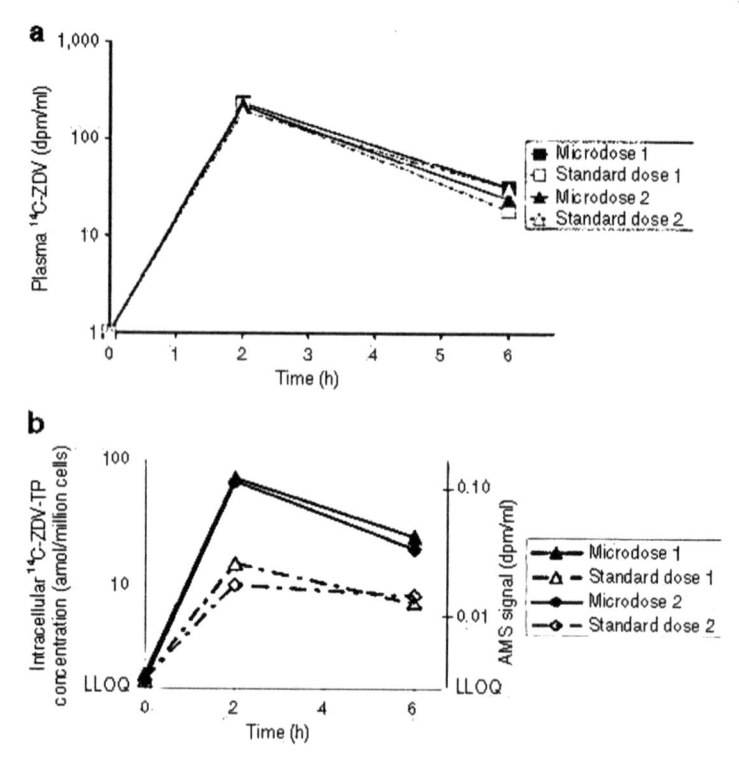

Figure 11.3 Comparison of plasma and intracellular 14C-labelled ZDV-TP concentrations after microdosing (100 μg ZDV/0.74 MBq) (solid line) and after a microdose combined with a 300 mg unlabelled ('standard') dose of ZDV (dashed line). (a) Plasma[14]C concentrations in two subjects. (b) Intracellular ZDV-TP concentrations in two subjects are plotted as amol/million cells determined by LC/MS/MS (left axis) or dpm/ml as detected by AMS (right axis).The numbers 1 and 2 refer to study subjects 1 and 2. AMS, accelerator mass spectrometry; amol, attomole; TP, triphosphate; ZDV, zidovudine (azidothymidine) (reproduced from Chen et al.[54]).

either 300 mg AZT/20 μCi ^{14}C-AZT or 100 μg AZT/20 μCi ^{14}C-AZT. Some of the results from this study are shown in Figure 11.3. Figure 11.3a shows total plasma concentration of parent drug after AZT microdose administration using AMS bioanalysis. Figure 11.3b shows AZT-TP concentrations in white blood cells using either AMS or LC/MS bioanalysis. Interestingly the AZT-TP concentrations are higher when measured using AMS than LC/MS. This was pointed out in Chen et al.[54] as an overestimation of AZT-TP when measured by AMS but it is just as likely that the LC/MS determination is an underestimate. The latter method relies on internal standards for quantitation while the AMS analysis requires no standard and is fully quantitative. Similar differences have been seen by the author in other studies where AMS and

LC/MS have been compared, as well as in studies where AMS has been compared with ELISA.[55]

The studies described in this section all support the concept that AMS has utility in measuring intracellular drug concentrations. It is interesting to speculate what other areas of drug development might lend themselves to intracellular drug concentration measurement. There seems to be no scientific reason, for example, why anti-infective drug concentrations could not be measured in target bacteria at infection sites, thus answering one of the key questions in this area, does the drug reach its bacterial target?

11.7 Special Applications of Microdosing

11.7.1 Use in Paediatric Studies

Most marketed drugs, as well as those under development, are focused on adult diseases and developing appropriate drug therapies. Drug treatment in the paediatric population, who comprise approximately 15% of the total European population, is all too often based on adult dosages. Knowledge of PK is pivotal in understanding a drug's safety and efficacy. PK data is essential in determining therapeutic dosages and avoiding adverse events. Absence of PK information can lead to (i) under-dosing, resulting in a lack of therapeutic benefit, or (ii) overdosing, resulting in adverse drug reactions. In adults, before a new drug can be marketed, extensive studies of its metabolism are conducted *in vitro*, in animal models and in volunteers and patients. This contrasts with the situation in paediatrics where there is widespread 'off-label' prescribing such that the administered drug has neither been tested for efficacy nor safety in this population. It has been estimated that between 11% and 80% of medicines prescribed to children in hospital are unlicensed for their age group.[56,57] We know that the ontogeny of many enzymes central to drug PK is determined by the age of the child.[58] As a consequence, there are specific drug metabolism issues for the paediatric population. Paediatric drug metabolism and PK may differ from adults for a variety of reasons including (i) differing fat content, decreased muscle, increased water, increased volume of distribution for water-soluble drugs; (ii) decreased concentrations of albumin; (iii) immature cytochrome P-450 subtypes and Phase II conjugation enzymes especially glucuronyl transferase and the unknown ontogeny of these enzymes as the child develops; and (iv) decreased glomerular filtration. These factors can lead to significantly different PK parameters in pre-term, neonates, infants, children and adolescents compared with adults. PK data has been difficult to obtain in paediatrics for a variety of reasons including (i) legal/ethical considerations; (ii) recruitment; (iii) study design; (iv) drug formulation; (v) sparse sampling; and (vi) small collected blood volumes causing analytical difficulties.[59]

There is a long history of adult drug PK studies where the above problems have been resolved. In the paediatric arena, however, none of these hurdles has been routinely overcome, which has led the EMA[60] and the US FDA[61] to publish a number of guidance documents to encourage drug developers to focus more attention on paediatric medicines. The EU has been particularly active, requesting that all applications for new Marketing Authorisations include a Paediatric Investigation Plan (PIP); applications without such a plan or waiver will be automatically refused. A major difficulty in paediatric clinical research is measuring drug and metabolite concentrations in the extremely small blood volume samples that are obtained especially in the 0–2 year old age group. The most sensitive analytical technology currently used for paediatric studies is LC/MS, which can measure parent drug in the pg/mL range. GC/MS or LC/MS has been used to study paediatric PK for a number of drugs,[62-64] but insensitivity in analysing drug concentrations in small volume samples is a significant barrier to their use. On the other hand AMS can measure drugs in the atto- to zeptogram mL^{-1} range (10^{-18} to 10^{-21} g mL^{-1}), thereby providing up to a million-fold increase in sensitivity over traditional PK techniques. AMS analysis does require the administered drug to be ^{14}C-labelled, but this handicap is overcome by the sensitivity of AMS, which allows the administration of radioactive doses that approach background levels of ^{14}C.

An EU-funded project known as PAMPER[65] is investigating the utility of AMS and microdosing in pre-term to 2 year-old infants to gain a better understanding of PK in this age group. AMS should have some benefits compared with other analytical techniques such as LC/MS including:

- analytical sensitivity;
- small sample size for analysis (no more than 10 µL plasma), permitting samples to be taken at multiple time points from the same baby without issues relating to the total blood volume taken;
- ability to use an intensive sampling regime;
- ng doses can be measured;
- generic analysis method irrespective of the drug structure and physico-chemical properties.

Three model drugs have been selected for study: (i) paracetamol, a very widely used analgesic and antipyretic in the paediatric population; (ii) spironolactone, a diuretic used mostly in conditions characterised by hyperaldosteronism such as congestive heart failure, liver disease and nephritic syndrome; and (iii) midazolam, a benzodiazepine drug used for sedation.

The human metabolism of paracetamol is catalysed by a combination of Phase I (mainly CYP3A4) and Phase II (glucuronosyl and sulphotransferase) metabolising enzymes. The precise ontogeny of these metabolising enzymes is not known but since the balance of metabolism between these various pathways is critical to paracetamol's potential toxic effects, understanding this balance better will assist in correct dosing.[66]

Each of the PAMPER drugs will be dosed at microdose levels in pre-term to 2 year-old infants, microlitre blood samples obtained at various times after drug administration and these analysed for parent drug and metabolites as appropriate using AMS bioanalysis. To date some 30 infants have been recruited and the plasma samples are currently being analysed by AMS.

A review of the use of AMS bioanalysis in paediatric studies has been published together with some preliminary information on the metabolism of ursodiol.[67] The results from the much more comprehensive PAMPER study will provide procedures and methods to microdose development drugs.

11.7.2 Use of AMS in Fluxomics

Understanding the dynamic changes of molecules within a cell, organ or tissue and how these are up- and down-regulated in health and disease, as well as the impact of drugs on these changes, forms the basis of the science of fluxomics.[68] The science is extremely broad in nature, ranging from single-cell studies[69] to whole individuals. Typically stable isotopes or deuterium are used for such studies but ^{14}C offers the advantage of a low background and high sensitivity, thus providing a re-examination of studies that might have failed previously using the former methods.

In the context of microdosing and AMS bioanalysis, consideration should be given to thinking not only about administration of drugs but also:

- ^{14}C-labelled enzyme substrates to examine substrate product relationships in the presence and absence of a drug designed to modulate the enzyme;
- ^{14}C-labelled ligand to study receptor/ligand binding in the presence and absence of drug; drug-induced alterations in plasma kinetics could indicate that competition between the ligand and drug is occurring at the receptor target;
- ^{14}C-labelled small molecule, peptide, oligonucleotide, protein, or other biopolymer to label an endogenous pool; once labelled the dynamics of the pool could be studied in the presence or absence of the investigational drug.

11.7.3 Microdosing of Protein Therapeutics

Approximately 50% of drugs currently under development are biologics and yet there have been no publications to date of human microdose biologics studies with AMS bioanalysis. Biologics have specific PK issues that most small molecules do not have, chief of which is the phenomenon of target-mediated disposition (TMD).[70] TMD is most often associated with drugs that have a high affinity to their target, which is reflected in the overall PK of the drug. Most notable are a dose- and time-dependent change in the apparent volume of distribution and a relatively long terminal elimination phase. Interestingly warfarin, one of the drugs in the CREAM trial, demonstrates TMD owing to its having a high-affinity, low-capacity binding site for its probable target vitamin K oxidoreductase (VKOR).[71] This results in PK non-linearity between a warfarin microdose and a therapeutic dose.[19,72]

Bearing in mind that TMD might be key to the overall PK profile of a biologic one could take advantage of this to determine the role of TMD for any particular antibody. In a preliminary study, TMD was investigated of a ^{14}C-labelled anti-CEA antibody in control and tumour-bearing mice.[73] ^{14}C-labelling the antibody was achieved *via* formylation, which is likely to be a general labelling procedure for studies using AMS bioanalysis of biologics. Clearance of the ^{14}C-labelled antibody was 2.28 mL h^{-1} in xenografted mice (LS147T colon cancer cells) compared with 0.98 mL h^{-1} in control mice; the mice received just 30 µg of labelled antibody containing approximately 100 Bq of ^{14}C. This study demonstrated that it was possible using chemical adduction of a biologic to study its PK with AMS. An earlier study[55] had demonstrated that PK could be measured of a biologically ^{14}C-labelled monoclonal antibody CAT-192 (metelimumab); ^{14}C-labelling in this study was achieved by adding ^{14}C-labelled amino acids to the growth medium of CHO cells expressing the recombinant monoclonal antibody. Clearance was estimated to be 7.3 × 10^{-4} mL h^{-1} g measured by ELISA and 4.6 × 10^{-4} mL h^{-1} g determined by AMS. The LLOQ was some 15-fold lower using AMS compared with ELISA – this could have been increased by using a higher specific radioactivity antibody.

These studies show that with the appropriate labelling methodology, it should be possible to consider human microdosing PK studies of a biologic with AMS bioanalysis. The methodology is likely to be most suitable for soluble therapeutic proteins. A recent announcement described the first human microdosing study of a biologic, namely human recombinant alkaline phosphatase (0.5 nmol dosed) being developed for immunomodulation; ^{14}C-labelling of the recombinant protein was *via* formylation.[74] The feasibility of the methodology used had been previously reported with erythropoietin labelled with ^{129}I.[75]

11.7.4 Drug–Drug Interaction Studies

The possible interaction of a development drug with another drug(s) with which it may be co-administered (polypharmacy) may lead to drug–drug interactions (DDIs) resulting in an alteration in PK of the development drug.[76,77] Interactions can occur at any step in the drug metabolism process, *e.g.* at the level of cytochrome P-450, efflux or uptake transporters, *etc.* resulting in an alteration of the AUC and hence giving rise to potential overdosing and toxicity. Both the FDA[7] and the EMA[6] have recently published guidance documents on measuring DDIs in humans which oblige companies to conduct *in vitro* and *in vivo* studies including human studies. Many DDIs are caused through competition between drugs at their site of metabolism or active transport. A cocktail of probe substrates is often used, each of which is metabolised by one or more of the major enzymes involved in drug metabolism. In a human cassette microdosing study[78] the metabolism of four probe substrates, namely midazolam (CYP3A4), tolbutamide (CYP2C9), caffeine (CYP1A2) and fexofenadine (Pgp efflux transporter) administered in a cocktail of 25 µg per probe was studied before and after 7 days' administration of ketoconazole (CYP3A4, CYP2C9 and Pgp inhibitor) and fluvoxamine (CYP1A2 and CYP2C9). The pharmacokinetics of each compound administered as a microdose were significantly altered after the administration of ketoconazole and fluvoxamine, showing statistically significant ($p < 0.01$) 12.8-, 8.1- and 3.2-fold increases in the area under the plasma concentration–time curve from time zero to infinity (AUC$_\infty$) for midazolam, caffeine and fexofenadine, respectively. A 1.8-fold increase (not statistically significant) in AUC$_\infty$ was observed for tolbutamide. While these findings are of interest, they provide no methodology that can be used to meet regulatory requirements. These make it clear that DDI studies should be conducted at therapeutic doses only. What is of interest from a scientific perspective in this study was the apparent ability to discern slow (CYP2C9*1/*2 and CYP2C9*1/*3) and normal (CYP2C9*1/*1) metabolisers of tolbutamide. While the number of subjects was small, the demonstration of polymorphism at a microdose is of interest. This result contrasts with a study examining polymorphism of the SLCO2B1 where polymorphism at a microdose was not seen.[79] Perhaps polymorphism in the CYP enzymes can be demonstrated at microdoses but not with ABC transporter substrates. In the latter case, deviation from linear PK may be more common although fexofenadine, a Pgp substrate, does show linear PK.[21] A report has been published on endeavouring to predict non-linearity of an orally dosed drug by plotting AUC from dose-ranging studies for a series of compounds against the dose normalised to the *in vitro* K_m for CYP3A4.[80] Although this may be a useful approach, oral absorption from the gut is complex, dependent as it is on active and passive transport processes, gut-metabolising enzyme activity, *etc.*

11.7.5 Absolute Bioavailability Determination

While not strictly a microdose study, because an extravascular therapeutic dose is administered concomitantly with an intravenous microdose, the determination of absolute bioavailability is currently probably the most practised application of microdosing. Absolute bioavailability studies have traditionally been conducted utilising a crossover study design in which the AUC$_\mathrm{oral}$ has been obtained followed by AUC$_\mathrm{IV}$. These values are then used to calculate absolute bioavailability (F). There are a number of issues with this approach, which have been reviewed[81,82] and will not be exhaustively covered here, but these include day-to-day variability as the oral and intravenous dosings are separated by time, formulation issues with the intravenous dose, a requirement to conduct intravenous toxicology in two species, one of which must be a non-rodent, even though the drug may be for oral use only, *etc.* The concept of administering an intravenous dose of stable isotope-labelled drug alongside an oral therapeutic dose was reported in the 1970s.[83] Blood samples were analysed by LC/MS for parent and stable isotope-labelled drug. While this study design provides better absolute bioavailability data than using a crossover design, the use of stable isotopes such as ^{13}C or deuterium has its own problems. These methods work on the presumption that the isotope-labelled and unlabelled molecules behave similarly from a metabolism standpoint and that there is no kinetic isotope effect. The vibration frequency of chemical bonds is dependent on the mass of the bonded isotope. Heavier isotopes require more energy for their bonds to be broken and so reaction rates are slowed. This contrasts with the situation between ^{14}C and ^{12}C where the kinetic isotope effect is so small that no metabolic isotope effect can be discerned. Hence studies utilising ^{14}C-labelled drugs to measure absolute bioavailability are preferred to the use of heavier isotopes. ^{13}C has issues relating to its natural abundance (1.1% of ^{12}C), such that the background signal severely impacts on assay sensitivity. Contrast this with ^{14}C where the natural background is approximately 1×10^{-12} ^{12}C and hence ^{14}C studies are so much more sensitive than stable isotope or ^{13}C studies. In 2005 the first absolute bioavailability study[84] was reported in which an intravenous microdose of ^{14}C-nelfinavir was co-administered with an oral therapeutic dose; plasma concentrations of nelfinavir were measured by LC and of ^{14}C-nelfinavir by AMS. Nelfinavir plasma steady-state levels of drug were achieved by daily oral dosing of nelfinavir for 10 days; at the end of this period an intravenous microdose of ^{14}C-nelfinavir was administered so that a comparison of the absolute bioavailability could be made after single and multiple dosing of cold drug. Bioavailability of nelfinavir was 0.88 on day 1, suggesting good absorption of nelfinavir in the gastrointestinal tract. A decrease in nelfinavir bioavailability was seen from 0.88 (day 1) to 0.47 (day 11) and this was attributed to metabolic enzyme induction of CYP3A4 after repeated dosing. This study design has applicability when it is thought metabolising enzyme induction or inhibition takes place. Other studies using an intravenous microdose protocol with AMS analysis have been reported of drugs in which data was submitted as part of a regulatory package. These include saxagliptin (Onglyza) and dapagliflozin, drugs used for the treatment of type II diabetes,[85] dabrafenib, a BRAF inhibitor[86] in cancer patients and vismodegib, a hedgehog pathway inhibitor in healthy female subjects.[87]

There are a number of advantages to measuring absolute bioavailability with an intravenous ^{14}C-labelled drug administered concomitantly with an extravascular therapeutic dose:

- Superior study design with no day-to-day variability issues. Timing of when the intravenous microdose is administered is important because it should be around the t_{max} of blood concentration.
- No formulation issues as the intravenous dose is a microdose.
- No requirement to conduct intravenous preclinical toxicology or local irritancy testing.
- Intravenous dose can be administered as part of a Phase I study design such as in a food interaction or DDI study.

- Intravenous dose can be administered in a multiple ascending dose study design to establish enzyme induction or inhibition.
- Patient studies can be conducted in a normal hospital environment owing to the small radioactive dose of drug being administered.

On the other hand there are some practical issues that need to be considered such as:

- Ensuring that the microdose of drug does not alter the pool size of drug administered extravascularly.
- Ensuring that the drug is given at the optimum time in relation to t_{max}.
- Binding tests to ensure that none of the intravenous dose of drug binds to infusion lines, containers, *etc*. It is very important to know the exact dose of drug being administered as the final absolute bioavailability value depends on dose normalisation.
- Accurate knowledge of the parent drug extraction efficiency from blood. The author has experience of absolute bioavailability studies in which values of F considerably greater than 100% were obtained; this is scientifically impossible. It was found after careful review of the data and experimental methodology that variable extraction efficiencies were responsible.
- Knowing with high accuracy the dose of drug administered, its specific radioactivity and the radiolabelled dose.

This absolute bioavailability method provides the pharmacokineticist with accurate measures of F, V and CL, permitting a better understanding of the drug's metabolism properties which cannot be obtained by oral dosing alone. The data helps the drug formulator to know whether formulation can improve bioavailability, particularly if the extravascular dose is administered as a solution. In this situation the true solution kinetics of the drug are being measured.

11.8 Conclusions

Microdosing to obtain human PK data early on in drug development offers advantages over traditional approaches of extensive preclinical metabolism studies to predict human PK. On the basis that the best model for humans are humans, microdosing permits the drug developer a reduction in preclinical testing and a sharper focus on humans. When the microdosing concept was originally proposed there was much scepticism that it offered any scientific benefits and that microdosing PK data would not correlate with therapeutic dose data. The large number of validation studies and microdose studies on development drugs has demonstrated that these concerns were misplaced. Predictivity looks to be in the range of 70–80%, a number that is unlikely to change from now. We have a good idea of the types of molecules that show non-linearity, such as drugs with high-affinity, low-capacity binding sites or where the rate-determining step in metabolism might be through ABC transporters. Drugs that are given at high therapeutic doses might also be more likely to show non-linear PK. On the other hand, as we now attempt to design drugs with a high affinity for their target, which are not extensively metabolised and where we seek once a day dosing, then microdosing should be appropriate for these molecules. The potency of today's drugs is higher than drugs developed years ago. Therapeutic doses of drugs have consistently reduced decade by decade, resulting in therapeutic doses that are likely to lie on the linear portion of the PK curve.

In contrast to the above benefits, the pharmaceutical industry has not widely adopted microdosing in their critical development plans. Reasons for this are given as microdosing studies are too expensive, delay the drug development process, require the synthesis and formulation of ^{14}C-labelled drugs, there are too few AMS providers and those that do provide this service are small companies with very limited budgets. It may well be in the future that microdosing is much more widely used, but using analytical techniques other than AMS such as LC/MS/MS. Most big pharma companies have their own equipment and so are keen to use this where possible. On the other hand the use of ^{14}C-labelled drugs offers benefits over unlabelled drugs, which in the author's opinion outweigh some of the reasons advanced for not conducting microdose studies with AMS bioanalysis. The costs of AMS analysis have dropped over the past few years, down to around US$100 per sample, a sum similar to what one would expect to pay for a high-resolution mass spectrometer analysis. This is likely to reduce still further as LC hyphenation to AMS becomes a reality and the cost and size of AMS instruments reduces even further. Furthermore ethical and environmental considerations, as well as public and political pressure, will become more prominent, forcing pharmaceutical companies to build microdosing into their drug development programmes.

The new paradigm for human PK prediction at therapeutic doses will encompass *in vitro* metabolism studies with human hepatocytes or microsomes to estimate clearance together with microdose PK data. These PK parameters will be entered into an appropriate *in silico* PBPK model to calculate PK at therapeutic doses after single and multiple dose administration. This paradigm will enable the goal of introducing new drugs to treat life-threatening disease faster, more safely and more cheaply than current methods can achieve.

References

1. J. A. DiMasi, R. W. Hansen and H. G. Grabowski, *J. Health Econ.*, 2003, **22**, 151.
2. http://www.forbes.com/sites/matthewherper/2012/02/10/the-truly-staggering-cost-of-inventing-new-drugs/.

3. J. W. Scannell, A. Blanckley, H. Boldon and B. Warrington, *Nat. Rev. Drug Discovery,* 2012, **11**, 191.
4. J. A. DiMasi, L. Feldman, A. Seckler and A. Wilson, *Clin. Pharmacol. Ther.,* 2010, **87**, 272.
5. FDA, U., *Innovation, stagnation. Challenge and Opportunity on the Critical Path to New Medical Products.* US Department of Health and Human Services, Food and Drug Administration, Center for Drug Evaluation and Research, 2004.
6. EMA *Guideline on the investigation of drug interactions.* 2012.
7. FDA, U., *Guidance for Industry. Drug Interaction Studies - Study Design, Data Analysis, Implications for Dosing and Labeling Recommendations.* US Department of Health and Human Services, Food and Drug Administration, Center for Drug Evaluation and Research, 2012.
8. FDA, U., *Guidance for Industry. Safety Testing of Drug Metabolites.* US Department of Health and Human Services, Food and Drug Administration, Center for Drug Evaluation and Research, 2008.
9. R. Vuppugalla, P. Marathe, H. He, R. D. Jones, J. W. Yates, H. M. Jones, C. R. Gibson, J. Y. Chien, B. J. Ring, K. K. Adkison, M. S. Ku, V. Fischer, S. Dutta, V. K. Sinha, T. Björnsson, T. Lavé, P. Poulin. *J. Pharm. Sci.,* 2011. Apr 7. doi: 10.1002/jps.22551.
10. P. Poulin, R. D. O. Jones, H. M. Jones, C. R. Gibson, M. Rowland, J. Y. Chien, B. J. Ring, K. K. Adkison, M. S. Ku, H. He, R. Vuppugalla, P. Marathe, V. Fischer, S. Dutta, V. K. Sinha, T. Bjornsson, T. Lave and J. W. T. Yates, *J. Pharm. Sci.,* 2011, **100**, 4127.
11. I. Kola and J. Landis, *Nat. Rev. Drug Discovery,* 2004, **3**, 711.
12. R. C. Garner., *Accelerating Drug Development by Accelerator Mass Spectrometry - a New Ultrasensitive Analytical Method.* European Pharmaceutical Contractor, 1999.
13. B. C. Cupid, T. J. Lightfoot, D. Russell, S. J. Gant, P. C. Turner, K. H. Dingley, K. D. Curtis, S. H. Leveson, K. W. Turteltaub and R. C. Garner, *Food Chem. Toxicol.,* 2004, **42**, 559.
14. T. Lightfoot, J. M. Coxhead, B. C. Cupid, S. Nicholson and R. C. Garner, *Mutat. Res.,* 2000, **472**, 119.
15. K. H. Dingley, S. P. Freeman, D. O. Nelson, R. C. Garner and K. W. Turteltaub, *Drug Metab. Dispos.,* 1998, **26**, 825.
16. E. A. Martin, K. Brown, M. Gaskell, F. Al-Azzawi, R. C. Garner, D. J. Boocock, E. Mattock, D. W. Pring, K. Dingley, K. W. Turteltaub, L. L. Smith and I. N. White, *Cancer Res.,* 2003, **63**, 8461.
17. K. W. Turteltaub, K. H. Dingley, K. D. Curtis, M. A. Malfatti, R. J. Turesky, R. C. Garner, J. S. Felton and N. P. Lang, *Cancer Lett.,* 1999, **143**, 149.
18. P. Williams, P. Zannikos, I. Weaner, R. J. Stubbs, S. P. Van Marle, R. C. Garner and Hutchison, *J., Clin. Pharmacol. Ther.,* 2002, **71**, 75.
19. G. Lappin, W. Kuhnz, R. Jochemsen, J. Kneer, A. Chaudhary, B. Oosterhuis, W. J. Drijfhout, M. Rowland and R. C. Garner, *Clin. Pharmacol. Ther.,* 2006, **80**, 203.

20. G. Lappin, Y. Shishikura, R. Jochemsen, R. J. Weaver, C. Gesson, J. Brian Houston, B. Oosterhuis, O. J. Bjerrum, G. Grynkiewicz, J. Alder, M. Rowland and C. Garner, *Eur. J. Pharm. Sci.*, 2011, **43**, 141.
21. G. Lappin, Y. Shishikura, R. Jochemsen, R. J. Weaver, C. Gesson, B. Houston, B. Oosterhuis, O. J. Bjerrum, M. Rowland and C. Garner, *Eur. J. Pharm. Sci.*, 2010, **40**, 125.
22. X. J. Zhou, R. C. Garner, S. Nicholson, C. J. Kissling and D. Mayers, *J Clin. Pharmacol.*, 2009, **49**, 1408.
23. A. Madan, Z. O'Brien, J. Wen, C. O'Brien, R. H. Farber, G. Beaton, P. Crowe, B. Oosterhuis, R. C. Garner, G. Lappin and H. P. Bozigian, *Br. J. Clin. Pharmacol.*, 2009, **67**, 288.
24. EMEA, Position *Paper On Non-Clinical Safety Studies To Support Clinical Trials With A Single Microdose.* 2004. CPMP/SWP/2599/02/Rev 1.
25. FDA, U., *Guidance for Industry. Exploratory IND Studies.* US Department of Health and Human Services, Food and Drug Administration, Center for Drug Evaluation and Research, 2006.
26. Agency, E.M., *ICH Topic M 3 (R2) Non-clinical Safety Studies for the Conduct of Human Clinical Trials and Marketing Authorization for Pharmaceuticals.* 2009.
27. J. Barker and R. C. Garner, *Rapid Commun. Mass Spectrom.*, 1999, **13**, 285.
28. S. R. Dueker, T. Vuong le, P. N. Lohstroh, J. A. Giacomo and J. S. Vogel, *Adv. Drug Delivery Rev.*, 2011, **63**, 518.
29. R. C. Garner, I. Goris, A. A. Laenen, E. Vanhoutte, W. Meuldermans, S. Gregory, J. V. Garner, D. Leong, M. Whattam, A. Calam and C. A. Snel, *Drug Metab. Dispos.*, 2002, **30**, 823.
30. T. Vuong le, J. L. Ruckle, A. B. Blood, M. J. Reid, R. D. Wasnich, H. A. Synal and S. R. Dueker, *J. Pharm. Sci.*, 2008, **97**, 2833.
31. S. K. Balani, N. V. Nagaraja, M. G. Qian, A. O. Costa, J. S. Daniels, H. Yang, P. R. Shimoga, J. T. Wu, L. S. Gan, F. W. Lee and G. T. Miwa, *Drug Metab. Dispos.*, 2006, **34**, 384.
32. N. Yamane, Z. Tozuka, M. Kusama, K. Maeda, T. Ikeda and Y. Sugiyama, *Pharm. Res.*, 2011, **28**, 1963.
33. K. Maeda and Y. Sugiyama, *Adv. Drug Delivery Rev.*, 2011, **63**, 532.
34. R. Boyd and R. Lalonde, *Clin. Pharmacol. Ther.*, 2007, **81**, 24.
35. R. C. Garner, *Bioanal*, 2010, **2**, 429.
36. K. Maeda, Y. Ikeda, T. Fujita, K. Yoshida, Y. Azuma, Y. Haruyama, N. Yamane, Y. Kumagai and Y. Sugiyama, *Clin. Pharmacol. Ther.*, 2011, **90**, 575.
37. Y. Sugiyama, *Drug Metab. Pharmacokinet.*, 2009, **24**, 127.
38. Z. Tozuka, H. Kusuhara, K. Nozawa, Y. Hamabe, I. Ikushima, T. Ikeda and Y. Sugiyama, *Clin. Pharmacol. Ther.*, 2010, **88**, 824.
39. T. Tachibana, M. Kato, J. Takano and Y. Sugiyama, *Curr. Drug Metab.*, 2010, **11**, 762.
40. M. Rowland, *J. Pharm. Sci.*, 2012, **101**, 4067.
41. T. P. Hill, *Cancer Med. Sci.*, 2012, **6**, 248.
42. C. Kurihara, *Adv. Drug Delivery Rev.*, 2011, **63**, 503.

43. R. D. Combes, T. Berridge, J. Connelly, M. D. Eve, R. C. Garner, S. Toon and P. Wilcox, *Eur. J. Pharm. Sci.*, 2003, **19**, 1.
44. G. Watts, *Br. Med. J.*, 2007, **334**, 182.
45. Y. Minamide, Y. Osawa, H. Nishida, H. Igarashi and S. Kudoh, *J. Sep. Sci.*, 2011, **34**, 1590.
46. A.-F. Aubry, *Bioanal*, 2011, **3**, 1819.
47. N. Yamane, T. Takami, Z. Tozuka, Y. Sugiyama, A. Yamazaki and Y. Kumagai, *Drug Metab. Pharmacokinet.*, 2009, **24**, 389.
48. H. Kusuhara, S. Ito, Y. Kumagai, M. Jiang, T. Shiroshita, Y. Moriyama, K. Inoue, H. Yuasa and Y. Sugiyama, *Clin. Pharmacol. Ther.*, 2011, **89**, 837.
49. N. Yamane, A. Igarashi, M. Kusama, K. Maeda, T. Ikeda and Y. Sugiyama, *Drug Metab. Pharmacokinet.*, 2013, **28**, 187.
50. N. A. Hosea, W. T. Collard, S. Cole, T. S. Maurer, R. X. Fang, H. Jones, S. M. Kakar, Y. Nakai, B. J. Smith, R. Webster and K. Beaumont, *J. Clin. Pharmacol.*, 2009, **49**, 513.
51. A. Rostami-Hodjegan and G. T. Tucker, *Nat. Rev. Drug Discovery*, 2007, **6**, 140.
52. G. M. Grass and P. J. Sinko, *Adv. Drug Delivery Rev.*, 2002, **54**, 433.
53. G. Lappin, M. J. Boyce, T. Matzow, S. Lociuro, M. Seymour and S. J. Warrington, *Eur. J. Clin. Pharmacol.*, 2013, **69**, 1673.
54. J. Chen, R. C. Garner, L. S. Lee, M. Seymour, E. J. Fuchs, W. C. Hubbard, T. L. Parsons, G. E. Pakes, C. V. Fletcher and C. Flexner, *Clin. Pharmacol. Ther.*, 2010, **88**, 796.
55. G. Lappin, R. C. Garner, T. Meyers, J. Powell and P. Varley, *J. Pharm. Biomed. Anal.*, 2006, **41**, 1299.
56. P. H. Caldwell, S. B. Murphy, P. N. Butow and J. C. Craig, *Lancet*, 2004, **364**, 803.
57. C. Pandolfini and M. Bonati, *Eur. J. Pediatr.*, 2005, **164**, 552.
58. R. N. Hines, *Int. J. Pharm.*, 2013, **452**, 3.
59. S. Abdel-Rahman, M. D. Reed, T. G. Wells and G. L. Kearns, *Clin. Pharmacol. Ther.*, 2007, **81**, 483.
60. E. Parliament., *Regulation (EC) No 1901/2006 of the European Parliament and of the Council*. 2006.
61. U. FDA., *Best Pharmaceuticals for Children Act*. 2002.
62. R. J. Meesters, J. J. van Kampen, M. L. Reedijk, R. D. Scheuer, L. J. Dekker, D. M. Burger, N. G. Hartwig, A. D. Osterhaus, T. M. Luider and R. A. Gruters, *Anal. Bioanal. Chem.*, 2010, **398**, 319.
63. P. Patel, H. Mulla, S. Tanna and H. Pandya, *Arch. Dis. Child*, 2010, **95**, 484.
64. M. F. Suyagh, P. L. Kole, J. Millership, P. Collier, H. Halliday and J. C. McElnay, *J. Chromatogr. B Analyt. Technol. Biomed. Life Sci.*, 2010, **878**, 769.
65. http://www.priomedchild.eu/fileadmin/cm/wetenschap_en_innovatie/priomedchild/PRIOMEDCHILD_ultrasensitive_rev2.pdf.
66. J. G. Bessems and N. P. Vermeulen, *CRC Crit. Rev. Toxicol.*, 2001, **31**, 55.
67. L. T. Vuong, A. B. Blood, J. S. Vogel, M. E. Anderson and B. Goldstein, *Bioanal*, 2012, **4**, 1871.

68. D. Mueller E. Heinzle. *Curr. Opin. Biotechnol.*, 2012.
69. F. Mannello, D. Ligi and M. Magnani, *Exp. Rev. Proteom.*, 2012, **9**, 635–648.
70. W. Wang, E. Q. Wang and J. P. Balthasar, *Clin. Pharmacol. Ther.*, 2008, **84**, 548.
71. J. Oldenburg, M. Watzka, S. Rost and C. R. Müller, *J. Thromb Haemost.*, 2007, **5**, 1.
72. G. Levy, D. E. Mager, W. K. Cheung and W. J. Jusko, *J. Pharm. Sci.*, 2003, **92**, 985.
73. G. Lappin, T. Oxley, S.-C. Chen, J. Balthasar http://www.xceleron.com/wp-content/uploads/2011/11/TMD-of-14C-Anti-CEA-using-AMS-T3359.pdf, 2012.
74. http://www.tno.nl/downloads/pb_2013_26_en_supplement.pdf.
75. R. J. Lamers, A. F. de Jong, J. M. López-Gutiérrez and J. Gómez-Guzmán, *Bioanal*, 2013, **5**, 53.
76. H. Yu and D. Tweedie, *Drug Metab. Dispos.*, 2013, **41**(3), 536.
77. T. D. Bjornsson, J. T. Callaghan, H. J. Einolf, V. Fischer, L. Gan, S. Grimm, J. Kao, S. P. King, G. Miwa, L. Ni, G. Kumar, J. McLeod, R. S. Obach, S. Roberts, A. Roe, A. Shah, F. Snikeris, J. T. Sullivan, D. Tweedie, J. M. Vega, J. Walsh and S. A. Wrighton, *Drug Metab. Dispos.*, 2003, **31**, 815.
78. M. Croft, B. Keely, I. Morris, L. Tann and G. Lappin, *Clin. Pharmacokinet.*, 2012, **51**, 237.
79. I. Ieiri, Y. Doi, K. Maeda, T. Sasaki, M. Kimura, T. Hirota, T. Chiyoda, M. Miyagawa, S. Irie, K. Iwasaki and Y. Sugiyama, *J. Clin. Pharmacol.*, 2012, **52**, 1078.
80. T. Tachibana, M. Kato and Y. Sugiyama, *Pharm. Res.*, 2012, **29**, 651.
81. G. Lappin, M. Rowlands and R. C. Garner, *Exp. Opin. Drug Metab. Toxicol.*, 2006, **2**, 1.
82. M. E. Arnold and F. LaCreta, *Bioanal*, 2012, **4**, 1831.
83. J. M. Strong, J. S. Dutcher, W. K. Lee and A. J. Atkinson Jr, *Clin. Pharmacol. Ther.*, 1975, **18**, 613.
84. N. Sarapa, P. H. Hsyu, G. Lappin and R. C. Garner, *J. Clin. Pharmacol.*, 2005, **45**, 1198.
85. D. W. Boulton, S. Kasichayanula, C. F. Keung, M. E. Arnold, L. J. Christopher, X. S. Xu and F. Lacreta, *Br. J. Clin. Pharmacol.*, 2013, **75**, 763.
86. C. L. Denton, E. Minthorn, S. W. Carson, G. C. Young, L. E. Richards-Peterson, J. Botbyl, C. Han, R. A. Morrison, S. C. Blackman and D. Ouellet, *J. Clin. Pharmacol.*, 2013, **53**, 955.
87. R. A. Graham, B. L. Lum, G. Morrison, I. Chang, K. Jorga, B. Dean, Y. G. Shin, Q. Yue, T. Mulder, V. Malhi, M. Xie, J. A. Low and C. E. Hop, *Drug Metab. Dispos.*, 2011, **39**, 1460.

Subject Index

absolute bioavailability determination, and microdosing, 259–261
Absorption, Distribution, Metabolism, Excretion and Toxicity (ADMET)
    impact of early human pharmacokinetics prediction, 116–117
    optimisation loop, 113–116
    overview, 110–113
    prediction tools, 117–126
        blood-brain barrier challenge, 117–121
        mechanisms of human toxicity, 121–125
        power of multiparametric screening in, 125–126
    science of, 112–113
    in vitro and *in vivo*, 126–128
accelerator mass spectrometry (AMS), 243
access policies, of human cells and tissues, 12–13
A549 cell, and respiratory system models, 71
adipose tissue, and body-on-a-chip technology, 143–144
ADME, and PCTS, 43–47
ADMET. *See* Absorption, Distribution, Metabolism, Excretion and Toxicity (ADMET)
Adverse Events Reporting System (AERS), 215
AERS (Adverse Events Reporting System), 215
AHR (airway hyper-responsiveness), 50
air–liquid interface (ALI) vs. submerged conditions, 77–78
airway hyper-responsiveness (AHR), 50
airway physiology and pharmacology, and PCTS, 51–53
ALI (air–liquid interface) vs. submerged conditions, 77–78
AMS (accelerator mass spectrometry), 243
analytical technologies, used for microdosing, 247–249
annotation, of human cells and tissues, 13–14
anti-inflammatory kinase inhibitors, 95–105
    JAK kinase inhibitors, 95–101
        analysis of tofacitinib in BioMAP Systems, 98–101
        background, 95–96
        JAK3 and tofacitinib, 97–98
    SYK kinase inhibitors, 101–105
        analysis of fostamatinib in BioMAP Systems, 103–105
        background, 101–102
        SYK and fostamatinib, 102–103

BBB (blood–brain barrier), 149
BDL (bile duct ligated) rats, 49

BEAS-2B cell, and respiratory system models, 71
bile duct ligated (BDL) rats, 49
biological data representation, and *in silico* analysis, 199
biological efficacy or toxicity, *in silico* analysis, 205–207
biological variability, *in silico* models and simulations, 230–234
BioMAP Systems, 90–95
    fostamatinib analysis in, 103–105
    tofacitinib analysis in, 98–101
blood–brain barrier (BBB), 149
    and ADMET, 117–121
blood surrogate, and body-on-a-chip technology, 150–151
body-on-a-chip technology
    functional single-organ *in vitro* models, 136–151
        adipose tissue, 143–144
        blood surrogate, 150–151
        central nervous system (CNS), 144–149
        gastrointestinal (GI) tract, 139–141
        heart/cardiac tissue, 137
        kidney, 143
        liver, 141–142
        lung, 137–139
        overview, 136–137
        peripheral nervous system (PNS), 144–149
        skeletal muscle, 150
    microfluidics, 151–153
    multi-organ *in vitro* models, 133–136
    overview, 132–133
brain networks, 145
brain uptake index (BUI), 118
BUI (brain uptake index), 118

Calu-3 cell, and respiratory system models, 70
cancer research, PCTS, 54–57
carcinoma-derived cells, and respiratory system models, 70–71
cardiac electrophysiology, computational
    overview, 220
    simulating interactions of medicines, 226–230
    single cardiomyocyte models, 220–223
    whole organ heart models, 223–226
cardiotoxicity, and iPSC-based disease models, 177–178
cardiovascular disease models, iPSC-based, 169–172
case studies, *in silico* analysis
    prediction of human adverse events, 215–216
    regulatory, 212–215
cell lines, and respiratory system models, 70
    A549 cell, 71
    BEAS-2B cell, 71
    Calu-3 cell, 70
    carcinoma-derived cells, 70–71
    16HBE14o- cells, 71–72
    immortalised cells, 71–72
cells and tissues, respiratory system models, 70–79
    cell lines, 70
        A549 cell, 71
        BEAS-2B cell, 71
        Calu-3 cell, 70
        carcinoma-derived cells, 70–71
        16HBE14o- cells, 71–72
        immortalised cells, 71–72
    ex vivo models, 74–76
        human isolated perfused lungs (hIPL), 75–76
        precision cut lung slices (PCLS), 75
        tissue explants, 75

Subject Index

primary and stem cells, 72–74
  primary cells, 72–73
  stem cells, 73–74
cell/seeding density, 76
central nervous system (CNS), and body-on-a-chip technology, 144–149
CFTR (cystic fibrosis conductance regulator), 71
chemical database search, and *in silico* analysis, 199–201
chemical representation, and *in silico* analysis, 196–198
chemical scaffolds, in molecular fragment descriptors, 204
chemical structures and associated data (*in silico* analysis), 196–199
  biological data representation, 199
  calculated data, 199
  chemical representation, 196–198
  overview, 196
chemotherapeutics, and PCTS, 55–57
chronic obstructive pulmonary disease (COPD), 50
clinical efficacy, and human isolated tissues, 30–31
clustering, and *in silico* analysis, 205–206
CNS (central nervous system), and body-on-a-chip technology, 144–149
co-culture conditions, in human-based respiratory models, 78–79
common chemical scaffolds, in molecular fragment descriptors, 204
Common Rule, 3
complex primary human cell systems
  anti-inflammatory kinase inhibitors, 95–105
    JAK kinase inhibitors, 95–101

    SYK kinase inhibitors, 101–105
  and kinase inhibitors, 89–90
  anti-inflammatory, 95–105
  and target-based drug discovery, 88–89
  for translational biology, 90–95
    BioMAP Systems, 90–95
computational cardiac electrophysiology
  overview, 220
  simulating interactions of medicines, 226–230
  single cardiomyocyte models, 220–223
  whole organ heart models, 223–226
connection table, 196
consent coverage, in human cells and tissues research, 8
continuous *vs.* static flow conditions, 78
contract research organisation (CRO), 22
COPD (chronic obstructive pulmonary disease), 50
CREAM study, 244
CRISPR-Cas9 system, 173
CRO (contract research organisation), 22
cystic fibrosis conductance regulator (CFTR), 71

DDIs (drug-drug interaction studies), 114, 258–259
decision trees, and *in silico* analysis, 207
DILI (drug-induced liver injury), 121
DMPK (drug metabolism and pharmacokinetics), 111
dorsal root ganglion (DRG) assay, 27
DRG (dorsal root ganglion) assay, 27
drug discovery
  approaches to, 164

drug discovery (*continued*)
　integration of iPSCs in, 179–181
　and iPSC-based disease models, 179–181
　　challenges, 179–180
　　future directions, 180
　　personalised medicine, 180–181
drug–drug interaction studies (DDIs), 114, 258–259
drug efficacy assessment, 173–176
　drug screens on, 176
　examples of drug testing to validate, 175–176
　target-based screening *vs.* phenotypic screening, 173–175
drug-induced liver injury (DILI), 121
drug metabolism and pharmacokinetics (DMPK), 111
drug screens, on drug efficacy assessment, 176
drug targeting, and microdosing, 252–255
drug testing examples, in drug efficacy assessment, 175–176
dynamic organ culture systems, 42

EBiSC (European bank for induced pluripotent stem cells), 182
ECG (electrocardiogram), 225
ECM (extracellular matrix), 76
ECs (epithelial cells), 55
EFS (electrical field stimulation), 51–53
electrical field stimulation (EFS), 51–53
electrocardiogram (ECG), 225
embryonic stem cells, 165–166
Environmental Protection Agency (EPA), 69
EPA (Environmental Protection Agency), 69
epithelial cells (ECs), 55

ethical considerations, in microdosing, 247
ethics and biological samples of, human cells and tissues, 1–2
EUMAPP study, 244
European bank for induced pluripotent stem cells (EBiSC), 182
European Medicines Agencies (EMA), 113
extracellular matrix (ECM), 76
extrinsic apoptosis pathway, 55
ex vivo models, and respiratory system, 74–76
　human isolated perfused lungs (hIPL), 75–76
　precision cut lung slices (PCLS), 75
　tissue explants, 75

FBS (foetal bovine serum), 71
FDA (Food and Drug Administration), 69
fibrotic diseases, and PCTS, 47–50
fluxomics, and microdosing, 257
foetal bovine serum (FBS), 71
Food and Drug Administration (FDA), 69
fostamatinib
　analysis in BioMAP Systems, 103–105
　and SYK kinase inhibitors, 102–103
FRAME (Fund for the Replacement of Animals in Medical Experiments), 67
functional single-organ *in vitro* models, 136–151
　adipose tissue, 143–144
　blood surrogate, 150–151
　central nervous system (CNS), 144–149
　gastrointestinal (GI) tract, 139–141
　heart/cardiac tissue, 137

Subject Index

kidney, 143
liver, 141–142
lung, 137–139
overview, 136–137
peripheral nervous system (PNS), 144–149
skeletal muscle, 150
functional tissues, and human isolated tissues, 32–34
Fund for the Replacement of Animals in Medical Experiments (FRAME), 67

gastrointestinal diseases, and human isolated tissues, 28–29
gastrointestinal (GI) tract, 139–141
general culture conditions, in human-based respiratory models, 76–79
    co-culture, 78–79
    static *vs.* continuous flow, 78
    submerged *vs.* ALI, 77–78
genome editing tools, 173
genotoxicity, and ADMET, 121–122
GFP (green fluorescent protein), 121
GI (gastrointestinal) tract, 139–141
glandular and skin tissues, human isolated tissues, 28–29
green fluorescent protein (GFP), 121
γ-secretase modulators (GSMs), 175

HAEpC (human airway epithelial cells), 72–73
hand held coring tool (Vitron), 41
HCS (high-content screening), 125
HCV (hepatitis C virus) infection, and PCTS, 57
Health and Human Services (HHS), 3
Health Insurance Portability and Accountability Act (HIPPA) of 1996, 3
heart/cardiac tissue, and body-on-a-chip technology, 137
hepatitis C virus (HCV) infection, and PCTS, 57

hepatotoxicity, and iPSC-based disease models, 178
HHS (Health and Human Services), 3
high-content screening (HCS), 125
high-content toxicology, and ADMET, 124–125
hIPL (human isolated perfused lungs), 75–76
HIPPA (Health Insurance Portability and Accountability Act) of 1996, 3
HipSci (human induced pluripotent stem cells initiative), 183
hormone replacement therapy (HRT), 34
HRT (hormone replacement therapy), 34
human adverse events prediction case study, 215–216
human airway epithelial cells (HAEpC), 72–73
human-based respiratory models
    cells and tissues, 70–79
        A549 cell, 71
        BEAS-2B cell, 71
        Calu-3 cell, 70
        carcinoma-derived cells, 70–71
        cell lines, 70
        ex vivo models, 74–76
        16HBE14o- cells, 71–72
        immortalised cells, 71–72
        isolated perfused lungs, 75–76
        precision cut lung slices, 75
        primary cells, 72–73
        stem cells, 73–74
        tissue explants, 75
    general culture conditions, 76–79
        co-culture, 78–79
        static *vs.* continuous flow, 78
        submerged *vs.* ALI, 77–78
    inhalation toxicology, 67–68
    overview of, 66–67

human-based respiratory models (*continued*)
   and safety assessment, 68
   trends and technology, 69
   in vitro toxicology, 68–69
human cells and tissues
   and access policies, 12–13
   annotation, 13–14
   ethics and biological samples of, 1–2
   legal issues, 2–8
      consent coverage, 8
      human participation in research, 3–4
      ownership of human samples, 4–8
   practical issues regarding, 8–12
      accessing material from diagnostic archive, 11–12
      obtaining, with consent for research, 9–11
human induced pluripotent stem cells initiative (HipSci), 183
human isolated perfused lungs (hIPL), 75–76
human isolated tissues
   applications, 29–34
      development of personalised medicines, 32–33
      predicting clinical efficacy, 30–31
      predicting safety and toxicology risks, 31–32
   experimental approaches to, 23–29
      and gastrointestinal diseases, 28–29
      organoculture systems and precision-cut tissue slices, 25–28
      perfusion myographs, 25
      skin and glandular tissue, 28–29
      tissue baths and wire myographs, 23–25
   overview, 17–20
   sourcing, storing and transporting, 21–22
human organs and tissues, and PCTS, 40–41
human participation, in cells and tissues research, 3–4
human pharmacokinetics prediction, and ADMET, 116–117
human phase 0 studies (microdosing). *See* microdosing (human phase 0 studies)
human stem cells, and drug discovery
   advances in, 165–167
   embryonic stem cells, 165–166
   existing approaches, 164
   iPSC-based disease models, 167–173
      cardiovascular, 169–172
      genome editing tools, 173
      neurological and psychiatric, 167–169
   overview, 162–164
   reprogramming somatic cells, 166–167
human toxicity mechanisms, and ADMET, 121–125
   genotoxicity, 121–122
   high-content toxicology, 124–125
   mitochondrial toxicity, 122–123
   phospholipidosis, 123–124
   reactive metabolites formation, 123

ICH (International Committee on Harmonisation), 213
IDM (indicator diffusion method), 118
IMI (Innovative Medicines Initiative), 181

Subject Index 273

immortalised cells, and respiratory system models, 71–72
immune thrombocytopenic purpura (ITP), 102
immunoreceptor tyrosinebased activation motif (ITAM), 102
indicator diffusion method (IDM), 118
IND (Investigational New Drug), 112
induced pluripotent stem (iPS) cells. See iPSC (induced pluripotent stem cells)-based disease models
inhalation toxicology, and respiratory system, 67–68
Innovative Medicines Initiative (IMI), 181
in silico analysis
    biological efficacy or toxicity, 205–207
    calculating molecular fragment descriptors, 201–205
        common chemical scaffolds, 204
        fragment types, 201–202
        scaffolds associated with data, 204–205
        structural alerts, 203–204
    case study
        prediction of human adverse events, 215–216
        regulatory, 212–215
    chemical structures and associated data, 196–199
        biological data representation, 199
        calculated data, 199
        chemical representation, 196–198
    overview, 194–196
    quantitative structure-activity relationship (QSAR) modelling, 208–212
        applying models, 210–211
        building models, 208–210
        model validation, 211
    searching chemical databases, 199–201
    structural features, 205–207
        classification based on substructures, 205
        clustering, 205–206
        decision trees, 207
in silico models and simulations
    computational cardiac electrophysiology, 220–230
        simulating interactions of medicines, 226–230
        single cardiomyocyte models, 220–223
        whole organ heart models, 223–226
    investigating variability, 230–234
    overview, 219–220
    validation of, 234–236
Institutional Review Boards (IRBs), 3
interactions of medicines simulations, 226–230
International Committee on Harmonisation (ICH), 213
intestinal fibrosis, PCTS in, 49–50
Investigational New Drug (IND), 112
in vitro and *in vivo* models, and ADMET, 126–128
in vitro toxicology, 68–69
iPSC (induced pluripotent stem cells)-based disease models
    cardiovascular, 169–172
    and drug discovery, 179–181
        challenges, 179–180
        future directions, 180
        personalised medicine, 180–181
    and drug efficacy assessment, 173–176
        drug screens on, 176
        examples of drug testing to validate, 175–176
        target-based screening *vs.* phenotypic screening, 173–175

iPSC (induced pluripotent stem cells)-based disease models (*continued*)
    emerging resources of, 181–183
        EBiSC (European bank for induced pluripotent stem cells), 182
        HipSci (human induced pluripotent stem cells initiative), 183
        StemBANCC (stem cells for biological assays of novel drugs and predictive toxicology), 181–182
    examples of, 172
    genome editing tools, 172
    neurological and psychiatric, 167–169
    overview, 167
    toxicity testing using, 177–179
        cardiotoxicity, 177–178
        hepatotoxicity, 178
        neurotoxicity, 178–179
IRBs (Institutional Review Boards), 3
ITAM (immunoreceptor tyrosinebased activation motif), 102
ITP (immune thrombocytopenic purpura), 102

Janus kinases (JAK) inhibitors, 95–101
    analysis of tofacitinib in BioMAP Systems, 98–101
    description, 95–96
    JAK3 and tofacitinib, 97–98

kidney, and body-on-a-chip technology, 143
kinase inhibitors, 89–90
    Janus (JAK), 95–101
        analysis of tofacitinib in BioMAP Systems, 98–101
        description, 95–96
        JAK3 and tofacitinib, 97–98
    SYK, 101–105
        analysis of fostamatinib in BioMAP Systems, 103–105
        description, 101–102
        SYK and fostamatinib, 102–103

legal issues, in human cells and tissues research, 2–8
    consent coverage, 8
    human participation in research, 3–4
    ownership of human samples, 4–8
liver
    and body-on-a-chip technology, 141–142
    fibrosis, PCTS in, 48–49
lungs, and body-on-a-chip technology, 137–139

Material and Data Transfer Agreements (MDTAs), 13
MDTAs (Material and Data Transfer Agreements), 13
MEA (multi-electrode array) technology, 145
mechanisms of human toxicity, 121–125
    genotoxicity, 121–122
    high-content toxicology, 124–125
    mitochondrial toxicity, 122–123
    phospholipidosis, 123–124
    reactive metabolites formation, 123
medicine interactions simulations, 226–230
microdosing (human phase 0 studies)
    analytical technologies used for, 247–249
    applications of, 249–252

Subject Index

description, 243–244
and drug targeting, 252–255
ethical considerations in, 247
overview, 241–243
regulatory authorisation, 245–247
scientific considerations, 244–245
special applications of, 255–261
    absolute bioavailability determination, 259–261
    in drug-drug interaction studies (DDIs), 258–259
    in protein therapeutics, 257–258
    use in fluxomics, 257
    use in paediatric studies, 255–257
microfluidics, and body-on-a-chip technology, 151–153
mitochondrial toxicity, and ADMET, 122–123
molecular fragment descriptors, and *in silico* analysis, 201–205
    common chemical scaffolds, 204
    fragment types, 201–202
    predefined features, 202–203
    scaffolds associated with data, 204–205
    structural alerts, 203–204
MRP (multi-drug-resistance protein), 117
multi-drug-resistance protein (MRP), 117
multi-electrode array (MEA) technology, 145
multi-organ *in vitro* models, and body-on-a-chip technology, 133–136
multiparametric screening, in ADMET, 125–126
myelination, 147–148
myeloproliferative diseases, 96

myographs, and human isolated tissues
    perfusion, 25
    wire, 23–25

NASH (non-alcoholic steatohepatitis), 48
National Cancer Registry Service (NCRS), 14
National Cancer Research Institute (NCRI), 12
National Centre for the Replacement, Refinement and Reduction of Animals in Research (NC3R), 67
National Research Council (NRC), 68
National Research Ethics Service (NRES), 4
NCEs (new chemical entities), 112
NCRI (National Cancer Research Institute), 12
NC3R (National Centre for the Replacement, Refinement and Reduction of Animals in Research), 67
NCRS (National Cancer Registry Service), 14
neurological disease models, iPSC-based, 167–169
neuromuscular junctions (NMJ), 145–147
neurotoxicity, and iPSC-based disease models, 178–179
new chemical entities (NCEs), 112
NHBE (normal human bronchial epithelial) cells, 72
NHNE (normal nasal epithelial) cells, 72
NMJ (neuromuscular junctions), 145–147
NOAEL (no observed adverse effect level), 199
non-alcoholic steatohepatitis (NASH), 48
non-steroidal anti-inflammatory drugs (NSAIDs), 175

no observed adverse effect level (NOAEL), 199
normal human bronchial epithelial (NHBE) cells, 72
normal nasal epithelial (NHNE) cells, 72
NRC (National Research Council), 68
NRES (National Research Ethics Service), 4
NSAIDs (non-steroidal anti-inflammatory drugs), 175

obstructive lung diseases, and PCTS, 50–54
OECD (Organisation for Economic Co-operation and Development), 69
Office of Human Research Protection (OHRP), 3
OHRP (Office of Human Research Protection), 3
optimisation loop, ADMET, 113–116
Organisation for Economic Co-operation and Development (OECD), 69
organoculture systems, and precision-cut tissue slices, 25–28
ownership, of human cells and tissues samples, 4–8

paediatric studies, use of microdosing in, 255–257
PAMPER project, 256
pathophysiology, and PCTS, 53–54
PCIS (precision-cut intestinal slices), 44
PCLS (precision cut lung slices), 75
PCTS. See precision-cut tissue slices (PCTS)
PDGF (platelet-derived growth factor), 47
perfusion myographs, and human isolated tissues, 25
peripheral nervous system (PNS), and body-on-a-chip technology, 144–149

personalised medicine
    and drug discovery, 180–181
    and human isolated tissues, 32–33
PGD (pre-implantation genetic diagnosis), 167
pharmacology, and PCTS, 51–53
phenotypic screening vs. target-based screening, drug efficacy assessment, 173–175
phospholipidosis, and ADMET, 123–124
physiology, and PCTS, 51–53
planar patch-clamp array chips, 145
platelet-derived growth factor (PDGF), 47
PNS (peripheral nervous system), and body-on-a-chip technology, 144–149
precision-cut intestinal slices (PCIS), 44
precision cut lung slices (PCLS), 75
precision-cut tissue slices (PCTS)
    in ADME and toxicology, 43–47
    applications, in disease models, 47–57
        cancer research, 54–57
        fibrosis in liver and intestine, 47–50
        obstructive lung diseases, 50–54
        viral infection research, 57
    and organoculture systems, 25–28
    overview, 38–40
    and pathophysiology, 53–54
    physiology and pharmacology, 51–53
    and sources of human organs and tissues, 40–41
    techniques to prepare, 41–43
predefined features, of molecular fragment descriptors, 202–203
prediction of human adverse events case study, 215–216

Subject Index

prediction tools, ADMET, 117–126
  blood–brain barrier challenge, 117–121
  mechanisms of human toxicity, 121–125
  power of multiparametric screening in, 125–126
pre-implantation genetic diagnosis (PGD), 167
primary cells, and respiratory system models, 72–73
  human airway epithelial cells (HAEpC), 72–73
  normal human bronchial epithelial (NHBE) cells, 72
  normal nasal epithelial (NHNE) cells, 72
primary human cell systems, complex
  anti-inflammatory kinase inhibitors, 95–105
    JAK kinase inhibitors, 95–101
    SYK kinase inhibitors, 101–105
  and kinase inhibitors, 89–90
    anti-inflammatory, 95–105
  and target-based drug discovery, 88–89
  for translational biology, 90–95
    BioMAP Systems, 90–95
protein therapeutics, microdosing in, 257–258
psychiatric disease models, iPSC-based, 167–169

QSAR modelling. *See* quantitative structure-activity relationship (QSAR) modelling
quantitative structure-activity relationship (QSAR) modelling, 208–212
  applying models, 210–211
  building models, 208–210
  model validation, 211

REACH (Registration, Evaluation, Authorisation and Restriction of Chemicals regulation), 67
reactive metabolites formation, and ADMET, 123
RECs (Research Ethics Committees), 11
Registration, Evaluation, Authorisation and Restriction of Chemicals regulation (REACH), 67
regulatory authorisation, in microdosing, 245–247
regulatory *in silico* case study, 212–215
reprogramming somatic cells, 166–167
Research Ethics Committees (RECs), 11
respiratory system. *See also* human-based respiratory models
  and inhalation toxicology, 67–68
  overview, 66–67

safety and toxicology risks, and human isolated tissues, 31–32
safety assessment, and human-based respiratory models, 68
'scheduled purposes', 3
SCID (severe combined immune deficiency), 96
science, of ADMET, 112–113
scientific considerations, in microdosing, 244–245
SCNT (somatic cell nuclear transfer), 166
seeding/cell density, 76
sensory systems, 148–149
severe combined immune deficiency (SCID), 96
signal transducers and activators of transcription (STATs), 95
single cardiomyocyte models, 220–223
16HBE14o- cells, and respiratory system models, 71–72

skeletal muscle, and body-on-a-chip technology, 150
skin and glandular tissues, human isolated tissues, 28–29
somatic cell nuclear transfer (SCNT), 166
sourcing, human isolated tissues, 21–22
static vs. continuous flow conditions, 78
STATs (signal transducers and activators of transcription), 95
StemBANCC (stem cells for biological assays of novel drugs and predictive toxicology), 181–182
stem cells, and respiratory system models, 73–74
stem cells for biological assays of novel drugs and predictive toxicology (StemBANCC), 181–182
storing, human isolated tissues, 21–22
Strategic Tissue Repository Alliance Through Unified Methods (STRATUM), 7
STRATUM (Strategic Tissue Repository Alliance Through Unified Methods), 7
structural alerts, and molecular fragment descriptors, 203–204
structural features, and *in silico* analysis, 205–207
    classification based on substructures, 205
    clustering, 205–206
    decision trees, 207
submerged vs. ALI conditions, 77–78
substructures classification, and *in silico* analysis, 205
SYK kinase inhibitors, 101–105
    analysis of fostamatinib in BioMAP Systems, 103–105
    description, 101–102
    SYK and fostamatinib, 102–103

target-based drug discovery, challenges of, 88–89
target-based screening vs. phenotypic screening, drug efficacy assessment, 173–175
target product profile (TPP), 114
TGF-β (tumour growth factor-beta), 17
tissue baths, and human isolated tissues, 23–25
tissue explants, and respiratory system, 75
tissue-specific promoter (TSP), 56
tofacitinib
    analysis in BioMAP Systems, 98–101
    and JAK kinase inhibitors, 97–98
toxicity/efficacy, biological, *in silico* analysis, 205–207
toxicity testing, and iPSC-based disease models, 177–179
    cardiotoxicity, 177–178
    hepatotoxicity, 178
    neurotoxicity, 178–179
toxicology, and PCTS, 43–47
TPP (target product profile), 114
translational biology, complex primary human cell systems for, 90–95
    BioMAP Systems, 90–95
transporting, human isolated tissues, 21–22
TSP (tissue-specific promoter), 56
Tufts Center for the Study of Drug Development, 110
tumour growth factor-beta (TGF-β), 17

viral infection research, and PCTS, 57
Vitron (hand held coring tool), 41

whole organ heart models, 223–226
wire myographs, and human isolated tissues, 23–25